T0132669

Physiology of Stressed Crops

Volume IV

Osmoregulation and Protection

U.S. Gupta
Professor (Adj)
Department of Crop & Soil Sciences
University of Georgia
Athens
USA

Enfield (NH) Jersey Plymouth

SCIENCE PUBLISHERS
An Imprint of Edenbridge Ltd., British Isles
Post Office Box 699
Enfield, New Hampshire 03748
United States of America

Website: *http://www.scipub.net*

sales@scipub.net (marketing department)
editor@scipub.net (editorial department)
info@scipub.net (for all other enquiries)

Library of Congress Cataloging-in-Publication Data

Gupta, U.S., 1940-
Physiology of stressed crops/U.S. Gupta
p. cm.
Includes bibliographical references and index.
Contents: v. IV. Osmoregulation and Protection
ISBN 1-57808-440-7
1. Crops--Effect of stress on. 2. Crops--Physiology I. Title.
SB112.5.G87 2004
632'.1--dc22

2004052486

ISBN 1-57808-440-7 [10 digits]
978-1-57808-440-1 [13 digits]

Published by Science Publishers, NH, USA
An Imprint of Edenbridge Ltd.
Printed in India

PREFACE

To cope with the abiotic stress-induced osmotic problems, one of the ways plants adapt is either increasing uptake of inorganic ions from the external solution or by *de novo* synthesis of organic compatible solutes acting as osmolytes. These two processes differ significantly in their time scales. Immediate changes in ion fluxes are believed to provide quick (within a few minutes) osmotic adjustment while a fine "tuning" by means of biochemical synthesis of compatible solutes has a scale of hours or days. The membrane transport processes play a crucial role in plant osmoregulation. Accumulation of inorganic ions is sufficient for osmotic adjustment under salinity stress. For osmotic adjustment at least two mechanisms have been suggested, one, by changes in cell volume by mechanosensitive receptors on the plasmamembrane and two, by intracellular osmosensing mechanisms affecting the degree of cytosol hydration. Most abiotic stresses alter plant metabolism, for instance, hormone metabolism (discussed in Vol. I of this series), nutrient metabolism, uptake and utilization (discussed in Vol. II of this series), allelochemical synthesis, exudation and sensitivity (discussed in Vol. III of this series), generation of compatible solutes, osmoregulants and protectants (discussed in this volume), membrane system, permeability, integrity and transport (to be discussed in Vol. V), and *de novo* synthesis of stress responsive and/or stress tolerant induced- proteins (to be discussed in Vol. VI).

Of the osmoregulants and protectants discussed in this volume, trehalose, fructans, ectoine and citrulline, which are generated in different species, in osmotically ineffective amounts, do mitigate the stress effects on cells/ plants and improve productivity. These are better termed as "protectants" rather than "osmoregulants" , with which the former have been clubbed in this volume for reasons of convenience. Many attempts to improve stress tolerance by enhancing osmolyte accumulation are based on the assumption that maintenance of cell turgor is a critical limitation to productivity under stresses that cause dehydration. To date, attempts to confer stress tolerance by increases in osmolyte levels are not consistent with this premise. The scope of improving crop stress tolerance by way of increasing the osmoregulating ions or metabolites does not appear to be bright, simply because of the energetic reasons (productivity decreases), but that of "protectants" is, and in fact by involving genetic engineering techniques, there are several pieces of encouraging research (discussed in this volume) showing significant improvement in stress tolerance and in turn productivity. This line of research should be vigorously followed.

In this volume, the most recent advances in this field have been discussed, some methodological hints for conducting research by beginners, and lead articles, have been included. I feel there is great scope for conducting more research in this area which will definitely, go a long way in increasing crop productivity in stressful environments which as of now are hardly being economically exploited. This volume should be a must for all scientists devoted to alleviating human hunger and solving the food problem.

Constructive criticism and suggestions by readers for further improvement of this volume will be highly appreciated. The patience and encouragement by my wife, Hiramani Devi, during the preparation phase of this volume is duly acknowledged. I also thank Sandra Goins Kirton for her help during the last phases of its preparation.

U.S. Gupta

CONTENTS

Preface iii

1. Introduction to Osmoregulation 1
   Introduction 1
   Non-osmotic (Specific) Effects of Some Stress-induced Metabolites 4
   Preconditioning 6
   Osmoregulation and Environment 7
   Inheritance of Osmotic Adjustment 7
   Improvement Aspects 8
   References 10
   Suggested Readings 12

2. Inorganic Osmolytes 13
   Introduction 13
   Drought Preconditioning 15
   Energetic Efficiency 17
   The Dominating Role of Potassium Ion 17
   In stomatal movement, cortex and epidermis, root steler cells, root hairs, phloem loading, vacuolar tonoplast, chloroplast, N-metabolism, electrical properties, and genotypic variability, K^+–uptake mechanism.
   Sodium : Potassium Uptake 24
   The Indirect Role of Calcium 25
   Ion Uptake as Affected by Different Types of Stresses 27
   Drought, salinity, acidity, heat stress, cold stress, radiation, and oxidative stress.
   Improvement Work 37
   References 38
   Suggested Readings 42

3. Sugars – The Key Osmolytes 43
   Introduction
   Freeze-induced Dehydration 45
   Drought-induced Dehydration 48
   Salinity-induced Dehydration 57

	Sugars and Dehydration-induced ROS	57
	Toxic Ion (arsenic) Stress	58
	High Light-intensity Stress	59
	Heat Stress	59
	Sucrose and Stress Metabolism	61
	References	66
	Suggested Readings	68
4.	Complex Sugars — Cyclitols	69
	Introduction	69
	Cold Stress	75
	Salinity Stress	76
	Drought Stress	80
	Nutrient Stress	81
	The Stress of ROS	81
	Improvement Work	83
	References	85
	Suggested Readings	87
5.	Proline	88
	Introduction	88
	Proline Biosynthesis in Stressed Plants	88
	Localization and Transport	92
	Stress-induced Proline Accumulation and Stress Mitigation	93
	Salinity Stress	93
	Drought Stress	95
	Cold and Freezing Stress	97
	Improvement	97
	References	101
	Suggested Reading	103
6.	Glycinebetaine	104
	Introduction	104
	Biosynthesis in Stressed Plants	105
	GB-induced Mechanism of Stress Tolerance	106
	Subcellular Localization of GB	108
	Stress-induced GB Accumulation and Stress Mitigation	108
	Drought Stress	108
	Salinity Stress	109
	Cold Stress	110
	Heat Stress	113
	Oxidative Stress	115
	Improvement	116
	Variability	116

Genetics/Inheritance 117
Metabolic Engineering 118
Practical Application and Future Perspective 123
References 123
Suggested Reading 126

7. Polyamines 127
Introduction 127
Biosynthesis 128
Polyamines and Root Growth 130
Stress Tolerance 133
Salinity Stress 134
Drought Stress 137
Cold Stress 138
Heat Stress 139
Mineral Ion Stress 140
UV-B Stress 140
Traumatic Stress 141
Improvement 141
References 143
Suggested Readings 146

8. Trehalose 147
Introduction 147
Biosynthesis and Sugar Metabolism 147
Increased Production by Unstressed Plants 149
Trehalose vs Stress Tolerance 152
Salinity Stress 155
Drought Stress 156
Improvement 159
Chloroplst vs Nuclear Engineering 159
Future Perspective 160
References 160
Suggested Reading 162

9. Fructan 163
Introduction 163
Biosynthesis 164
Abiotic Stress Tolerance 168
Drought 168
Cold Stress 170
Hypoxia 173
Mineral Nutrient Stress 174
Salinity Stress 174
Improvement 175

References 177
Suggested Reading 179

10. Ectoine 180
Introduction 180
Biosynthesis 182
Transport 184
Engineering for Ectoine Production and Improvement 185
References 188
Suggested Reading 189

11. Citrulline 190
Introduction 190
Biosynthesis 190
Mechanism of Action 193
Genetic Basis 194
References 195
Suggested Reading 195

12. ROS and Antioxidants 196
Introduction 196
ROS Generation and ROS Stress 196
The Plant Defence Mechanism : Generation of ROS-degenerating Enzymes and Antioxidants 198
Enzymatic Defence 199
Non-enzymatic Defence 201
Stress induced ROS-detoxification by Antioxidants 204
Drought 205
Salinity 209
Cold Stress 213
Heat Stress 215
Hypoxia/Anoxia 216
Heavy Metal Stress 218
References 221
Suggested Readings 224

Appendix 225
Author Index 237
Subject Index 242

1

INTRODUCTION TO OSMOREGULATION

Introduction

Abiotic stress factors such as drought, salinity, and extremes of temperature are the major limiting factors of crop production. As the plants under these situations experience constant fluctuations in water availability, they have to some extent evolved adaptive features to search for and absorb water through their root systems, to cope with the excessive transpirational water loss, and adjust their physiology and metabolism for continued growth and survival. Plants respond to these stresses at morphological, anatomical, cellular, and molecular levels. Tolerance to dehydration depends on the ability of their cells to maintain membrane integrity and prevent protein denaturation. When cells are subjected to slow osmotic stress,compatible*solutes (osmolytes/osmoprotectants) are accumulated resulting in maintenance of a higher turgor potential at a given leaf water potential. Thus, zero turgor will occur at a lower water potential in the osmotically adjusted tissues. The accumulation of osmolytes is termed osmotic adjustment or osmoregulation. Osmolyte accumulation represents a mechanism for maintaining a positive turgor potential at low water potential. Maintenance of a positive turgor potential is generally required for stomata opening and the tissue/plant expansion growth.

The osmolytes are often termed as compatible solutes because even at higher concentrations, these solutes do not inhibit the activity of enzymes; rather they protect enzymes and membranes against the deleterious effects of destablizing ions such as Na^+ and Cl^-. Accumulation of such solutes in response to water stress is presumably a metabolic adaptation developed by the stress-tolerant species. These osmoprotectant solutes synthesized in response to stress are localized in cytoplasm; but the inorganic

* Metabolites that serve as compatible solutes differ among plant species and include polyhydroxylated sugar alcohols, amino acids and their derivatives, tertiary sulfonium compounds, and quarternary ammonium compounds. The major role of these metabolites is to serve as organic osmolytes with compatible properties at high concentrations; such osmolytes increase the ability of cells to retain water without disturbing normal cellular functions.

osmoprotectant ions such as K^+, Na^+ and Cl^- preferentially accumulate into the vacuole. Thus, these compatible solutes and ions lead to turgor maintenance for the cell under osmotic stress. The osmoprotectants can protect proteins via a chaperone-like action on protein folding, e.g. glycinebetaine may act as a signal molecule that elicits the expression of genes associated with stress tolerance. The osmoprotectants are, as a rule, not species-specific, so that they function to protect organisms that do not naturally over-produce them. Such a protection has been observed when glycinebetaine, proline, or polyvols are supplied in the medium or when they are synthesized endogenously as a result of metabolic engineering (Rontein et al., 2002). Crop species and varieties also differ with respect to the type of osmolytes accumulated, and the osmolyte that plays a major role in osmotic adjustment is species dependent. Osmotic adjustment results in increased water retention and turgor maintenance. The compatible osmolytes accumulate, replace water in some biochemical reactions and associate with lipids or proteins and prevent membrane damage, dissociation of protein complexes, or inactivation of enzymes (Zhang et al., 1999). Such adapted plants develop deeper roots or higher root length density in soil, leading to greater soil moisture extraction (Tangpremsri et al., 1991). Osmoregulation occurs primarily in expanding organs such as roots, leaves, hypocotyls and inflorescences, where it plays an important role in maintaining the inflow of water during expansion growth. When expansion is complete, the capacity for osmoregulation is gradually lost. The recently fully expanded leaves of determinate species such as wheat, generally have a greater capacity for osmoregulation than those formed earlier. Consequently during water deficits, lower leaves wilt at higher water potentials than the upper leaves, even though the latter may receive a higher radiation load. Reasons for the change in osmoregulation with leaf position and age are influenced by factors affecting supply of assimilates to the leaf such as a declining rate of photosynthesis of older leaves and demand for assimilates and other component solutes in other parts of the plant.

Osmotic adjustment of root tips during drought is important for continued root growth, water and nutrient absorption, CO_2 assimilation, growth and yield. Root osmotic adjustment is reflected in leaf osmotic adjustment. It may be noted here that at moderate moisture stress, increased abscisic acid (ABA) content promotes root growth but inhibits shoot growth, stimulates floral sterility and senescence, stomatal closure and the expression of genes controlling the contrasting growth behavior of root and shoot at moderately low moisture potentials. The leaf osmotic adjustment further maintains leaf turgor and growth, photosynthesis, transpiration, and thus increasing moisture and nutrient uptake and delaying the production of ABA and floral abortion (Ludlow and Muchow, 1990).

In young expanding organs, osmoregulation is entirely dependent on import of solutes from the photosynthetic organs. However, in expanded leaves, increase in sugars and amino acids results from a difference between the rate of CO_2 fixation and the rate of metabolite transport. Changes in the potassium (K^+) level also contribute to osmoregulation in accompaniment with changes in sugars and amino acids. Organic acids e.g., malate and citrate, and nitrate and chloride ions also contribute to osmoregulation. Accumulation of osmolytes is not a passive accumulation of solutes which may not be utilized in growth processes, since the osmolyte over-producing transgenic plants perform better under water-stress conditions. Thus the osmotic adjustment in plants is an ancient evolutionary trait, but there is no simple osmolyte, rather a host of osmolytes in different crop species perform the osmoprotective function. Osmolytes are also valuable storage compounds for generating reducing power and a source of carbon and/or nitrogen contributing to relief from stress. Preferential accumulation of a particular osmolyte in a given plant depends on the availability of its precursors under adverse conditions.

High concentrations of most compatible solutes confer protection against oxidative damage by scavenging free radicals in addition to their roles in maintenance of osmotic equilibrium without perturbing macromolecule-solvent interactions (Akashi et al., 2001). For instance, mannitol increases resistance to oxidative stress in tobacco (Shen et al., 1997). Concentrations of compatible solutes, e.g. mannitol, are higher in the cytoplasm, whereas in the vacuole and apoplast, electrolytes (both organic and inorganic) make the major contribution to osmotic potential (Hare and Cress, 1997). Thus the accumulation of osmolytes and electrolytes plays a critical role in mitigation of water stress caused by drought, salinity, and cold; and the active osmolyte overproduction activity of certain enzymes involved in the biosynthesis are increased following imposition of stress. For instance, the transgenic P5C ($\Delta^{\cdot}$ -pyrroline-5-carboxylase) increased proline accumulation and osmoregulation in *Vigna aconitifolia* (Kavi Kishore et al., 1995) but the transgenic tobacco plants which produce twice as much proline as comparable controls after water deprivation display improved growth following drought and salinity stress, yet do not undergo osmotic adjustment upon soil drying. Thus increasing the osmolyte accumulation has been proposed as a strategy to enhance at least one aspect of stress tolerance; provided it is not accompanied by yield reduction under optimal growth conditions. Even a marginal improvement in stress tolerance resulting from such manipulations may have profound economic benefit in large-scale farming (Hare et al., 1996).

NON-OSMOTIC (SPECIFIC) EFFECTS OF SOME STRESS-INDUCED METABOLITES

Apart from the compatible osmolytes (osmoprotectants) being produced under abiotic stress, in some plants there are some other metabolites synthesized in small amounts, far less to be osmotically effective, but are very effective in imparting stress tolerance otherwise. Such metabolites cannot be included in the group of stress-induced hormones (discussed in Vol.1), proteins (to be discussed in a later volume), or osmolytes (being discussed in this volume), for instance—trehalose, fructan, ectoine, and citrulline – and the work done on their roles in stress tolerance is not enough to warrant a separate volume. Since these have often been loosely discussed by some authors with compatible solutes, I decided to discuss the roles of these metabolites here, and modified the title of the volume from "Osmoregulation" to "Osmoregulation and Protection", so as to include these chemical mediators.

The stress stimulated osmolytes result in cell osmotic adjustment at relatively much higher concentrations than the other stress-produced chemical mediators, named above, that are effective at relatively much lower concentrations. The protectants of the latter group offer protection to cell membranes and to the macromolecules. Thus the multi-gene transformed transgenics with osmoprotectants and the other chemical mediators with the membrane/nucleotides-, proteins/enzyme-protection ability, are likely to have greater stress tolerance ability.

Trehalose stabilizes the dehydrated enzymes, proteins, and lipid membranes efficiently, as well as protects biological structures from damage during desiccation. Engineering of trehalose over-production in rice has been achieved (Garg et al., 2002) which accumulates increased amount of trehalose and shows higher levels of tolerance to salt, drought, and low temperature stresses as compared with the non-transformed plants. The transgenic plants develop vigoros root system and maintain a higher level of selectivity for K^+ over Na^+ uptake in roots and Na^+ exclusion from shoots, compared with the non-transformed-stressed plants. The transgenic plants have improved photosynthesis and thus the plant growth, under drought conditions by limiting the photooxidative damage (Garg et al., 2002). The transgenic plants also exhibit higher carbohydrate level suggesting that trehalose may be involved in sugar sensing and modulating carbon metabolism.

Fructans have a much stronger effect on different lipid systems than other (poly)saccharides, which is related to their hydrophobic properties (Vereyken et al., 2001). Fructans are able to utilize the liquid-crystalline lamellar phase, which is consistent with a drought protecting role in plants. Fructan increases mobility of acyl chains, but immobilizes the lipid-head group region. Most likely, fructans insert between the head groups of lipids, thereby spacing the acyl chains. This results in a lower phase transition temperature. Location of interaction with the lipid head group is different for the inulin-type fructan compared to the levan-type fructan, since inulin shows interaction with the lipid phosphate group, whereas levan does not. Further, the water-stress protective effect of fructans is probably induced by membrane-fructan interaction which prevents lipid condensation and phase transition to take place.

Jebbar et al. (1992) report that ectoine accumulates in *Escherichia coli* cells proportionately to the osmotic strength of the medium, and is not metabolized. Ectoine also confers protection to cell membranes in transgenic plants at a very low level, rather than conferring an osmolyte effect (Nakayama et al., 2000). Ectoine synthesis allows maintenance of the normal permeability of cell membrane even upon exposure to hyper-osmotic shock. Further, Nakayama et al. suggest that the multigene introduction system, in which accumulation of osmolytes such as proline and mannitol for intracellular osmotic adjustment, and ectoine production to protect the cell membrane from initial damage, are simultaneously achieved, would be a powerful strategy of metabolic engineering to grow useful salt- and drought-tolerant crops.

The highly drought-tolerant wild watermelon plant, a native of Botswana, accumulates citrulline which acts as a hydroxyl radical scavenger (Akashi et al., 2001). Its effectiveness as hydroxyl radical scavenger is greater than those of mannitol and proline. Thus citrulline that accumulates in the drought-tolerant watermelon leaves increases the antioxidative potential of the cells and protects them from oxidative damage, and appears even more important than ascorbate and glutathione. Citrulline effectively protects DNA and the metabolic enzymes from oxidative injuries. Yokota et al. (2002) conclude that citrulline contributes to oxidative stress tolerance under drought conditions as a novel hydroxyl radical scavenger.

The reactive oxygen species (ROS)/ free radicals (O^-, H_2O_2, OH^-) are formed as by-products in electron transport chains of chloroplasts, mitochondria, and plasma membrane. Several abiotic stresses like drought, salinity, cold, intense light, and toxic metals; biotic stresses like wounding, and pathogenesis; and chemical stresses like some herbicides, produce ROS which affect cytoplasmic macromolecules, lipids, proteins, and membranes causing considerable subcellular damage. To prevent the damage caused by free radicals, plants have evolved complex protective mechanisms which include antioxidant enzymes such as superoxide dismutase, catalases, ascorbic peroxidase, and glutathione reductase, and the free radical scavengers such as ascorbate, glutathione, carotenoids, tochopherol, and citrulline (discussed in the last chapter of this volume).

PRECONDITIONING

The degree of osmoregulation is affected by both, the rate of stress, and stress preconditioning. For instance, plants which have been previously exposed to water stress exhibit an improved capacity to tolerate subsequent periods of water stress through increases in solute levels, which may be of two types. One, a decrease in the osmotic potential at full turgor, and two, a change in the degree of turgor maintenance during the period of water stress subsequent to preconditioning stress. Solutes that accumulate during the preconditioning stress are not fully dissipated immediately after rewatering. Hence the rate of preconditioning stress may affect the magnitude of decrease in the osmotic potential. For instance, the magnitude of the change in osmotic potential differs with leaf position on plants after the plants are subjected to a series of pre-stress cycles in which the minimum water potential reaches –2.0 MPa. A decrease in osmotic pressure of 0.2 MPa occurs in young emerging leaves, but a small increase in osmotic pressure occurs in the older fully expanded leaves.

Osmolytes which accumulate as a result of pre-stressing eventually dissipate if no further water stress occurs, and the rate of dissipation depends on the stress-hardening conditions. For instance, the slow rates of pre-stressing or more than one pre-stress leads to incomplete dissipation in rice plants after 15 days, while plants exposed to one rapid-pre-stress cycle showed complete recovery of osmotic pressure within this period (Steponkus et al., 1982). Similarly, the osmotic pressure of sorghum leaves exposed to a single stress of –1.5 MPa recovers to the level of control plants in approximately 12 days (Jones and Rawson, 1979). The differences in water potential with regard to field-grown sorghum below –0.6 MPa, and continued solute accumulation in the higher osmoregulating type to an osmotic potential level of –2.2 MPa compared with –1.5 MPa in the lower osmoregulating type was observed. The relative water content (RWC) at a water potential of –2.5 MPa was 0.82 compared with 0.70 for sorghum cultivars with high- and low-osmoregulation, respectively. Further, Karami et al. (1980) observed that the cotton genotypes differed with respect to leaf morphology; the high osmoregulating line having a super-okra type and the low a normal type. During the diurnal stress cycle, turgor was fully maintained at 0.5 MPa in the super-okra type, while it declined from 0.3 MPa at 6 AM to zero at 2 PM in normal type. Turgor maintenance in older cotton leaves is greater in water-stressed plants than in previously unstressed plants. However, in younger fully expanded leaves, where the degree of turgor maintenance was the same as that of the hardened older leaves, there was no response to pre-stressing other than a small change in osmotic pressure at full turgor (Ackerson and Hebert, 1981).

OSMOREGULATION AND ENVIRONMENT

Apart from the tissue osmotic pressure, atmospheric CO_2 concentration and light intensity also affect osmoregulation. Sionit et al.(1981) observed that increase in CO_2 concentration from the ambient (350 µl/L) led to a marked improvement in the maintenance of leaf turgor in wheat plants exposed to water stress, with a decrease in osmotic pressure of 0.2 MPa at a water potential of –1.7 MPa. Reduction in photosynthetically active radiation (PAR) by leaf shading from 1500 $\mu Em^{-2}\ s^{-1}$ to 30-50 $\mu Em^{-2} s^{-1}$ on one day led to a reduction in reducing sugars in sorghum leaves, but in unshaded plants, osmotic pressure decreased and the reducing sugars content increased (Acevedo et al., 1979).

By dissipating excess photochemical energy generated, photosynthetic membranes are protected against light-induced damage when CO_2 assimilation is limited. An enhanced photorespiratory capacity is associated with increased tolerance of high-intensity light, while a diminished capacity for photorespiration increases susceptibility to photo-oxidative damage induced by high light intensity.

As most of the solutes in plant cells contributing to osmotic adjustment originate from the photosynthetic products, one can easily understand how atmospheric CO_2 concentration and PAR affect osmotic potential. Thus it is also understandable that the lower leaves in a crop canopy which receive low levels of irradiance show low levels of osmoregulation leading to rapid senescence than the leaves positioned at higher levels during the periods of water stress.

INHERITANCE OF OSMOTIC ADJUSTMENT

The capacity to adjust osmotically when the plant experiences water stress is an inherited trait. Osmotic adjustment in wheat is conditioned by alternative alleles at a single locus, with high response being recessive (Morgan, 1991). The gene appears to condition primarily a difference in potassium accumulation; with amino acid accumulation as a possible secondary and dependent response (Morgan, 1992). Analysis of a single chromosome substitution series of Chinese Spring indicated that genes coding for this response are located on chromosome 7A (Morgan, 1991). However, Galiba et al.(1992) reported that genes controlling osmotic adjustment in wheat are primarily located on chromosome 5A and 5D based on their studies on stress-induced free amino acid accumulation profiles. The genes identified by Galiba et al. regulate free amino acid accumulation and thus differ from the genes identified by Morgan (1991).

Grumet and Anderson (1986) developed two barley isopopulations differing in glycinebetaine accumulation. The high glycinebetaine concentration-population maintained an osmotic potential of 0.1 MPa less than the low glycinebetaine population at all salinity levels studied. Difference in osmotic pressure between the two populations was not accounted for by glycinebetaine *per se*. The authors suggest that the glycinebetaine levels in barley are controlled by osmoregulatory genes that have pleiotropic effects and whose alleles confer various osmotic pressure levels by influencing solute levels as a whole.

In sorghum, inheritance of osmotic adjustment to water stress was investigated in a series of generations derived from the three possible biparental crosses between two inbred lines with a high capacity for osmotic adjustment, and one with a low capacity (Basnayake et al., 1995). Analysis of segregation ratios identified two independent major genes for high osmotic adjustment. In the same year, Yang et al. (1995) working with maize, developed a series of near-isogenic F_8 pairs of glycinebetaine-containing and glycinebetaine deficient lines. The pairs of lines differed for alternative alleles at a single locus; the wild-type allele conferring glycinebetaine accumulation was designated *Bet1*. This gene affected glycinebetaine biosynthesis at the level of choline oxidation and was located near the centromere on the short arm of chromosome 3 (Rhodes et al., 1993).

IMPROVEMENT ASPECTS

The strategies used to create more tolerant plants through manipulation of osmotic adjustment and its components include approaches through conventional breeding, genetic engineering and quantitative trait loci (QTL) mapping. The conventional breeding methods are expensive and the progress is slow. Selection for improved osmotic adjustment by measuring osmotic potential in segregating populations is tedious and is, therefore, considered impractical by most plant breeders except at the stage of selecting parents for initial crossing.

A number of genes encoding osmolyte biosynthesis have been transformed into model plant species such as tobacco. The expression of a bacterial gene *mtlD* allowing mannitol accumulation conferred salinity tolerance to transgenic tobacco plants (Tarczynski et al., 1993). Since then many plants expressing or over-expressing biosynthesis of various osmolytes such as proline, glycinebetaine, ononitol, mannitol, and the so called osmolytes, trehalose and fructan, have been generated. Introduction of osmolyte production makes transgenic plants tolerant to water-scarcity stress (drought, salinity, or cold). It is argued that osmolytes may play a more complex role in conferring drought tolerance than simply contributing to osmotic adjustment.

Morgan and Tan (1996) using information from their conventional genetic studies on osmotic adjustment, selected 14 recombinant inbred lines of hexaploid wheat for QTL mapping. These lines formed two groups with contrasting osmotic adjustment capacity. Analysis of linkage with RFLP loci suggested a probable position for osmotic adjustment on the short arm of chromosome 7A. The study of Lilley et al.(1996) on 52 rice recombinant inbred lines, a major QTL explaining $1/3^{rd}$ of the phenotypic variation for osmotic adjustment, was identified at the RG1 region of chromosome 8.

Zhang et al. (1999) state that starting with the BC_1F_1 generation, plants are screened with the markers previously identified as being linked with QTLs for high osmotic adjustment capacity. Introgression of QTLs for high osmotic adjustment capacity is accomplished by repeatedly backcrossing one genotype carrying high osmotic adjustment genes into another genotype that has other desirable properties but low osmotic adjustment capacity (recurrent parent). In each generation plants carrying the targetted QTLs are selected with flanking markers for backcrossing. This process is repeated until BC_6F_1 generation when plants have 99.2 percent of genome of the recurrent parent. Plants carrying the targetted QTLs are selfed to produce lines homozygous for those QTLs. Lines homozygous for the QTLs of interest are advanced one more generation for seed multiplication.

For crop breeders, the most useful application of the identified QTLs is to form marker-assisted selection aimed at accumulating favorable alleles and breaking their possible linkage with undesirable loci. Marker-assisted selection can increase the efficiency and accuracy of selection, especially for traits that are difficult to phenotype. For marker-assisted selection to be effective, marker(s) should be closely linked with the desired trait. To achieve this goal, a series of near-isogenic lines should be genotyped with markers in the QTL region and compared for the phenotypic expression of the trait being mapped. By identifying the chromosomal region unique to those segments which cause phenotypic effect, one can potentially allow placement of a QTL to a very small interval on the map (Zhang et al., 1999). Although it could be useful in crop improvement, the application of marker-assisted selection has not been rigorously evaluated in many crop species, especially for abiotic stress resistance improvement. Applying marker-assisted selection to a QTL of major effect on osmotic adjustment will make it easier for osmotic adjustment to be incorporated into a breeding program (Zhang et al., 1999). However, in view of the complex physiological and biochemical impacts that osmotic stress has on plant cells, it seems unlikely that changing the expression of only one of the osmotic expression genes will have a major effect. It is expected that the successful release of stress-tolerant crops will require large-scale metabolic engineering of not only genes encoding osmolytes, but also the genes for antioxidants, water channel proteins, *lea* proteins, etc., because osmotic adjustment is just part of a complex web of adaptive strategies that plants have evolved (Bray, 1997).

One of the approaches to improve stress tolerance in crop plants is to transfer the genes for adaptive traits from the tolerant plants or even the lower organisms to commercial crop plants. Now the available genetic transformation technology enables us to achieve gene transfer in a precise and, to some extent, predictable manner. Metabolic traits, especially pathways with a few enzymes, are better characterized genetically and more amenable to such manipulations than the structural and developmental traits. Thus the metabolic engineering is the directed improvement of cellular properties through the modification of specific biochemical reactions or the introduction of new ones, with the use of recombinant DNA technology. However, availability of the precursor to synthesize the osmoprotectant could limit the amount of osmoprotectant synthesized in a transgenic host. Also the negative physiological consequences of diverting the precursor to the osmoprotectant away from the primary metabolism has to be kept in mind. However todate, most of the transgenic plants have rarely been subjected to the rigorous assessments for their supposedly improved stress-tolerance ability.

Since the water stress caused by drought, salinity or cold leads to oxidative stress, manipulations aimed at improving oxidative stress tolerance have resulted in salinity tolerance of tobacco (Roxas et al., 1997). Some of the traits when engineered together with osmoprotectant synthesis, are expected to enhance whole plant stress tolerance. This can be done by reiterative engineering or crossing and selecting transgenic plants engineered for different traits. For example, manipulation of genes involved in ion transport together with osmoprotectant synthesis is expected to increase the cell's ability to withstand salinity stress (Serrano et al., 1999). Further, understanding the function and regulation of other genes involved in water and ion transport is expected to provide important tools for engineering salinity and osmotic stress tolerance in plants.

Production of the osmoprotectant is also regulated by the stress itself. Expression of some of the genes in the stress signal transduction cascade is mediated by the plant growth regulator abscisic acid (ABA). But some other genes act independent of ABA. Components of the same signal transduction pathway are also shared by various stress factors such as drought, salinity and cold (Shinozaki and Yamaguchi-Shinozaki, 1999).

REFERENCES

Acevedo, E.; E. Fereres; T.C. Hsiao and D.W. Henderson, 1979. Diurnal growth trends, water potential, and osmotic adjustment of maize and sorghum leaves in the field. Plant Physiol. 64: 76-80.

Ackerson, R.C. and R.R. Hebert, 1981. Omoregulation in cotton in response to water stress, translocation, and ultrastructure. Plant Physiol. 67: 484-88.

Akashi, K.; C. Miyake and A. Yokota, 2001. Citrulline, a novel compatible solute in drought-tolerant wild watermelon leaves, is an efficient hydroxyl radical scavenger. FEBS Lett. 508: 438-42.

Basnayake, J.; M. Cooper; M.M. Ludlow; R.G. Henzel and P.J. Snell, 1995. Inheritance of osmotic adjustment to water stress in three grain sorghum crops. Theo. Applied Genet. 90: 675-82.

Bray, E.A. 1997. Plant responses to water deficit. Trends in Plant Science 2:48-54.

Galiba, G.; L. Simonsarkadi; G. Kocsy; et al. 1992. Possible chromosal location of genes determining the osmoregulation of wheat. Theo. Appl. Genet. 85: 415-18.

Garg, A.K.; J.K. Kim; T.G. Owens et al. 2002. Trehalose accumulation in rice plants confers high tolerance levels to different abiotic stresses. Proc. Nat. Acad. Sci. 99: 15898-903.

Grumet, R. and A.D. Anderson, 1986. Genetic evidence for an osmoregulatory function of glycinebetaine accumulation in barley. Aust. J. Plant Physiol. 13: 353-64.

Hare, P.D. and W.A. Cress, 1997. Metabolic implications of stress-induced proline accumulation in plants. Plant Growth Regul. 21: 79-102.

Hare, P.D.; S. du Plessis; W.A. Cress and J. van Staden, 1996. Stress-induced changes in plant gene expression : prospects for changing agricultural productivity in South Africa. South African J. Sci. 92: 431-39.

Jebbar, M.; R. Talibart; K. Gloux et al. 1992. Osmoprotection of *Escherichia coli* by ectoine : uptake and accumulation characteristics. J. Bacteriology 174 (15): 5027-35.

Jones, M.M and H.M. Rawson, 1979. Influence of rate of development of leaf water deficits upon photosynthesis, leaf conductance, water use efficiency, and osmotic potential in sorghum. Physiol. Plant. 45:103-11.

Karami, E.; D.R. Krieg and J.E. Quisenberry, 1980. Water relations and carbon-14 assimilation of cotton with different leaf morphology. Crop Sci. 20: 421-26.

Kavi Kishore, P.B.; Z. Hong; G.H. Miao; C.A.A. Hu and D.P.S. Verma, 1995. Overexpression of Δ^{-} -pyrroline-5-carboxylate synthetase increases proline production and confers osmotolerance in transgenic plants. Plant Physiol. 108: 1387-94.

Lilley, J.M.; M.M. Ludlow; S.R. Mc Couch and J.C. O'Toole, 1996. Locating QTL for osmotic adjustment and dehydration tolerance in rice. J. Exp. Bot. 47: 1427-36.

Ludlow,M.M and R.C. Muchow, 1990.A critical evaluation of traits for improving crop yields in water-limited environments. Adv. Agron. 43: 107.

Morgan, J.M. 1991. A gene controlling differences in osmoregulation in wheat. Aust. J. Plant Physiol. 18: 249-57.

Morgan, J.M. 1992. Osmotic components and properties associated with genotypic differences in osmoregulation in wheat. Aust. J. Plant Physiol. 19: 67-76.

Morgan J.M. and M.K. Tan, 1996. Chromosomal location of a wheat osmoregulation gene using RFLP analysis. Aust. J. Plant Physiol. 23: 803-06.

Nakayama, H.; K. Yoshida; H. Ono et al. 2000. Ectoine, the compatible solute of *Halomonas elongata,* confers hyperosmotic tolerance in cultured tobacco cells. Plant Physiol. 122: 1239-47.

Rhodes, D.; W.J. Yang; Y. Samaras et al. 1993. Map locations of genes confering glycinebetaine and trigonelline accumulation in maize. Plant Physiol. 102 S, 160.

Rontein, D.; G. Basset and A.D. Hanson, 2002. Metabolic engineering of osmoprotectant accumulation in plants. Metab. Eng. 4: 49-54.

Roxas, V.P; R.K. Smith,Jr; F.R. Allen and R.D. Allen, 1997. Overexpression of glutathione S-transferase/glutathione peroxidase enhances the growth of transgenic tobacco seedlings during stress. Nature Biotechnol. 15: 988-91.

Serrano, R.; J.M. Mulet; G. Rios; J.A. Marquez et al. 1999. A glimpse of the mechanisms of ion homeostasis during salt stress. J. Exp. Bot. 50: 1023-36.

Shen, B.; R.G. Jensen and H.J. Bohnert, 1997. Increased resistance to oxidative stress in transgenic plants by targeting mannitol biosynthesis to chloroplasts. Plant Physiol. 113: 1177-83.

Shinozaki, K. and K. Yamaguchi-Shinozaki, 1999. Molecular responses to drought stress. pp 11-28. In: Molecular Responses to Cold, Drought, Heat, and Salt Stress in Higher Plants. (eds) K. Shinozaki and K. Yamaguchi-Shinozaki, R.G.. Landes Co. Austin, TX.

Sionit, N.; B.R. Strain; H. Hellmers and P.J. Kramer, 1981. Effects of atmospheric CO_2 concentration and water stress on water relations of wheat. Bot. Gaz. 142: 191-96.

Steponkus, P.L.; K.W. Shahan and J.M. Cutler, 1982. Osmotic adjustment in rice. pp 181-94. In: Drought Resistance in Crops with Emphasis on Rice. Los Banos, Leguna, Philippines :IRRI.

Tangpremsri, T.; S. Fukai; K.S. Fischer and R.G. Henzell, 1991. Genotypic variation in osmotic adjustment in grain sorghum. II. Relation with some growth attributes. Aust. J. Agric. Res. 42: 759-67.

Tarczynaski, M.C.; R.G. Jensen and H.J. Bohnert, 1993. Stress protection of transgenic tobacco by production of the osmolyte mannitol. Science 259: 508-10.

Vereyken, I.J.; V. Chupin; R.A. Demel; S.C. Smeekens et al. 2001. Fructans insert between the head groups of phospholipids. Biochem. Biophys. Acta. 1510: 307-20.

Yang, W.J.; A. Nadoska-Orczyk; K.V. Wood; D.T. Hahn et al. 1995. Near-isogenic lines of maize differing for glycinebetaine. Plant Physiol. 107: 621-30.

Yokota,A.; S. Kawasaki; M. Iwano and C. Nakamura, 2002. Citrulline and DRIP-1 protein (ArgE Homologue) in drought tolerance of wild watermelon. Ann. Bot. 89: 825-32.

Zhang, J.; H.T. Nguyen and A. Blum, 1999. Genetic analysis of osmotic adjustment in crop plants. J. Exp. Bot. 50(332): 291-302.

SUGGESTED READINGS

Hare, P.D.; W.A. Cress and J. van Staden, 1998. Disecting the roles of osmolyte accumulation during stress. Plant, Cell Environ. 21: 535-53.

Rontein, D.; G. Basset and A.D. Hanson, 2002. Metabolic engineering of osmoprotectant accumulation in plants. Metab. Eng. 4: 49-56.

Xiong, L. and J. K. Zhu, 2002. Molecular and genetic aspects of plant responses to osmotic stress. Plant, Cell Environ. 25: 131-39.

2

INORGANIC OSMOLYTES

Introduction

Several abiotic stresses including drought, salinity, and low temperature limit crop growth and productivity by imposing osmotic stress on plants. To overcome the stress, plant cells adjust their osmotic potential to prevent water stress, either by enhancing the uptake of inorganic ions from the soil, or by *de novo* synthesis of compatible solutes (sugars, amino acids, polyvols, quarternary amines) acting as osmolytes. The almost immediate changes in ion fluxes provide quick (a few minutes) osmotic adjustment while osmotic adjustment by synthesis of compatible solutes takes hours or days. Thus the role of membrane transport processes is of crucial importance in osmoregulation. Accumulation of inorganic ions in salt-stressed maize (Cerda et al., 1995) and barley (Huang and Redmann, 1995) genotypes has been noted to be sufficient for osmotic adjustment. There are two mechanisms by which plants sense the osmotic conditions, (a) change in cell volume could be sensed by mechanosensitive receptors on the plasma membrane, and (b) intracellular osmosensing mechanisms detect the degree of cytosol hydration. Shabala et al. (2000) made non-invasive measurements of H^+, K^+, Ca^{2+}, and Cl^- fluxes from broad bean (*Vicia faba)* leaf mesophyll in response to hyperosmosis, and noted the changes in plasma membrane potential. They provided data to support that the influx of K^+ and Cl^- observed in response to the hyperosmotic treatment (mannitol 150 mM) provides an adequate osmotic adjustment in broad bean mesophyll cells.

Further, Shabala and Lew (2002) observed that immediately after the onset of hyperosmotic stress (100/100 mM mannitol/sorbitol treatment), the cell turgor dropped from 0.65 to about 0.25 MPa. The turgor recovery started within 2-10 minutes after the treatment and was accompanied by a significant (30- 80 nmol m^{-2} s^{-1}) increase in uptake of K^+, Cl^-, and Na^+ by root cells. In most cells, almost complete (> 90% of initial values) recovery of the cell turgor was observed within 40-50 minutes after the onset of stress. The rate of recovery varies between different plant organs and cells. Some of them have a lag of about 5- 15 minutes, while in others the process starts almost immediately. The quickness of turgor recovery implies that this process is mediated by mechanisms other than biosynthesis of compatible solutes. Shabala and Lew present evidence to show that the rapid recovery is mediated by the increased uptake mechanisms of major osmotica (inorganic ions) present in the growth medium. They found that the uptake of K^+ was rapid and the flux normally stabilized at a new steady level, a few minutes after mannitol treatment and then exhibited a pronounced fluctuation around the new base line. Chloride and Na^+ flux were usually more delayed (up to 20- 30 minutes).

Net deposition rate of the total osmotica, cations, anions, and sugars (mmol kg^{-1} H_2O h^{-1}) increases from the base of the leaf to the most actively elongating location and then decreases near the end of the elongation zone (Hu and Schmidhalter, 1998). Contributions of the cations, anions, and sugars to osmotic adjustment varies with distance from the wheat-leaf-base, and are about 21- 30, 15- 21, and 13 percent, respectively, in the elongation zone. The absorbed ions are released from the xylem parenchyma into the xylem apoplast by simultaneous flow of cations and anions through channels, following electrical gradients set up by the ion uptake processes in the cortex, and possibly, the release and reabsorption of ions on their way to the xylem (Wegner and Raschke, 1994).

The calculations of Shabala et al. (2000) made on ion-specific mechanisms of osmoregulation in *Vicia faba* mesophyll cells show that influx of K^+ and Cl^- observed in response to hyperosmotic treatment (150 mM mannitol) provide an adequate osmotic adjustment, which suggests that the activity of plasma membrane transporters for these ions should be targeted to improve osmotolerance. The potassium rectifier channels are the primary targets in the mechanism of osmotic stress. Anion channels are known to play an important role in cell turgor and osmoregulation, and there are several reports suggesting that at least some of these anion channels are mechanosensitive.

Membrane potential measurements in root cells of *Arabidopsis* by Spalding et al. (1999) demonstrated that the *AKT1* component of the wild-type K^+ permeability was between 55 and 63 percent when external K^+ was between 10 and 1,000 µM, and NH_4^+ was absent. NH_4^+ specifically inhibited the non-*AKT1* component, apparently by competing for K^+ binding sites on the transporter(s). This inhibition by HN_4^+ had significant consequences for *akt1* plants : K^+ permeability, $^{86}Rb^+$ fluxes into roots, seed germination, and seedling growth rate of the mutant were each similarly inhibited by NH_4^+. Wild-type plants were much more resistant to NH_4^+. Thus, *AKT1* channels conduct the K^+ influx necessary for the growth of seedlings in conditions that block the non-*AKT1* mechanism. In contrast to the effects of NH_4^+, Na^+ and H^+ significantly stimulated the non-*AKT1* portion of the K^+ permeability. These results indicate that the *AKT1* channel is an important component of the K^+ uptake apparatus supporting growth, even in the "high-affinity" range of K^+ concentrations. In the absence of *AKT1* channel activity, an NH_4^+–sensitive, Na^+/H^+–stimulated mechanism can suffice.

High-affinity K^+ uptake is an essential process for plant nutrition under K^+-limiting conditions (Angeles Martinez-Cordero et al., 2005). Pepper (*Capsicum annuum)* plants grown in the absence of NH_4^+ and starved of K^+ show an NH_4^+ -sensitive high-affinity K^+ uptake that allows plant roots to deplete external K^+ to values below 1 µM. When plants are grown in the presence of NH_4^+, high-affinity K^+ uptake is not inhibited by NH_4^+ . Although NH_4^+ -grown plants deplete external K^+ below 1 µM in the absence of NH_4^+, when 1 mM NH_4^+ is present they do not deplete external K^+ below 10 µM. A K^+ transporter of the HAK family, CaHAK1, is very likely mediating the NH_4^+ -sensitive component of the high-affinity K^+ uptake in pepper roots. *CaHAK1* is strongly induced in the roots that show the NH_4^+-sensitive high-affinity K^+ uptake and its induction is reduced in K^+-starved plants grown in the presence of NH_4^+. The NH_4^+ -insensitive K^+ uptake may be mediated by an AKT1-like K^+ channel.

Su et al. (2001) characterized transcripts for three K^+ channel homologs in the AKT/KAT subfamily (Shaker type) from the common ice plant (*Mesembryanthemum crystallinum*), with a focus on their expression during salt stress (up to 500 mM NaCl). *Mkt1* and 2, *Arabidopsis* AKT homologs, and *Kmt1*, a KAT homolog, are members of small gene families with two to three isoforms each. *Mkt1* is root specific; *Mkt2* is found in leaves, flowers, and seed capsules; and *Kmt1* is expressed in leaves and seed capsules. *Mkt1* is present in all cells of the root, and in leaves a highly conserved isoform is present in all cells with highest abundance in the vasculature. MKT1 for which antibodies were made is localized to the plasma membrane. Following salt stress, MKT1 (transcripts and protein) is drastically down-regulated, *Mkt2* transcripts do not change significantly, and *Kmt1* is strongly and transiently (maximum at 6 hour) up-regulated in leaves and stems. The detection and stress-dependent behavior of abundant transcripts representing subfamilies of K^+ channels provides information about tissue specificity and the complex regulation of genes encoding K^+ uptake systems in a halophytic plant.

Drought preconditioning

Jiang and Huang (2001) tried to integrate the effects and responses of drought pre-conditioning,* heat tolerance, osmotic adjustment and root growth acceleration of *Poa pratensis* . Compared with non-preconditioned plants, the drought-preconditioned plants had 17 and 48 percent higher

*Plants were subjected to two 14-day cycles of soil drying (drought preconditioning) and rewatering (14 days in the first cycle and 17 days in the second cycle).

osmotic adjustment at 14 and 21 days of heat stress (35°C/30°C day/night), respectively. Total ion (K^+, Ca^{2+}, Na^+, Mg^{2+},Cl^-, and P) concentration of cell sap increased during heat stress and was 11 and 16 percent higher in preconditioned plants than in nonpreconditioned plants. The concentration of K^+ accounted for 59 to 65 percent of the total ion solutes in both groups of plants during heat stress. The soluble carbohydrate content of leaves increased during heat stress and was about 21 and 44 percent higher in drought-preconditioned plants at 14 and 21 days, respectively (Table 2.1). Root growth in the deeper soil layer (40 – 60 cm) was accelerated in drought preconditioned plants as a result of increased ABA synthesis. The drought preconditioned plants had 100 percent higher root dry weight than the non-preconditioned plants in the 20 – 40 cm soil layer.

Table 2.1 Ion concentrations in cell sap of drought-preconditioned (Pre) and non-preconditioned (Non) Kentucky Bluegrass exposed to 7, 14, and 21 days of heat stress (After Jiang and Huang, 2001).

Days of heat stress	Treatment	← Ion concentration →						
		K	Ca	Na	Mg	Cl	P	Total ions
		← (mmol L^{-1}) →						
0	Non	108.6a*	8.7a	0.88a	9.1a	13.9a	14.4a	155.6a
	Pre	108.4a	7.5a	0.99a	7.6a	14.0a	12.6a	155.1a
7	Non	98.4b	10.1b	0.76a	10.6a	15.5a	18.5a	153.9b
	Pre	117.0a	13.3a	0.75a	11.2a	17.3a	19.4a	178.9a
14	Non	115.3b	17.5b	1.00b	13.8a	17.0b	21.8b	186.4b
	Pre	126.3a	21.9a	1.43a	13.8a	20.5a	26.4a	210.3a
21	Non	95.8b	14.1b	0.87b	11.4a	12.0a	23.6a	159.8b
	Pre	103.4a	18.3a	1.78a	12.5a	14.2a	26.5a	176.9a

- Means followed by the same letters within a column at a given day of treatment were not significantly different based on LSD test (P=0.05).

The drought preconditioned *Poa pratensis* plants showed about 8–19 percent higher level of potassium than the non-preconditioned plants during heat stress (Jiang and Huang, 2001). The accumulated K^+ is the major ion solute contributing to osmotic adjustment in sorghum (Premchandra et al., 1995) and other crops. The pattern of the increased level of Ca^{2+} is similar to that of K^+ but its concentration is less than K^+ in both preconditioned and nonpreconditioned plants. Both K^+ and Ca^{2+} regulate guard cell turgor and stomatal aperture, and the accumulation of K^+ and Ca^{2+} contribute to increased osmotic adjustment during heat stress following drought preconditioning. Accumulation of cytosolic free Ca^{2+} also has been observed during heat shock which alleviates heat injury. Application of Ca^{2+} to leaves also increases heat tolerance by regulating Ca^{2+} concentration of cell sap. Concentrations of P and Cl^- increase under heat stress, however, the accumulation of these two ions only shows up to a 14 day of heat stress (Jiang and Huang, 2000). The drought-preconditioned plants have about 43 percent and over 100 percent higher level of Na^+ on 14 and 21 days of heat stress, respectively, but its contribution to total ion accumulation is less than 1 percent. The results indicate that Na^+ ,P and Cl^- are less important for osmotic adjustment than the K^+ and Ca^{2+} accumulation.

Energetic efficiency

The energetic cost of osmotic adjustment using inorganic ions is much lower than that using the organic molecules synthesized in plant cells (Hu and Schmidhalter, 1998). The overall contribution of inorganic ions to the cell osmotic adjustment is estimated to be 50 – 60 percent for wheat plants. Cerda et al. (1995) concluded that accumulation of inorganic ions is sufficient for osmotic adjustment in salt-stressed maize genotypes, and that no single organic solute appears to be important in this process. The total contribution of K^+ and Cl^- uptake in the cell osmotic adjustment is -0.32 MPa. Nearly 84 percent of the total change in cell osmotic potential caused by hyperosmotic stress is thus compensated by the uptake of these two inorganic ions in one hour (Shabala et al., 2000).

There are at least two major advantages to plants by using inorganic ions for cell osmotic adjustment. One of them is the rapidity of turgor recovery and the other is the low energy cost. For the synthesis of compatible solute(s) in required amount, longer time is required in hours or even days. This is the crucial time for cell metabolism, especially if the stress is acute. Although synthesis of compatible solutes may have physiological significance for a slowly developing stress, uptake of inorganic ions is the only way to provide fast and efficient osmotic adjustment. The energetics of osmotic adjustment is another important issue to be considered. Generating enough organic solutes to achieve full osmotic adjustment under hyperosmotic conditions can be a costly exercise. The difference in ATP cost between active uptake and compartmentation of inorganic ions and synthesis of compatible solutes is approximately a factor of 10. It is reasonable to suggest, therefore, that the cheapest option (ion uptake) might be getting first preference (Shabala and Shabala, 2002). Thus it is important/urgent to identify genes that control important plant traits related to mineral nutrient acquisition and abiotic stress tolerance, in order to facilitate crop improvement for stress environments. The limited literature available has been discussed at the end of this chapter.

THE DOMINATING ROLE OF POTASSIUM ION

Potassium contributes up to 6 percent of plant dry weight and serves as a major inorganic osmolyte, thus is crucial for cell osmoregulation and turgor maintenance, cell expansion, and stomatal function. Potassium is a cofactor of several enzymes and thus, plays a key role in photosynthesis, protein synthesis, and oxidative metabolism. Potassium also affects phloem transport and provides charge balance during ion transfer across cellular membranes. The epidermal cells are virtually unable to produce organic solutes and rely heavily on inorganic ions, namely K^+ , for osmotic adjustment. However, in mesophyll cells, the total contribution of organic solutes is much higher (20 – 30 percent) than in epidermal cells, although K^+ remains the dominant osmoticum (Shabala et al., 2000; Gonzales et al., 2002).

Stomatal movement

Potassium ions greatly influence stomatal opening and closure. In plasma membrane of guard cells, two major types of potassium channels are present : voltage-dependent K^+-selective inward and outward rectifying channels. The K^+ –selective inward channels are activated by membrane hyperpolarization and mediate stomatal opening, whereas the K^+-selective outward rectifying channels are opened by voltages more positive than E_k and mediate stomatal closure. Similar inward and outward K^+ channels exist in both the cortex and stele of maize roots with respect to their selectivity, activation kinetics, and single channel conductance. However, although the same channel types appear to be present in both the cortex and stele of maize roots, the channels are regulated directly. This could reflect the fact that either different signal transduction pathways exist in the cortical and steler cells or that the channels themselves are different with respect to their sensitivity to signaling compounds (Roberts, 1998).

Decrease in stomatal aperture is accompanied by volume and turgor loss in stomatal guard cells bordering the pore. This turgor loss is achieved through net K^+ loss from guard cells. In turgid guard cells, vacuole is the dominant cellular compartment and, therefore, a high proportion of the K^+ released on closure must originate from the vacuolar lumen. The vacuolar K^+ release occurs independently of the identity of the closing stimuli and signal transduction pathway, a point exemplified for ABA-induced closure which can be mediated by Ca^{2+} –dependent or Ca^{2+} -independent (pH cyt-dependent) signalling (Allen et al., 1998). Vacuolar K^+ channels are involved in K^+ release and in this study two K^+ –selective channels were identified in *Vicia faba* guard cell vacuoles that differ in their regulation by cytosolic free Ca^{2+} and cytosolic pH.

In *Zea mays*, the stomatal complex consists of two flanking subsidiary cells in addition to the two guard cells with the pore in between. Two K^+ –selective channel types have been identified in the plasma membrane of subsidiary cells (Majore et al., 2002). These channels act as hyperpolarization-dependent K^+ uptake and depolarization-activated K^+ -release channels during stomatal movement. The plasma membrane of subsidiary cells and guard cells has to be inversely polarized in order to achieve the anti-parallel direction of K^+ fluxes between these cell types during stomatal movement.

Using whole-cell patch-clamp technique, Li et al. (1994) demonstrated that inward K^+ currents in guard cells are inhibited by nanomolar concentrations of okadaic acid or calyculin (inhibitors of protein phosphatases 1 and 2A), whereas outward K^+ currents are not affected. However, the same inhibitors enhance the magnitude of outward K^+ currents in mesophyll cells. These findings suggest that protein phosphatases 1 and/or 2A play different physiological roles in modulating the activity of K^+ channels in mesophyll cells and guard cells.

Cortex and epidermis

In the cortex and epidermis, inward K^+ channels of plant roots mediate low-affinity K^+ uptake from the soil solution. Water stress and ABA have no significant effect on the K^+ channels and the pathway for low affinity K^+ uptake. In the stele, K^+ channel activity underlying the time dependent outward current is significantly reduced by both water stress and ABA (Roberts, 1998). Root steler cells regulate the ionic composition of the transpiration stream. The outward K^+ channels in these cells represent a pathway for the efflux of K^+ from the root symplasm into the xylem apoplast. The inward K^+ channels may be involved in the reabsorption of K^+ from the xylem apoplast. Thus, water stress and ABA modify the permeability of the plasma membrane of root steler cells in favor of K^+ influx and against K^+ efflux. However, the regulation of K^+ channels by themselves only affects the capacity for passive transport of K^+. Potassium transport via ion channels also depends on membrane potential and cytosolic and extracellular K^+ concentrations (Roberts,1998). Thus, the plasma membrane hyperpolarization is necessary for increased passive K^+ uptake via K^+ channels in the steler cells.

Root steler cells

Several studies have shown that ABA and water stress promote K^+ accumulation by root steler cells; ABA reduces K^+ efflux from the steler symplasm without any effect on K^+ uptake by the root cortex (Pitman and Wellfare,1978). ABA inhibits ion efflux in low-salt roots, whereas in high-salt roots ABA promotes ion efflux. Potassium accumulation by roots is part of an important adaptation of the plant, allowing it to survive drying soils. An early response of plants to water stress is an inhibition of shoot growth while root growth is maintained. Thus the ions transported from the root could accumulate in the apoplast of leaves, lowering the extracellular osmotic potential and decreasing the water deficit of leaf cells.

Abscisic acid regulates K^+ channels in two ways; one, via expression and/or incorporation of channel proteins in the plasma membrane, and two, by regulating the activity of channel proteins after incorporation into the plasma membrane. Application of ABA to steler protoplasts decreases the outward current in approximately 55 percent of steler cells, consistent with ABA regulation of the activity of the outward K^+ channels after their incorporation into the plasma membrane (Roberts, 1998). It is noteworthy that ABA regulation of K^+ channel activity in maize root stele is opposite to that observed in guard cells.

Root hairs

Root hairs constitute an important site of nutrient uptake from the soil. Patch-clamp studies of Gassmann and Schroeder (1994) showed activation of a large inward current carried by K^+ ions into root hairs at membrane potentials more negative than –75 mV. This K^+ influx current is mediated by hyperpolarization-activated K^+–selective cations of $K^+ > Rb^+ \approx NH_4^+ > Na^+ \approx Li^+ > Cs^+$. The authors conclude that inward-rectifying K^+ channels in root hairs can function as both a physiologically important mechanism for low-affinity K^+ uptake and as regulators of membrane potential.

Phloem loading

Phloem loading with assimilates is accompanied by a significant increase in symplastic K^+ concentration, required to maintain electrical neutrality during vectorial H^+ transfer.The most abundant K^+ channels in the phloem tissue are AKT3 (Marten et al., 1999) and their homologues. Another major type of K^+ channel detected in minor veins is KAT2 involved in K^+ loading into the phloem sap.

Vascular tonoplast

In the vascular tonoplast, several types of K^+ permeable channels are present. The most abundant are slow-activating and fast-activating vascular channels. The slow-activating vascular channel is permeable to both mono- and di-valent cations and is activated by cytosolic Ca^{2+} and positive vacuolar voltage. The fast-activating vacuolar channel is selective for monovalent cations only,activated by positive voltages, and may be blocked by divalent cations (Shabala, 2003).

Chloroplast

In chloroplasts, the K^+ transport barrier is the inner membrane, which contains transporters for a selected number of low molecular weight substrates. However, the outer membrane contains specific pore-forming proteins and is permeable to substances with molecular weights of several kDa. Most of these "pores" are also able to conduct ions. In mitochondria, similar studies on K^+ transporters are still at an early stage.

N-metabolism

Potassium also plays a role in the transfer of nitrate from the roots to the shoots and leaves. Without adequate potassium, nitrate accumulates in the roots and a feedback mechanism to the root cells stops further nitrate uptake (Fig. 2.1). Consequently, nitrate remains in soil at risk to loss to the environment, either when leached into surface and ground water or

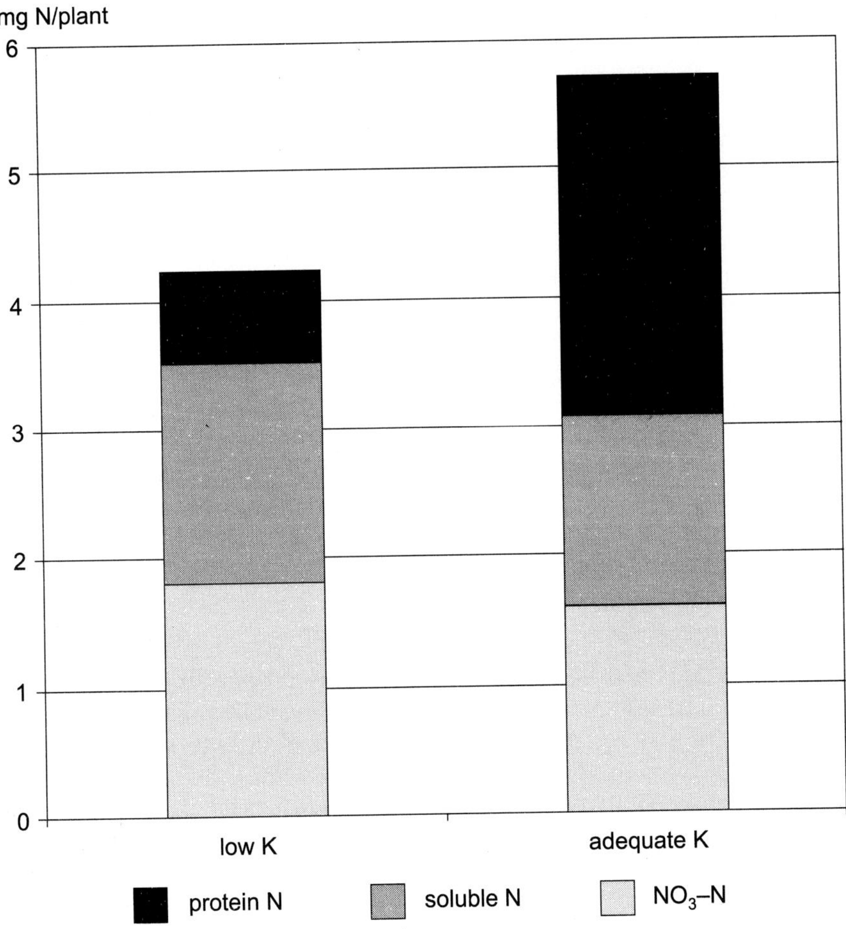

Fig. 2.1 N-metabolism of tobacco plants as affected by potassium supply (After Koch and Mengel, 1974).

denitrified and lost to the atmosphere as nitrogen gas or nitrous oxide. In other words, too little potassium decreases the efficiency with which N fertilizers are used. Nitrate in the plant is reduced first to amines and then incorporated into amino acids to ultimately form protein. With an inadequate potassium supply, the plant not only absorbs less nitrate from the soil solution, but less is converted into protein. Consequently, nitrate accumulates in the shoot and the protein content remains low (Fig. 2.1). Both factors, nitrate and protein content are important contributors to crop quality.

Further, potassium plays a major role in the transport of assimilates from the leaves to the roots for them to function properly and also to storage organs like tubers and grains. Thus, plants with too little potassium retain most of the low molecular assimilates, predominantly sugars, in the leaves and roots. As a consequence of the accumulation of low molecular weight carbohydrates in the leaves of plants with an inadequate potassium supply, it substantially decreases the efficiency of conversion of solar energy into assimilates and excessive electrons released from the chloroplasts lead to the formation of highly reactive oxygen radicals (ROS) (Cakmak, 2002).

Energy conversion, carbohydrate formation and translocation, N-metabolism and other metabolic processes in plants are controlled by enzyme systems, some 50 of which are activated by potassium. This again points to the versatile and multiple role, potassium plays in plants. In order to function properly, the metabolic processes require an appropriate hydration and acidity of plant cells. Potassium again is involved in both processes. It helps to adjust the osmotic potential in plant cells in the roots, to enable them to absorb water from the soil. In the whole plant, potassium activates the movement of water and solutes, it regulates stomatal opening and thus, intake of carbon dioxide, and the loss of water to the atmosphere; it affects the pH of plant cells.

Electrical properties

Dietrich et al.(1998) compared the electrical properties of an inward rectifying guard cell K^+ channel in *Vicia faba, Solanum tuberosum* and *Nicotiana tabacum* using patch-clamp method. The channel density in *S. tuberosum* was higher than in *V. faba* and *N. tabacum* while the activation and deactivation kinetics were faster in the latter two species. Among different monovalent cations, K^+ channels discriminated strongly against Na^+, Li^+, and Cs^+. Extracellular Ca^{2+} blocked the *V. faba* K^+ channel at concentrations greater than or equal to 1 mM but only affected its functional homologs in *S. tuberosum* and *N. tabacum* at higher concentrations and more-negative membrane potentials. Like the differences in Ca^{2+}-sensitivity, protoplasts from the 3 species differed remarkably in their response towards extracellular pH changes. The observed differences in cation-sensitivity and voltage-dependent kinetics between K channels reflect the diversification of guard-cell channels that may contribute to species-specific variations in the control of stomatal aperture.

Genotypic variability

Earlier, Morgan (1992) studied genotypic differences in osmoregulation (K^+) in water-stressed wheat breeding lines (sap collected from flag leaves of breeding lines with high and low osmoregulation). The lines showed either very low (0.1 MPa) or very high (1.35 MPa) levels of solute accumulation. These levels were similar to a single gene effect observed in other lines. High osmoregulation was largely (78 percent) due to K^+ accumulation, with amino acids the only other important contributor (22 percent). The osmotic potential due to the solutes' K^+, amino acids and sugars accounted for 90 percent of the sap osmotic potential. The major components were K^+ (42-53 percent) and amino acids (35-43 percent). In high lines, increases in the proportion of K^+ with increased stress were associated with changes in the relationship between sap osmotic potential and either tissue osmotic potential or that derived from the component solutes.

K^+-uptake mechanism

There is a rapid up-regulation of high-affinity K^+ uptake when the exogenous K^+ supply is interrupted. Under these conditions, root K^+ concentrations are rapidly depleted by K^+ translocation to the shoot and also by root growth (Walker et al., 1996). Likewise, when K^+ uptake is limited by pruning the root system or by restricting the K supply to certain regions of the root, high-affinity K^+ uptake increases to compensate for the reduced absorptive surface. The elevated rates of high-affinity K^+ influx observed in K-deprived plants are rapidly down-regulated when K^+ is supplied (Fernando et al., 1990). In both barley and wheat roots, a rapid and large up-regulation of *HKT1* mRNA levels resulted when K^+ was withdrawn from the growth media (Fig. 2.2; Wang et al., 1998). This effect was specific for K^+ ; withholding N caused a modest reduction of *HKT1* mRNA levels. Up-regulation of *HKT1* transcript levels in barley roots occurred within 4 hours of removing K^+. Resupply of 1 mM K^+ was sufficient to strongly reduce *HKT1* transcript levels. The rapid up-regulation of *HKT1* mRNA levels in response to K^+ withdrawal provides a potent tool to further investigate the functional and molecular basis of K^+ sensing in plants.

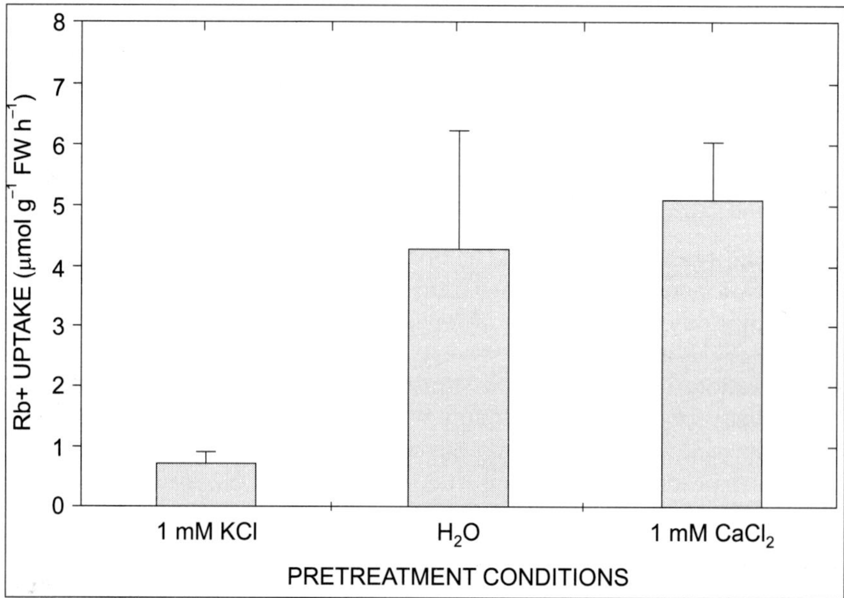

Fig. 2.2 High affinity K^+ (Rb^+) uptake in wheat roots in response to K^+ deprivation. High-affinity uptake was suppressed in roots grown in 1 mM KCl. In K^+–depleted roots grown hydroponically in H_2O or in 1 mM $CaCl_2$, high-affinity Rb^+ uptake was induced. Average uptake rates from 10 ($CaCl_2$) or 3(H_2O) replicate experiments are illustrated. The [Rb^+] was 10 mM. FW, Fresh weight. (After Wang et al., 1998).

Buschmann et al. (2000) investigated the effects of K^+ starvation on Na^+ and K^+ uptake mechanisms in the plasma membrane of wheat root cortex cells using the patch-clamp technique. Unexpectedly, K^+ starvation of wheat seedlings enhanced the magnitude and frequency of occurrence of time-dependent inward-rectifying K^+ channel currents. TaAKT1 (a cDNA coding for a wheat root K^+ channel homolog) mRNA levels were up-regulated in roots in response to withdrawal of K^+ from the growth medium. Furthermore, K^+ starvation caused an enhancement of instantaneous Na^+ currents.

SODIUM : POTASSIUM UPTAKE

Potassium is the most abundant cation in plants and is crucial for plant nutrition, growth, tropisms, enzyme homeostasis and osmoregulation. Potassium accumulation is rate-limiting for agricultural production, and its uptake from soil into roots is largely mediated by high affinity K^+ uptake. The transport mechanism of *HKT1* (a complementary DNA encoding a membrane protein) uses K^+ – H^+ co-uptake. Expression of *HKT1* is localized to specific root and leaf regions which represent primary sites for K^+ uptake in plants (Schachtman and Schroeder, 1994). The wheat root high affinity K^+ uptake transporter (*HKT1*) functions as a high affinity K^+ – Na^+ co-transporter (Rubio et al., 1995). High affinity K^+ uptake is activated by micromolar Na^+ concentrations; moreover, high-affinity Na^+ uptake is activated by K^+. However, at physiologically detrimental concentrations of Na^+ , K^+ accumulation mediated by *HKT1* is blocked and low-affinity Na^+ uptake occurs, which correlates to Na^+ toxicity in plants. Sodium uptake and Na^+ inhibition of K^+ accumulation indicate a possible role for *HKT1* in physiological Na^+ toxicity in plants.

Roots have at least two distinct pathways with high and low affinities to accumulate K^+. The system for high affinity K^+ uptake, which takes place against the electrochemical K^+ gradient, requires direct energization. Energization of K^+ uptake via Na^+ coupling has been observed in wheat and barley plants (Maathuis et al., 1996). Wheat and barley plants are able to sustain growth and K^+ uptake in absence of Na^+. Furthermore, the addition of Na^+ is either without effect or inhibits K^+ absorption. Thus, Maathuis et al. suggest that Na^+ –coupled K^+ transport has no or limited physiological relevance in wheat and barley, but in certain aquatic angiosperms and algae, this type of secondary transport energization plays a significant role. However, in red beet cultivars Klein Bol and Ruby Queen, Subbarao et al. (2000) observed that reducing potassium levels from 5.0 to 0.10 mM quadrupled the Na^+ uptake, and lamina sodium levels reached 20g kg^{-1} dry weight. Lamina potassium levels decreased from 60g kg^{-1} dry weight at 5.0 mM K to 4.0g kg^{-1} dry weight at 0.10 mM K. Ruby Queen and Klein Bol responded differently to these changes in Na and K status. Klein Bol showed a linear decline in dry matter production with a decrease in available K, whereas for cultivar Ruby Queen, growth was stimulated at 1.25 mM K and relatively insensitive to further decreases of K down to 0.10 mM.

THE INDIRECT ROLE OF CALCIUM

Ionic calcium stimulates accumulation of the osmolyte, proline, under extreme temperature (cold or heat)-stress in tomato cultured cells and seedlings (De et al., 1996). De et al. observed reduction in proline accumulation when tomato seedlings were pretreated with EGTA (a Ca^{2+} chelator) or $LaCl_3$ (0.05 mM)(a Ca^{2+} channel blocker) at 4°C for 4 or 8 hours (Fig. 2.3). The 8 hour pretreated seedlings showed more proline content than that pretreated for 4 hours. Pretreatment with EGTA and $LaCl_3$ most likely caused an initial Ca^{2+} chelating and channel-blocking effects and masked the Ca^{2+} -induced proline accumulation effect, which might have slightly recovered in time (4 to 8 hours) due to adaptation. On the other hand, when calcium ($CaCl_2$) was added, proline accumulation increased to a large extent. Further, in the contrasting experiment with heat shock (45°C for 4 or 8 hours) with tomato seedlings, almost similar trends of EGTA and $LaCl_3$ induced reduction in proline accumulation and $CaCl_2$ –induced enhancement of proline accumulation were observed (Fig. 2.4).

Under cold stress the membrane-bound Ca^{2+} is transported into the cytosol. Umashankar (1985) and Suresh et al. (1991) have also reported involvement of Ca^{2+} in moisture stress, NaCl stress and ABA induced accumulation of proline in *Eleucine coracana.* Calcium acts as a mediator in both high- and low-temperature-induced accumulation of proline. EGTA acts by chelating the divalent charges on calcium ion and $LaCl_3$ by competitive binding to calcium binding site (Umashankar, 1985) preventing

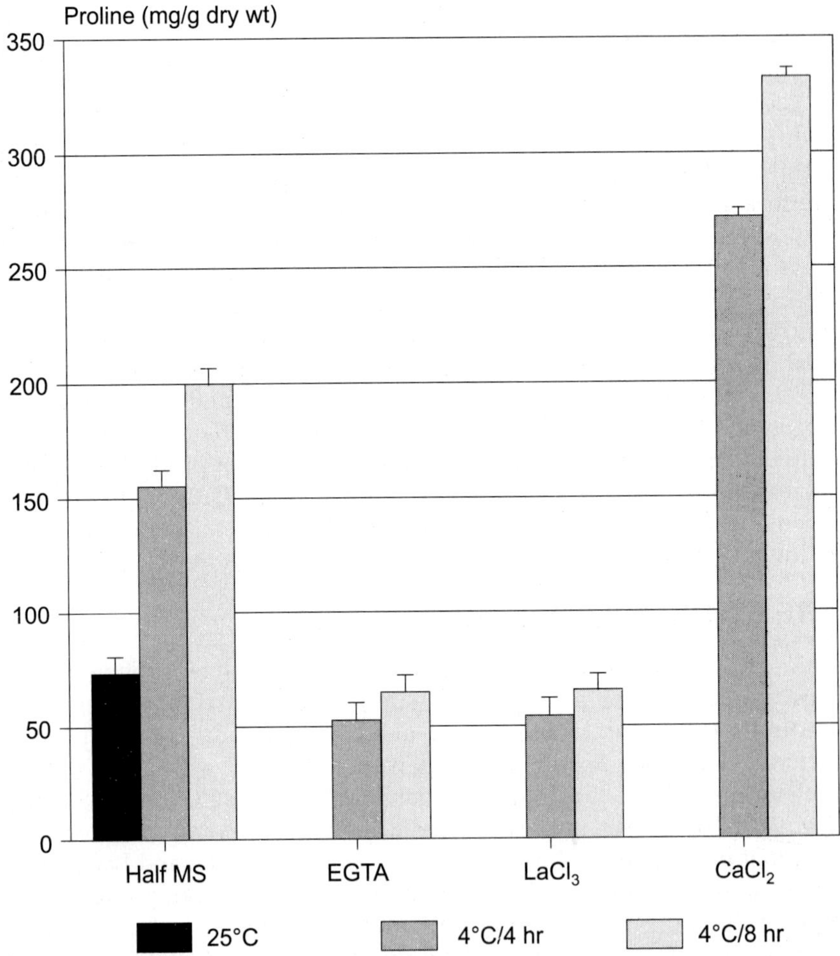

Fig. 2.3 Cold shock induced proline accumulation in seedlings of tomato under the influence of EGTA, $LaCl_3$ and $CaCl_2$. Error bars represent ± SE. (After De et al., 1996).

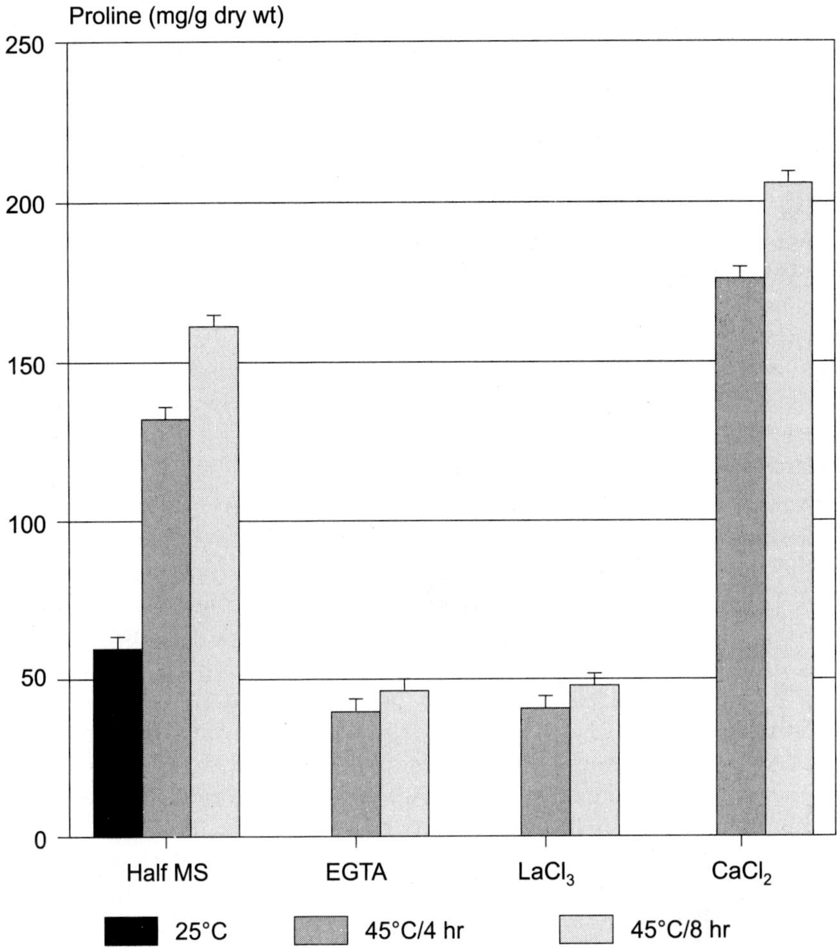

Fig. 2.4 Heat shock induced proline accumulation in seedlings of tomato under the influence of EGTA, $LaCl_3$ and $CaCl_2$. Error bars represent ± SE. (After De et al., 1996).

the entry of displaced Ca^{2+} into the cytoplasm of stressed cells. Liu and Zhu (1998) suggested that intracellular calcium signaling through a calcineurin-like pathway mediates the beneficial effect of calcium on plant salt tolerance.

ION UPTAKE AS AFFECTED BY DIFFERENT TYPES OF STRESSES

Drought

Under hyperosmotic stress (150 mM mannitol added after 5 min) to *Vicia faba* plants, net Ca^{2+} flux exhibited complex transit behavior (Shabala et al.,

2000, Fig. 2.5). in both K^+ –free variants (3 types of solutions used : A= 0.1 mM $CaCl_2$ + 1 mM KCl; B= 0.1 mM $CaCl_2$, and C = 0.1 mM $CaSO_4$), the net Ca^{2+} flux remained positive (inward directed, Fig. 2.5 B,C), while in the KCl + $CaCl_2$ bath, initial flux was followed by a noticeable flux starting at about 5 minutes after mannitol application and lasting for another 30 minutes. The most dramatic was the difference in the steady-state K^+ fluxes, with the net K^+ uptake in the "plus K^+ " solution (Fig. 6.A), and significant net K^+ eflux for both K^+–free variants (Fig. 2.6 B, C). In spite of this difference in the steady-state K^+ values, however, a large shift towards net K^+ influx in response to hyperosmotic stress was evident in all variants. In both $CaCl_2$ and $CaSO_4$ treatments, net K^+ flux quickly reached the zero level and oscillated around it (Fig. 2.5 B, C), while in the "plus K^+" solution a substantial net K^+ influx was observed (Fig. 2.5 A). Hyperosmotic stress also induced net Cl^- uptake into the mesophyll cells; a significant shift towards net Cl^- influx was observed.

Plants use complex signal transduction pathways to perceive and react to various abiotic stresses. As a consequence of this signaling, plants modify their metabolism to adapt themselves to the new conditions. Sadiqov et al. (2002) studied drought and salinity-induced proline accumulation and the roles of Ca^{2+} (10 mM) and indoleacetic acid (0.3 mM) in this response. Subjecting wheat seedlings to both drought (6% PEG) and salinity (150 mM $NaCl_2$) stress resulted in a dramatic increase in proline accumulation (7-fold). However, application of Ca^{2+} prevented the drought-induced proline accumulation indicating that these stress factors had different effects. Unlike salinity stress, Ca^{2+} prevented the drought-induced proline accumulation indicating that these stress factors use distinct signaling pathways to induce similar responses. Experiments with IAA supported this interpretation and suggested that Ca^{2+} and IAA participate in signaling mechanisms of drought-induced proline accumulation. Drought- and salt stress-induced proline accumulation was compared on salt resistant (cv Gerek 79) and salt sensitive (cv Bezostaya) wheat varieties. Although proline level of the resistant cultivar was 2-fold lower than that of the sensitive cultivar as control, relative proline accumulation was dramatically higher in case of the salt resistant cultivar under the stress condition.

Potassium channel activity in guard cells is known to be regulated by ABA (Luan, 2002). Biosynthesis of ABA in both root and shoot tissues is significantly enhanced by soil drying and/or salinity. Molecular and elecrophysiological techniques have identified a number of discrete transporters which are involved in the translocation of K^+ from the soil solution to the shoots of higher plants. Furthermore, ABA regulates K^+ channel activity in roots which suggests that ABA regulation of K^+ transport in roots is, at least in part, ion channel-mediated (Roberts and Snowman,

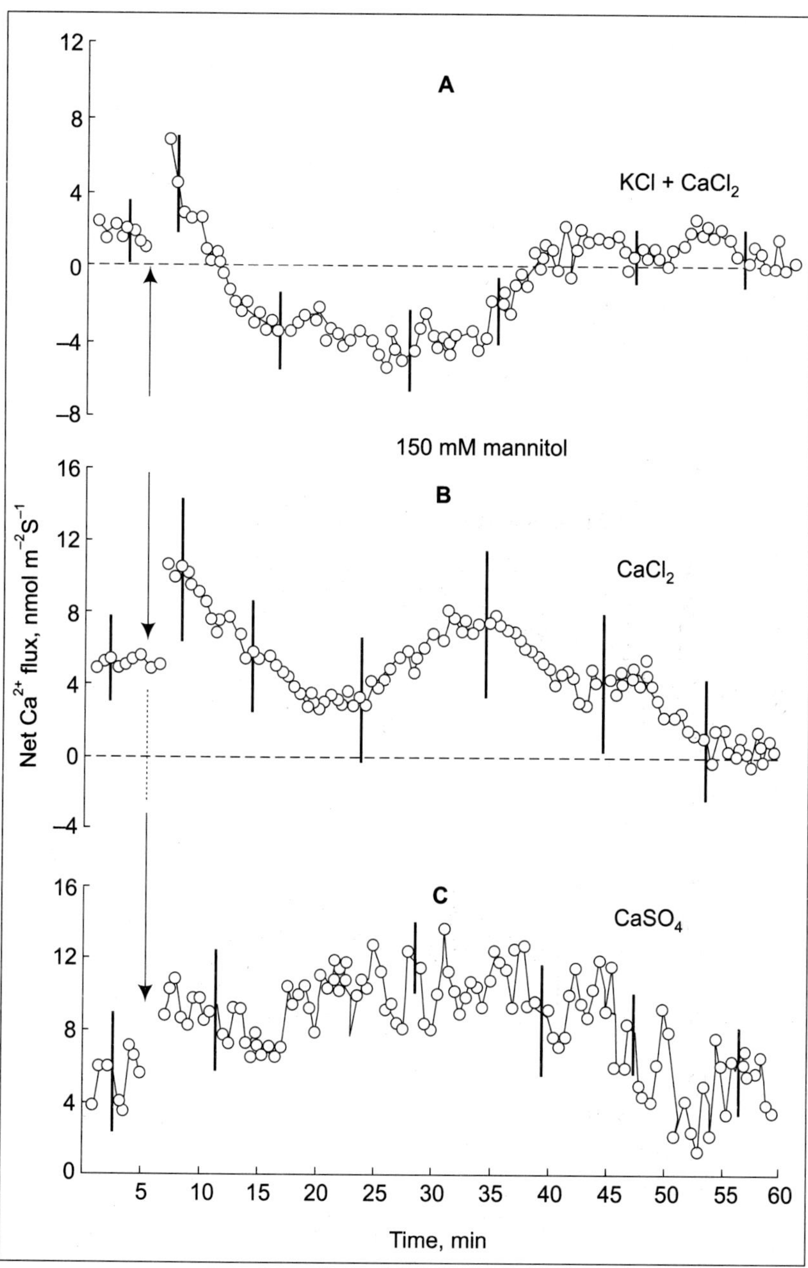

Fig. 2.5 Transient changes in the net Ca^{2+} flux measured from the bean mesophyll tissue in response to hyperosmotic stress (150 mM mannitol added at 5 min, A= 0.1 mM $CaCl_2$ + 1.0 mM KCl solution, B= 0.1 mM $CaCl_2$ solution, C=0.1 mM $CaSO_4$ solution). Number of replicates is 6, 6, and 5 for A, B, and C, respectively. (After Shabala et al., 2000).

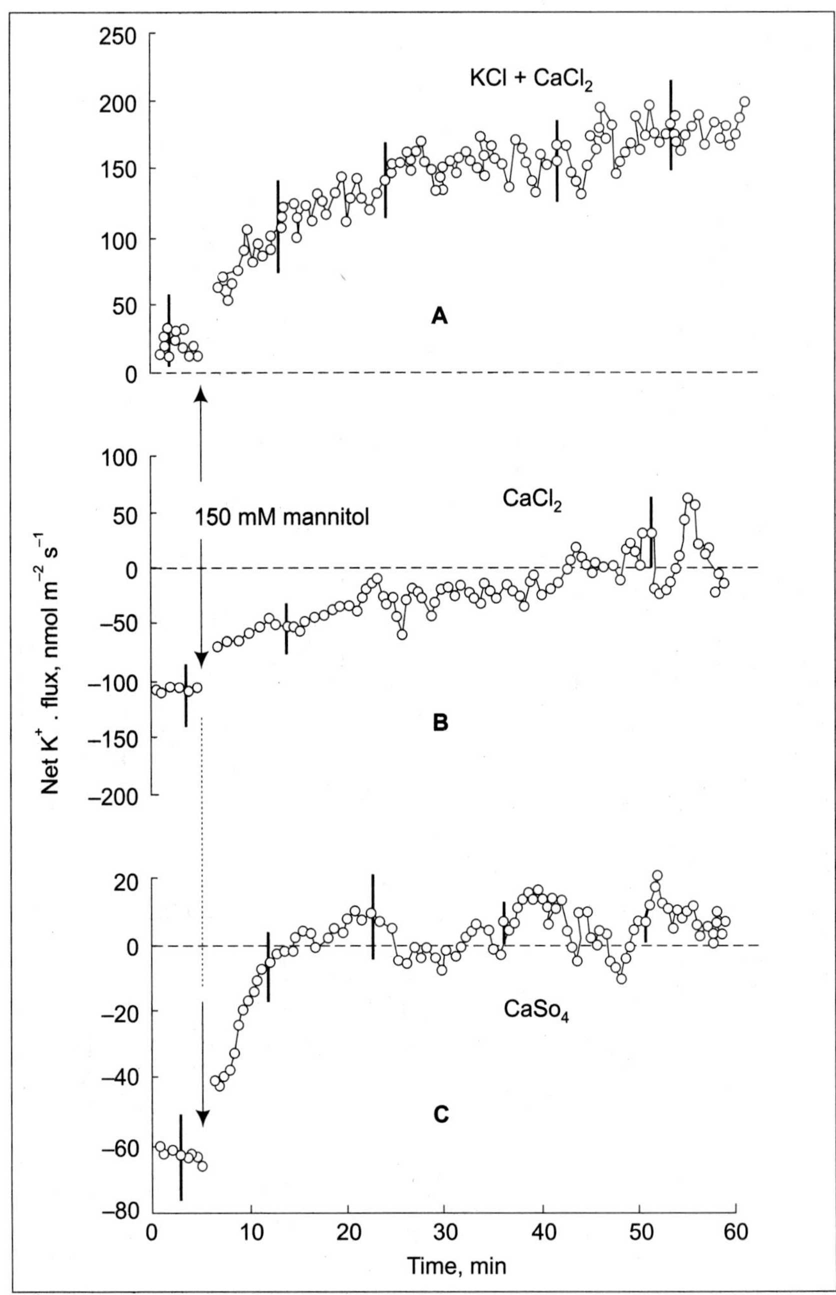

Fig. 2.6 Transient changes in the net K^+ flux measured from the bean mesophyll tissue in response to hyperosmotic stress. All the details are as in Fig. 2.5. Number of replicates is 8, 5, and 4 for A, B, and C, respectively. (After Shabala et al., 2000).

2000). Roberts and Snowman found that ABA regulated the membrane potential of root cells and that this regulation is consistent with the hypothesis that ABA-induced K^+ accumulation in roots is mediated by K^+ channels.

Salinity

Accumulation of inorganic ions, Na^+ , K^+ and Cl^-, in the leaf tissues of maize cultivars is considered sufficient for osmotic adjustment, and no single organic solute appeared important for osmoregulation (Cerda et al., 1995). The experiments of Wu et al. (1996) further demonstrate that K^+ acquisition is a critical process for salt tolerance of glycophytic plants. Li and Lin (2000) concluded that *Bruguiera gymnorrhiza* seedlings cultured at high salinity accumulated more metal ions, which enhanced osmotic adjustment of plants, and kept highly selective absorption of Na^+/K^+. These characteristics are the main physiological mechanisms of adaptation to high salinity, while suitable Ca^+/Na^+ ratio in soil is also one of the important causes of survival under high salinity. Carden et al. (2003) reported that the salt tolerant barley variety maintained a 10-fold lower cytosolic Na^+ than the more sensitive variety, although by the 8th day the two varieties were not significantly different. At this time, the more tolerant variety was better at maintaining root cytosolic K^+ in the high-NaCl background than was the more sensitive variety. Under salt stress, plants maintain a high concentration of K^+ and a low concentration of Na^+ in the cytosol. Plants do this by regulating the expression and activity of K^+ and Na^+ transporters and of H^+ pumps that generate the driving force for transport (Zhu, 2003). A protein kinase complex consisting of the myristoylated calcium-binding protein SOS_3 is activated by a salt-stress-elicited calcium signal. The protein kinase complex then phosphorylates and activates various ion transporters, such as plasma membrane Na^+/ H^+ antiporter SOS_1 . In the same year Fernandez-Ballester et al. (2003) reported that in *Citrus macrophylla* (root stock), Cl^- and Na^+ uptake rates were high at the beginning of the saline treatments, but low at the end of the 14-day experiment. In contrast, sour orange (*C. aurantium)* showed high uptake rates at the beginning and end of the experiment. In response to increasing salinity, root and shoot concentrations of Cl^- and Na^+ increased in sour orange, but not in *C. macrophylla*. Different loading characteristics of Cl^- and Na^+ were observed between young and old segments of the root system. In general, old root segments reached quasi-steady-states later than young root segments. Thus the commercial and the root stock orange species have different regulatory mechanisms for uptake and transport of Cl^- and Na^+ .

The thermodynamic calculations show that K^+ is actively transported into the vacuoles of both epidermal and mesophyll cells of salinized and non-salinized plants (Cuin et al., 2003). The cytosolic K^+ is similar in both root and leaf cells of barley plant, but the vacuolar K^+ in leaf cells is approximately twice than in root cells. This suggests differences in the regulation of vacuolar K^+ but not cytosolic K^+, in leaf and root cells. Addition of 100 mM NaCl in the root medium of barley plants caused an immediate reduction in elongation velocity of the growing leaf (Fricke, 2004). After 20–30 minutes, elongation velocity recovered suddenly to 40-50 percent of the prestress level. Bulk osmolality increased significantly in the proximal half of the elongation zone. Over the following 3 days, osmolality increases became significant in the distal half of the elongation zone, the adjascent, enclosed nonelongation zone and finally in the emerged portion of the blade. The developmental gradient and time course in osmolality increase along the growing leaf was reflected in the pattern of solute (Cl^-, Na^+ and K^+) accumulation in bulk tissue and epidermal cells. The partitioning of newly accumulated solutes between epidermis and bulk tissue changed with time. Even though the solute accumulation does not contribute to the sudden and partial growth recovery 20-30 minutes after exposure to salt, it does facilitate residual growth from one hour onwards. This is due to a high sink strength for solutes of the proximal part of the growth zone and its ability to accumulate solutes rapidly and at high rates.

Yang et al. (2003) also compared salt tolerance of three wild soybean (*Glycine soja*) populations. The main reason for the salt tolerance in salt-tolerant wild soybean was the accumulation of Na^+ and Cl^- in roots and stems and the higher selective transportation of K^+ from root to shoot. Salinity at a concentration of 30 mM NaCl caused a substantial reduction in chickpea plant height, number of leaves and the dry weights of the leaves, stems and roots. Biomass allocation to the leaves increased, predominantly at the expense of roots. The authors report substantial effects of salinity and ozone on chickpea growth and ion concentrations.

A higher K^+ uptake under normal and saline conditions is evident in alkali grass (*Puccinellia tenuiflora*) compared with that in wheat, and the authors (Peng et al., 2004) explain that the difference in uptake results from the higher K^+/Na^+ selectivity of the plasma membrane. High K^+ and low Na^+ existed in the endodermal cells of alkali grass roots, suggesting this to be the tissue where Cs^+ inhibition occurs. These results suggest that the potassium apoplastic barrier, the Casparian bands of the endodermis, lead to the lateral gradient of K^+ and Na^+ across root tissue, resulting not only in high levels of K^+ in the shoot but also a large Na^+ gradient between the root and the shoot. Tester and Davenport (2003) suggest that we require more knowledge of cell-specific transport processes and the consequences of manipulation of transporters and signalling elements in specific cell types.

Acidity

Grabov and Blatt (1997) examined the influence of cytosolic pH in controlling K^+ channel activity and its interaction with cytosolic-free Ca^{2+} concentration in stomatal guard cells of *Vicia faba.* Extreme acid loads evoke reversible increases in cytosolic-free calcium concentration, although the effect is generally delayed with respect to the time course of cytosolic pH changes and K^+ channel responses. Cytosoilc pH is considered as a second messenger capable of acting in parallel with, but independent of cytosolic-free calcium concentration in controlling the pH of channels. In the same year, Shabala et al. (1997) also measured the net fluxes of H^+ and Ca^{2+} around the elongation region of low salt maize roots using microelectrode ion flux estimation technique. At pH 5.2, two oscillatory components were found : fast 7-minute oscillations in H^+ flux were superimposed on slow oscillations of about 1.5 hour. Fast oscillations in Ca^{2+} flux showed a strong dependence on the H^+ oscillations and were normally leading in phase by about 1 to 1.5 minutes. Both oscillatory components were strongly affected by external pH values. The fast and slow oscillatory components of H^+ flux seem to relate to biophysical and biochemical mechanisms of intracellular pH homeostasis, respectively. The most important factor regulating K^+ channels in guard cells is pH; both apoplastic and cytosolic acidification lead to activation of inward K^+ currents in guard cells.

Heat stress

Ilan et al. (1995) used patch-clamp technique to examine the effect of temperature (13-36°C) on the depolarization-activated K^+ channels and on the hyperpolarization-activated channels in the plasma membrane of *Vicia faba* guard-cell protoplasts. The steady-state whole-cell conductance of both K^+ channel types increased with temperature up to 20°C. However, whereas the whole-cell conductance of the hyperpolarization-activated K^+ channels increased further and saturated at 28°C, that of the depolarization-activated K^+ channels decreased at higher temperatures. The mean number of available hyperpolarization activated K^+ channels was not affected significantly by temperature, but the mean number of available depolarization-activated K^+channels increased significantly between 13 and 20°C and declined drastically above 20°C. This may be the basic mechanism by which leaves of well-watered plants keep their stomata open during heat stress to promote cooling by transpiration.

Exposure of plants to elevated temperatures results in a complex set of changes in gene expression that induce thermo-tolerance and improve cellular survival to subsequent stress (Gong et al., 1998). Pretreatment of young tobacco (*Nicotiana plumbaginifolia*) seedlings with Ca^{2+} or ethylene glycol-bis (β-aminoethylether) NNN'N-tetra-acetic acid enhanced or diminished subsequent thermotolerance, respectively, compared with untreated seedlings, suggesting a possible involvement of cytosolic Ca^{2+} in heat-shock signal transduction. The seedlings responded to mechanical stimulation or cold-shock with cytosolic Ca^{2+} increases similar to untreated controls. These observations suggest that there may be specific pools of cytosolic Ca^{2+} mobilized by heat treatments or that the refractory period results from a temporary block in heat-shock perception or transduction. Use of the inhibitor suggests that heat-shock mobilizes cytosolic Ca^{2+} from both intracellular and extracellular sources.

Jiang and Huang (2001 b) reported that heat stress increased osmotic adjustment in both drought-preconditioned (subjected to two 14 day cycles of soil drying and rewatering 14 day/ys in the first cycle and 17 day/ys in the second cycle) and non-preconditioned *Poa pratensis* plants (Fig. 2.7). However, osmotic adjustment was significantly higher in drought preconditioned plants (0.23, 0.90, 0.42 MPa) than in non-preconditioned plants (0.13, 0.75, 0.22 MPa) at 7, 14 and 21 days of heat stress, respectively. The differences in osmotic adjustment between the two groups of plants increased with the extended period of heat stress. Under normal temperature conditions, no differences in ion concentrations of cell sap were observed between drought preconditioned plants. However, preconditioned plants had 18.9, 9.5 and 7.9 percent higher concentrations of K^+, and 31.7, 19.4 and 29.8 percent higher concentrations of Ca^{2+} than non-preconditioned plants (see Table 2.1) at 7, 14 and 21 days of heat stress, respectively. Significantly higher (about 43 percent at 14 d and 100 percent at 21 d) concentration of Na^+ was observed in drought-preconditioned plants than in non-preconditioned plants. The preconditioned plants showed 20.6 percent higher level of Cl^- and 21.1 percent higher P than non-preconditioned plants on the 14th day of heat stress, respectively. The total concentration of the six ion solutes combined was 16.2, 12.8, and 10.7 percent higher in the drought preconditioned plants than in non-preconditioned plants on 7, 14 and 21 days of heat stress, respectively. Potassium alone accounted for 59 to 65 percent of the total ion solutes in both groups of plants during heat stress.

Cold stress

At low root temperatures, K^+ uptake by roots is considerably decreased (Kafkafi, 1997). The same applies to nitrate and phosphate while the uptake of Na^+ and Cl^- is less affected by root temperature (Table 2.2). For example, when carnations were cultivated in open fields in Israel there was an increased percentage of stem-brittle after cool nights due to impaired K^+ uptake when the potassium supply with the irrigation water was low. Increasing potassium concentration in the irrigation water, and thus

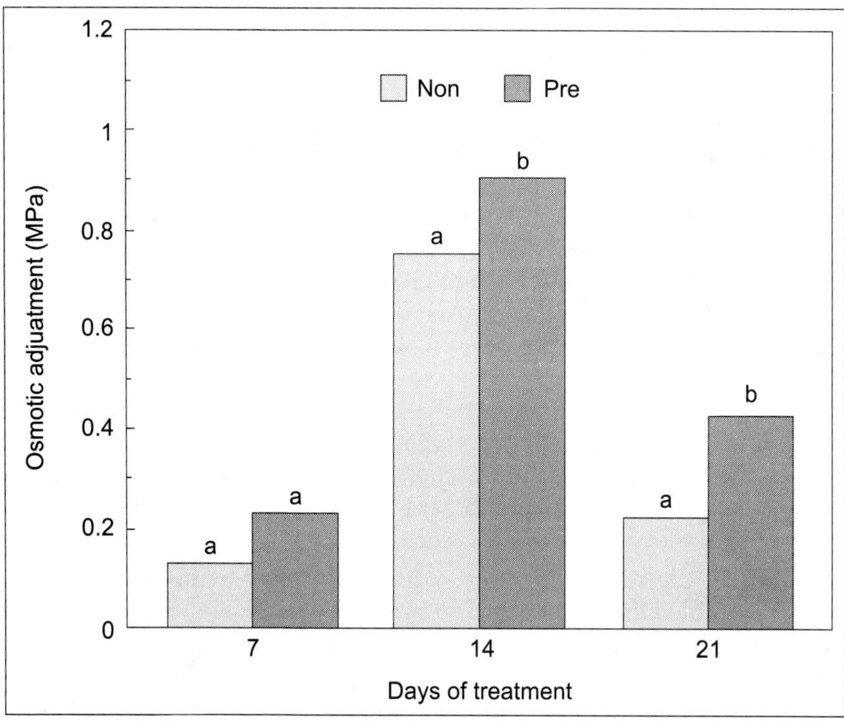

Fig. 2.7 Osmotic adjustment as affected by heat stress in drought-preconditioned and non-preconditioned Kentucky bluegrass. Means followed by the same letters within a column at a given day of heat stress treatment were not significantly different based on LSD test (P= 0.05). (After Jiang and Huang, 2001).

improving the K^+ uptake, reduced the incidence of frost damage. Similarly, Grewal and Singh (1980) found a fairly strong negative correlation between the leaf potassium content and the degree of frost damage in potatoes. The higher the potassium content of the plant, the lower the damage, and thus the larger the final tuber yield.

Table 2.2 Concentration of water-soluble cations and anions (meq/kg DM) in tomato shoots as affected by root temperature (After Kafkafi,1997).

Root temperature	Na	K	Cl	NO_3
Constant 12°C	76	682	197	19
Variable 12°C ± 4°C	81	793	222	61
Constant 17°C	71	1040	217	218
Variable 25°C ± 5°C	61	1279	278	454

Further, Cakmak (2003) draws a relation between impaired K^+ uptake at low temperatures and the photo-oxidative damage associated with potassium deficiency. This view is supported by the fact that NADPH oxidation and thus, the production of oxygen radicals is much larger at chilling than at normal temperatures.

Shabala and Shabala (2002) made a comparative study of low temperature effects on the activity of plasma membrane transporters of different ions in maize leaf and root tissues by non-invesive measurements of net ion fluxes using the ion-selective microelectrode technique. Kinetics of net H^+ , Ca^{2+} , K^+ , Na^+ , NH_4^+ and Cl^- fluxes were measured as the plant tissues recovered after short term (3 h) chilling stress. The major findings include : (a) the critical temperatures, under which recovery of the activity of plasma membrane transporters took place, were found to be the same for all ions measured and are likely to be associated with the phase transition of membrane lipids, (b) the most pronounced was the reduction in net uptake of K^+ and NH_4^+ , (c) chilling treatment caused a significant net influx of Cl^- and Na^+ in the leaf tissue, and (d) for the same species, critical temperatures for membrane transport procresses in roots were 2 to 2.5°C lower than in leaves.

To study the role of Ca^{2+} in cold acclimation, Monroy and Dhindsa (1995) examined the relationship between Ca^{2+} influx and accumulation of transcripts of two *cas* (cold acclimation specific) genes of alfalfa *cas 15* and *cas 18*. Whereas a decline in temperature from 25 to 15°C had little effect on the influence of extracellular $^{45}Ca^{2+}$, an increasing influx was observed when the temperature was lowered further. The influence of $^{45}Ca^{2+}$ at 4°C was nearly 15 times greater than at 25°C. The addition of calcium chelators or of Ca^{2+} channel blockers which have been shown to prevent cold acclimation, inhibited the influx of extracellular $^{45}Ca^{2+}$ as well as the expression of *cas* genes at 4°C. Addition of a calcium ionophore or a calcium channel antagonist to non-acclimated cells caused the influx of extracellular $^{45}Ca^{2+}$ and induced the expression of *cas* genes at 25°C. These results suggest that a cold-induced calcium influx plays an essential role in cold acclimation. They also isolated two sequences corresponding to Ca-dependent protein kinases. The transcript level of one of them was markedly upgraded at 4°C.

Radiation

Shabala and Newman (1999) by using ion-selective vibrating microelectrodes measured the kinetics of H^+ , Ca^{2+}, K^+ and Cl^- fluxes and the change in their concentrations caused by illumination near the mesophyll and attached epidermis of *Vicia faba*. These flux measurements were related to light-induced changes in the plasma membrane potential. The influx of Ca^{2+} was the main depolarising agent in electrical responses to light in the mesophyll. Changes in the net fluxes of H^+ , K^+ , and Cl^- occurred only after a significant delay of about 2 minutes, whereas light-stimulated influx of Ca^{2+} began within five seconds. In the absence of hydrogen flux, light caused an initial quick rise of external pH near the mesophyll and epidermal tissues. In the mesophyll this fast alkalization was followed by slower, oscillatory pH changes (5-15 min); in the epdermis, the external pH increased steadily and reached a plateau three minutes later.

Most guard cell K^+ channels are voltage-dependent and thus are coupled with light-induced stimulation of the H^+ATPase (Dietrich et al., 2001). Further, light-induced apoplastic acidification provides an additional mechanism for K^+ gating. Also, many guard cell K^+ channels are Ca^{2+} sensitive, and light-induced elevation in cytosolic free Ca^{2+} is a widely reported phenomenon.

IMPROVEMENT WORK

Huang and Redmann (1995) determined the ion relations, water content, leaf water potential, and osmotic adjustment for the cultivated barley cv Harrington and the wild barley grown under mixed SO_4 –salts with varied Ca^{2+} supply. The cultivated barley had lower Ca^{2+} concentrations than the wild barley, especially in the low calcium treatment. Although K/Na and Ca/Na ratios were higher in the control wild barley plants than in Harrington, they declined under salt stress, irrespective of calcium supply. Major osmotica in the wild barley leaves were K^+ , sugars, organic acids, and quarternary ammonium compounds, while in Harrington they were cations, including Na^+ , K^+ and Mg^{2+}, and anions such as PO_4^{2-} and SO_4^{2-}. The wild barley maintained better water status than Harrington under low calcium salt treatment. Supplemental calcium improved water status more in Harrington than in the wild barley. Lack of osmotic adjustment to salinity in the wild barley resulted from its ion exclusion. In the high calcium salt treatment, Harrington had improved water and ion relations, as well as positive turgor. Golldack et al. (2003) further reported that plant adaptation to salt stress involves significant reprograming of K^+ –channel gene expression, especially in leaves, and includes strong and progressive increase in the level of *AtKCl* transcript and a decrease in *AKT2* mRNA accumulation.

Kaya et al. (2001) showed that applying additional potassium under conditions of salt-stress substantially alleviated the stress symptoms. On the contrary, addition of sodium substantially reduced the dry matter production. Applying supplementary potassium together with phosphorus, however, resulted in increased dry matter production of cucumber and pepper varieties, chlorophyll and water content and uptake, and the membrane permeability was restored to the levels similar to those of the control plants. Rascio et al. (2001) identified a mutant of durum wheat by screening an M_4 population for genotypes with enhanced capacity for potassium accumulation in leaves. The mutant 422 accumulates about 5 mg/g dry weight more potassium than the wild-type and is less salt sensitive. Under severe stress imposed by salts and mannitol, the mutant germinates better than the wild-type. In soil containing increasing amounts of NaCl, mutant 422 had higher potassium concentration than the wild-type. The capacity to accumulate potassium could improve tissue hydration, because water content of the mutant 422 leaves was greater than the wild-type, and increased linearly in relation to leaf potassium concentration. In the same year, Zhang and Blumwald (2001) reported that the transgenic tomato plants over expressing a vacuolar Na^+/H^+ antiport were able to grow, flower, and produce fruit in presence of 200 mM NaCl. Although the leaves accumulated high Na^+ concentrations, the fruits displayed very low Na^+ content. The authors demonstrated that with a combination of breeding and transgenic plants it is possible to produce salt-tolerant crops with good quality fruits. Membrane potential measurements in root cells of tomato indicate that the gene *tss1* is affected in both NH_4^+–sensitive and –insensitive components of K^+ transport at low Ca^{2+} concentrations and that this defective transport is rescued by increasing the concentration of Ca^{2+} (Rubio et al., 2004). The results suggest that the *TSS1* gene product is part of a crucial pathway mediating the beneficial effects of Ca^{2+} involved in K^+ nutrition and salt tolerance.

Lin et al. (2004) produced an F_2 and an equivalent F_3 population of rice derived from a cross between a high salt-tolerant indica variety, Nona Bokra, and a susceptible elite japonica variety, Koshihikari, under salt stress. The damage to leaves was attributed to accumulation of Na^+ in the shoot by transport of Na^+ from the root to the shoot in extreme high concentrations. The two major QTLs with the very large effect, q SNC-7 for shoot Na^+ concentration and q SKC-1 for shoot K^+ concentration, explained 48.5 and 40.1 percent of the total phenotypic variance, respectively. The QTLs detected between the shoots and roots almost did not share the same map locations, suggesting that the genes controlling the transport of Na^+ and K^+ between the shoots and the roots are different.

REFERENCES

Allen, G.; A. Amtmann and D. Sanders, 1998. Calcium-dependent and calcium-independent K^+ mobilization channels in *Vicia faba* guard cell vacuoles. J. Exp. Bot. 49:305-18.

Angeles Martinez-Cordero, M.; V. Martinez and F. Rubio, 2005. High-affinity K^+ uptake in pepper plants. J. Exp. Bot. 56 (416): 1553-62.

Buschmann, P.H.; R. Vaidyanathan; W. Gassmann and J.I. Schroeder, 2000. Enhancement of Na^+ uptake currents, time-dependent inward-rectifying K^+ channel currents, and K^+ channel transcripts by K^+ starvation in wheat root cells. Plant Physiol. 122: 1387-98.

Cakmak, I. 2003. The role of potassium in alleviating detrimental effects of abiotic stresses in plants. In: Proc. IPI Congress on "Feed of the Soil to Feed the People : The Role of Potash in Sustainable Agriculture:. 8-10 Oct., 2002. Basel, Switzerland pp 325-43.

Carden, D.E.; D.J. Walker; T.J. Flowers and A.J. Miller, 2003. Single-cell measurements of the contributions of cytosolic Na^+ and K^+ on salt tolerance. Plant Physiol. 131: 676-83.

Cerda, A.; J. Pardines; M.A. Botella and V. Martinez, 1995. Effect of potassium on growth, water relations, the inorganic and organic solute contents on two maize cultivars grown under saline conditions. J. Plant. Nutr. 18: 839-51.

Cuin, T.A.; A.J. Miller; S.A. Laurie and R.A. Leigh, 2003. Potassium activities in cell components of salt-grown barley leaves. J. Exp. Bot. 54(383): 657-61.

De, B.; S. Bhattacharjee and A.K. Mukherjee, 1996. Short term heat shock and cold shock induced proline in *Lycopersicon esculentum* (Mill) cultured cell and seedlings. Indian J. Plant Physil., New Series 1:32-35.

Dietrich, P.; I. Dreyer; P. Wiesner and R. Hedrich, 1998. Cation sensitivity and kinetics of guard cell potassium channels differ among species. Planta 205: 277-87.

Dietrich, P.; D. Sanders and R. Hedrich, 2001. The role of ion channels in light-dependent stomatal opening. J. Expt. Bot. 52: 199-67.

Fernendez-Ballester, G.; F. Garcia-Sanchez; A. Cerda and V. Martinez, 2003. Tolerance of citrus root stock seedlings to saline stress based on their ability to regulate ion uptake and transport. Tree Physiol. 23: 265-71.

Fernando, M.; J. Kulpa; M.Y. Siddiqui and A.D.M. Glass, 1990. Potassium dependent changes in the expression of membrane-associated proteins in barley roots.1.Correlations with K^+ ($^{86}Rb^+$) influx and root K^+ concentration. Plant Physiol. 92; 1128-32.

Fricke, W. 2004. Rapid and tissue specific accumulation of solutes in the growth zone of barley leaves in response to salinity. Planta 219: 515-25.

Gassmann, D. and J.I. Schroeder, 1994. Inward-rectifying K^+ channels in root hairs of wheat (A mechanism for aluminum-sensitive low affinity K^+ uptake and membrane potential control). Plant Physiol. 105: 1399-1408.

Golladack, D.; F. Quigley; C.B. Michalowski; U.R. Kamasani et al. 2003. Salinity stress-tolerant and –sensitive rice (*Oriza sativa*) regulate *AKT1*-type potassium channel transcripts differently. Plant Mo. Biol. 51: 71-81.

Gong, M.; A.H. van der Luit; M.R. Knight and A.J. Trewavas, 1998. Heat-shock-induced changes in intracellular Ca^{2+} level in tobacco seedlings in relation to thermotolerance. Plant Physiol. 116: 429-37.

Gonzales, E.M.; C. Arrese-Igor; P.M. Aparicio-Tejo; M. Royuela et al. 2002. Solute hetrogeniety and osmotic adjustment in different leaf structures of semi-leafless pea (*Pisum sativum* L.) subjected to water stress. Plant Biol. 4: 558-66.

Grabov, A. and M.R. Blatt, 1997. Parallel control of the inward-rectifier K^+ channel by cytosolic free Ca^{2+} and pH in *Vicia* guard cells. Planta 201: 84-95.

Grewal, J.S. and S.N. Singh, 1980. Effect of potassium nutrition on frost damage and yield of potato plants on aluvial soils of the Punjab (India). Plant Soil 57: 105-10.

Hu, Y. and U. Schmidhalter, 1998. Spacial distributions of ions and sugars contributing to osmotic adjustment in the elongating wheat leaf under saline soil conditions. Aust. J. Plant Physiol. 25: 591-97.

Huang, J. and R.E. Redmann, 1995. Solute adjustment to salinity and calcium supply in cultivated and wild barley. J. Plant Nutr. 18: 137-89.

Ilan, N.; N. Moran and A. Schwartz, 1995. The role of potassium channels in the temperature control of stomatal aperture. Plant Physiol. 108: 1161-70.

Jiang, Y. and B. Huang, 2001 a. Effects of calcium on antioxidant activities and water relation associated with heat tolerance in two cool season grasses. J. Exp. Bot.52 (355): 341-49.

Jiang, Y. and B. Huang, 2001 b. Osmotic adjustment and root growth associated with drought preconditioning-enhaced heat tolerance in Kentucky Bluegrass. Crop Sci. 41: 1168-72.

Kafkafi, U. 1997. Impact of potassium in releiving plants from climatic and soil-induced stresses. In: Food Security in the Wana Region, the Essential Need for Balanced Fertilization,(ed) A.E. Johnston, 26-30 May,1997, Izmir, Turkey, pp 313-27.

Kaya, C.; H. Kirnak and D. Higgs, 2001. Effects of supplementary potassium and phosphorus on physiological development and mineral nutrition of cucumber and pepper cultivars grown at high salinity (NaCl). J. Plant Nutr. 24: 1457-71.

Koch, K. and K.K.. Mengel, 1974. The influence of the level of potassium supply to young tobacco plants (*Nicotiana tabacum* L.) on short-term uptake and utilization of nitrate nitrogen. J. Sci. Food Agric. 25: 465-71.

Li, Y. and P. Lin, 2000. (Impact of salinity on accumulation of several metal elements in *Bruguiera gymnorrhiza* seedlings and Ca effect). Ying Yong Sheng Tai Xue Bao11: 177-80.

Li, W.; S. Luan; S.L. Schrciber and S.M. Assmann, 1994. Evidence for protein phosphatase 1 and 2A regulation of K^+ channels in two types of leaf cells. Plant Cell Environ. 25: 229-37.

Lin, X.H.; M.Z. Zhu; M. Yano; J.P. Gao et al. 2004. QTLs for Na and K uptake of the shoots and roots controlling rice salt tolerance. Theo. Appl. Genet. 108: 253-60.

Liu, J. and J.K. Zhu, 1998. A calcium sensor homolog required for plant salt tolerance. Science 280 (5371): 1943-45.

Luan, S. 2002. Signalling drought in guard cells. Plant Cell Environ. 25: 229-37.

Maathuis, F.J.M.; D. Verlin; F.A. Smith; D. Sanders et al. 1996. The physiological relevance of Na^+ coupled K^+ - transport. Plant Physiol. 112: 1609-16.

Majore, I.; B. Wilhelm and I. Marten, 2002. Identification of K^+ channels in the plasma membrane of maize subsidiary cells. Plant Cell Physiol. 43: 844-52.

Marten, I.; S. Hoth; R. Deeken; P. Ache et al. 1999. *AKT3,* a phloem-localized K^+ channel, is blocked by protons. Proc. Nat. Acad. Sci., U.S.A. 96: 7581-86.

Monroy, A.F. and R.S. Dhindsa, 1995. Low-temperature signal transduction : induction of cold acclimation-specific genes of alfalfa by calcium at 25°C. Plant Cell 7: 321-31.

Morgan, J.M. 1992. Osmotic components and properties associated with genotypic differences in osmoregulation in wheat. J. Plant Physiol. 19: 67-76.

Peng, Y.H.; Y.F. Zhu; Y.Q. Mao; S.M. Wang et al. 2004. Alkali grass resists salt stress through high K^+ and an endodermis barrier to Na^+ . J. Exp. Bot. 55(398): 939-49.

Premchandra, G.S.; D.T. Ilahn; D. Rhodes and R.J. Joly, 1995. Leaf water relations and solute accumulation in two grain sorghum lines exhibiting contrasting drought tolerance. J. Exp. Bot. 46: 1833-41.

Pitman, M.G. and D. Wellfare, 1978. Inhibition of ion transport in excised barley roots by abscisic acid : relation to water permeability of the roots. J. Exp. Bot. 29: 1125-38.

Rascio, A.; M. Russo; L. Mazzucco; C. Platani et al. 2001. Enhanced osmotolerance of a wheat mutant selected for potassium accumulation. Plant Sci. 160: 441-48.

Roberts, S.K. 1998. Regulation of K^+ channels in maize roots by water stress and abscisic acid. Plant Physiol. 116: 145-53.

Roberts, S.K. and B.N. Snowman, 2000. The effect of ABA on channel mediated K^+ transport across higher plant roots. J. Exp. Bot. 51(350): 1585-94.

Rubio, F.; W. Gassmann and J.I. Schroeder, 1995. Sodium-driven potassium uptake by the plant potassium transporter *HKT1* and mutations conferring salt tolerance. Science 270 (5242): 1660-63.

Rubio,L.; A. Rosado; A. Linares-Rueda; O. Borsani et al. 2004. Regulation of K^+ transport in tomato roots by the *TSS1* locus. Implications in salt tolerance. Plant Physiol. 134: 452-59.

Sadiqov, S.T.; M. Akbulut and V. Ehmedov, 2002. Role of Ca^{2+} in drought stress signaling in wheat seedlings. Biochemistry (Mosc) 67(4): 491-97.

Schachtman D.P. and J.I. Schroeder, 1994. Structure and transport mechanism of a high affinity potassium uptake transporter from higher plants. Nature 370(649): 655-58.

Shabala, S. 2003. Regulation of potassium transport in leaves from molecular to tissue level. Ann. Bot. 92: 627-34.

Shabala, S. and R.R. Lew, 2002. Turgor regulation in osmotically stressed *Arabidopsis* epidermal cells. Direct support for the role of inorganic ion uptake as related by concurrent flux and cell turgor measurements. Plant Physiol. 129: 290-99.

Shabala, S. and I.A. Newman, 1999. Light-induced changes in hydrogen, calcium, potassium and chloride fluxes and concentrations from the mesophyll and epidermal tissues of bean leaves : understanding the ionic basis of light-induced bioelectrogenesis. Plant Physiol. 119:1115-24.

Shabala, S.; O. Babourina and I.A. Newman, 2000. Ion-specific mechanisms of osmoregulation in bean mesophyll cells. J. Exp. Bot. 51(348): 1243-53.

Shabala, S.; I.A. Newman and J. Morris, 1997. Oscillations in H^+ and Ca^{2+} ion fluxes around the elongation region of corn roots and effects of external pH. Plant Physiol. 113: 111-18.

Shabala S. and L. Shabala, 2002. Kinetics of net H^+ , Ca^{2+} , K^+ , Na^+ , and NH_4^+ , and Cl^- fluxes associated with post chilling recovery of plasma membrane transporters in *Zea mays* leaf and root tissues. Physiol. Plant. 114: 47-56.

Spalding, E.P.; R.E. Hirsch; D.R. Lewis; et al. 1999. Potassium uptake supporting plant growth in the absence of AKT1 channel activity. J. Gen. Physiol. 113: 909-18.

Suresh, P.; K.R. Lalitha; M. Udaykumar; et al. 1991. Involvement of calcium in ABA, KCl, NaCl and moisture stress induced accumulation of proline in finger millet, *Elusine coracana* . Indian J. Exp. Biol. 29: 355-58.

Su, H.; D. Golldack; M. Katsuhara; et al. 2001. Expression and stress-dependent induction of potassium channel transcripts in the common ice plant. Plant Physiol. 125: 604-14.

Subbarao, G.V.; R.M. Wheeler; G.W. Stutte; et al. 2000. Low potassium enhances sodium uptake in red-beet under moderate saline conditions. J. Plant Nutr. 23 (10): 1449-70.

Tester, M. and R. Davenport, 2003. Na^+ tolerance and Na^+ transport in higher plants. Ann. Bot. 91: 503-27.

Umashankar, R. 1985. Moisture stress induced loss of membrane integrity, accumulation of proline and their interrelationship. Ph. D. Thesis, Univ. Agric. Sci., Bangalore, India.

Walker, D.J.; R.A. Leigh and A.J. Miller, 1996. Potassium homeostasis in vacuolate plant cells. Proc. Natl. Acad. Sci., U.S.A. 93:10510-14.

Wang,T.B.; W. Gassman; F.Rubio; J.I. Schroeder et al. 1998. Rapid upregulation of *HKT1*, a high affinity potassium transporter gene, in roots of barley and wheat following withdrawal of potassium. Plant Physiol. 118: 651-59.

Watson, R.; J. Pritchard and M. Malone, 2001. Direct measurement of sodium and potassium in the transpiration stream of salt-excluding and non-excluding varieties of wheat. J. Exp. Bot. 52(362): 1873-81.

Wegner,I.H. and K. Raschke, 1994. Ion channels in the xylem parenchyma of barley roots (A procedure to isolate protoplasts from this tissue and a patch-clamp exploration of salt passage ways into xylem vessels. Plant Physiol. 105: 799-813.

Wu, S.J.; L. Ding and J.K. Zhu, 1996. *sos1*, a genetic locus essential for salt tolerance and potassium acquisition. J. Plant Cell 8: 619-27.

Yang, X.; W. Zhang; Q.Wang and Y. Liu, 2003. (Salt tolerance wild soybeans in Jiangsu and its relation with ionic distribution and selective transportation). Ying Yong Tai Xue Bao 14: 2237-40.

Zhang, H..X and E. Blumwald, 2001. Transgenic salt-tolerant tomato plants accumulate salt in foliage but not in fruit. Nat. Biotechnol. 19: 765-68.

Zhu, J.K. 2003. Regulation of ion homeostasis under salt stress. Curr. Opin. Plant Biol. 6: 441- 45.

SUGGESTED READINGS

Roberts, S.K. 1998. Regulation of K^+ channels in maize roots by water stress and abscisic acid. Plant Physiol. 116: 145-53.

Shabala, S. 2003. Regulating potassium transport in leaves from molecular to tissue level Ann. Bot. 92: 627-34.

Shabala, S.; O. Babourina and I. Newman, 2000. Ion-specific mechanisms of osmoregulation in bean mesophyll cells. J. Exp. Bot. 51(348): 1243-53.

Wang, T.B.; W. Gassman; F. Rubio; J.I. Schroeder and A.D.M. Glass, 1998. Rapid upregulation of *HKT1*, a high-affinity potassium transporter gene, in roots of barley and wheat following withdrawal of potassium. Plant Physiol. 118: 651-59.

3

SUGARS – THE KEY OSMOLYTES

Introduction

Under conditions of osmotic stress, sugars (mono- or di-saccharides) levels increase and they do contribute towards osmotic protection, but they also play several other important regulatory functions in stressed plants. The plastidic starch, a sugar reserve, is rapidly converted to sucrose under stress conditions which in turn inhibits further starch synthesis. Many tissues of stressed plants have increased demand for the rapidly metabolizable carbohydrates, and thus initiate inter-organ transport of sugars which performs some other regulatory functions in addition to the osmotic one. For instance, sucrose provides carbon for cellular metabolism only after its conversion to hexose phosphates. Sugars not only sustain growth of sink tissues, but they also affect sugar-sensing systems that regulate the expression, either positively or negatively, of a variety of genes involved in photosynthesis, respiration, sucrose and starch syntheses and degradation, and nitrogen metabolism (Koch, 1996). Phosphorylation of hexose sugars by hexokinase, is the most important mediator of sugar sensing. Hexoses affect the sugar-sensing system more directly than sucrose. During stress, the irreversible hexokinase reaction is an important determinant of whether hexoses (glucose, fructose and mannose) are required or accumulated, possibly for osmotic purposes.

Sugar accumulation at low temperature is an adaptive mechanism that helps to stabilize membranes during freeze-induced desiccation by preventing phase separation and membrane fusion. Sucrose also maintains the activity of key enzymes in the dehydrated state. Bruni and Leopold (1991) suggest that the ability of disaccharides to protect membranes and proteins against dehydration is linked to their capacity to form a glassy state that slows down the molecular motion and helps in minimizing the damaging interactions between cell components.

Sucrose functions as a typical osmoprotectant by stabilizing cellular membranes and maintaining turgor, and acts as water replacement for maintaining membrane phospholipids in the liquid-crystalline phase and to prevent structural changes in soluble proteins (Kerepesi and Galiba, 2000). In addition, sucrose may also serve as an intermediate energy source upon rehydration as an easily metabolizable reducing sugar.

At low temperatures, the photosynthetic energy-capture is reduced but to a lesser degree than the metabolic utilization processes. This leads to surplus-reducing potential in the form of pyridine nucleotides (NADH, NADPH) and ATP production, which is channeled into the CO_2 fixation processes. Active growth is almost always reduced or suspended resulting in decreased demand for the products of photosynthesis. Thus the production of photosynthate in excess of need, leads to reserve accumulation in the form of starch. However, in temperate perennials, starch content decreases and free saccharides exhibit a direct quantitative increase. In less hardy species such as *Solanum* and subtropical *Citrus* , starch accumulates during low temperature treatment but only in concert with a marked accretion of free sugars.

Accumlation of soluble sugars (sucrose, glucose, fructose) is strongly correlated to the acquisition of drought tolerance (Hoekstra and Buitink, 2001). The hydroxyl group of sugars substitute for water to maintain hydrophyllic interactions in membranes and proteins during dehydration. Thus, sugars interact with proteins and membranes through hydrogen-bonding, thereby preventing protein denaturation. Also, sugars are the major contributing factors to vitrification, i.e., the formation of biological glass in the cytoplasm of dehydrated cells (Leopold et al., 1994). These intracellular glasses, by virtue of their high viscosity, drastically reduce molecular movement, impede the diffusion of reactive compounds in the cell, and maintain the structural and functional integrity of macromolecules. It is by this property that glasses prolong the longevity of desiccated tissues by slowing down degradation processes during storage (Buitink et al., 1998). Thus the soluble sugar content appears to be a better marker for accelerating/ improving drought tolerant wheat genotypes (Kerepesi and Galiba, 2000).

Sucrose is accumulated in several plant tissues in response to drought, salinity, and low temperature, and plays a role in osmoregulation and cryoprotection. There are three enzyme systems directly involved in sucrose metabolism. The first one is sucrose phosphate synthase, a soluble enzyme located in the cytoplasm, which catalyzes the reaction UDPG +F-6-P $\longleftrightarrow$ sucrose phosphate + UDP. Sucrose phosphate synthase plays a major role in sucrose biosynthesis. The second enzyme is the system in invertase catalyzing an irreversible sucrose-cleaving reaction. In crop plants, there are invertase enzymes which are categorized as acidic or alkaline based on the pH optima of other reactions. The soluble alkaline invertase is present in cytoplasm. The third key enzyme of sucrose metabolism is sucrose synthase, which catalyzes a reversible reaction UDP-glucose + fructose ⟵⟶ sucrose + UDP. This is the main catalytic enzyme for sucrose cleavage and synthesis in many plant tissues (Wang et al., 2000).

During stress, sugars not only function as osmoprotectants and as substrates for growth, but are also important regulators of gene expression (Koch, 1996). There is a sugar sensing system that initiates changes in gene expression, which vary depending on the carbohydrate and levels of carbon sources. The genes associated with starch breakdown can also be up-regulated by carbohydrate depletion and repressed by high glucose levels (Krapp and Stitt, 1994).

FREEZE-INDUCED DEHYDRATION

During cold acclimation, coordinated alterations in the activity of various enzyme systems result in modification of the composition and properties of various cell organnels, particularly the cell walls. The alterations in cell wall composition occurring during hardening at chilling temperatures cause changes in their thickness and elasticity. Plants subjected to hardening conditions not only exhibit an increase in hardiness but also show a substantial parallel change in the osmotic potential of cell sap that resembles an osmotic adjustment response. The major component responsible for the change in osmotic potential of hardened plants is sugar, and sucrose is often the most abundant sugar. Undoubtedly , sucrose is one of the most universal osmolytes in higher plants. In keeping with the dehydrative effect caused by freezing, sucrose is universally associated with cold acclimation and freezing tolerance.

Accumulation of water-soluble oligosaccharides with a molecular weight of 3-5 kDa occurs during the first few hours of low temperature adaptation in the root cells of winter wheat seedlings (Zabotina et al., 1998). Oligosaccharides from the hardened seedlings inhibit elongation of wheat coleoptile segments showing its contribution to low temperature adaptation. Sucrose helps plant survive the severe desiccation imposed by extracellular ice formation by stabilizing membranes and membrane proteins, and also during the desiccation phase of the maturing seeds (Hoekstra et al., 1989; Chen and Buris, 1990). In order to confirm the contribution of sugars to cold hardening, Castonguay et al. (1995) further studied the carbohydrate composition of alfalfa (*Medicago sativa*) crown and freezing tolerance in cultivars of contrasting winter hardiness during their acclimation to low temperature. During hardening, freezing tolerance and crown levels of soluble sugars increased with a concomitant decrease in starch. The differences in freezing tolerance between a cold-tolerant and a cold-sensitive cultivar were closely associated with the accumulation of oligosaccharides raffinose and stachyose but were not related to the levels of sucrose.

Freezing tolerance and carbohydrate changes during cold acclimation of green type annual bluegrass (*Poa annua* L.) ecotypes were studied by Dionne et al. (2001). Sucrose levels in crowns of annual bluegrass markedly increased at temperatures below freezing and maximum sucrose concentration coincided with the maximum freezing tolerance. After two weeks of acclimation at 2°C, sucrose content slightly increased, but the incubation at subzero temperatures induced a marked rise in sucrose levels with a concomitant decrease in fructans. Fructose and glucose levels increased after exposure to freezing with the cold sensitive ecotype (Central Quebec) accumulating significantly higher levels than Coastal Maryland and Western Pennsylvania. The total carbohydrate nearly doubled in annual bluegrass acclimated to low, non-freezing temperatures but these levels noticeably declined in plants exposed to subfreezing temperatures as a result of fructan hydrolysis. The cold sensitive ecotype Coastal Maryland accumulated significantly higher levels of sucrose, glucose, and fructose than the other two ecotypes after acclimation to simulated winter conditions in an unheated greenhouse. Dionne et al. suggested that sucrose plays an importat cryoprotective role by stabilizing membranes and proteins, preventing adhesion of ice to critical cellular tissues during freezing, or by modifying ice crystal formation.

Considering the major shift in the carbohydrate status of plants at low temperature, Guy et al. (1992) studied the enzymology of carbohydrate metabolism of spinach (*Spinacia oleracea*) at low temperatures. Plants were grown at 25°C for three weeks and then transferred to a constant 5°C, and it was noted that sucrose, glucose and fructose accumulated to high levels during a 14-day period. Sucrose was the most abundant free sugar before, during, and after exposure to 5°C. Leaf sucrose phosphate synthetase activity significantly increased by the low temperature treatment, whereas the activities of sucrose synthase and invertase did not. Synthesis of the sucrose phosphate synthase subunit increased during and after low temperature exposure and paralleled an increase in the steady state level of the subunit.

Menendez et al. (2002) studied the cold sweetening of diploid potatoes and noted that the QTLs for glucose, fructose, and sucrose content are located on all potato chromosomes. QTLs are linked to genes encoding invertase, sucrose synthase 3, sucrose phosphate synthase, ADP-glucose phosphorylase, sucrose transporter 1, and a putative sucrose senser. Their results suggest that the allelic variants of enzymes operating in carbohydrate metabolic pathways contribute to the genetic variation in cold sweetening. When the Japanese potato tubers were stored at 4°C, three types of changes were observed among the cultivars (Matsuura-Endo et al., 2004): (a) increased levels of reducing sugars during storage, (b) a pattern similar to that of type 1, but with 4-6-fold lower levels of reducing sugars throughout storage, and (c) increased sucrose, but not reducing sugars. The activity of vacuolar acid invertase increased in type 1 cultivars, but in the type 2 and type 3 cultivars, the activities were very low during storage at 4°C. These results suggest that the activity of vacuolar acid invertase is related to the types of changes that occur in sugar content during the low temperature storage among the potato cultivars. However, working with two alfalfa cultivars, Castonguay et al. (1995) found that during hardening under environmentally controlled conditions, freezing tolerance and crown levels of soluble sugars increased significantly with a concomitant decrease in starch. Incubation of the prehardened plants at subzero temperatures (-2°C) increased freezing tolerance and promoted higher accumulations of sucrose, raffinose, and stachyose and further reduction in starch levels. Stachyose and raffinose accumulated later in fall than sucrose and reached higher levels in winter-hardy than in nonhardy cultivars. Differences in the maximum level of freezing tolerance between non-hardy and winter hardy cultivars are better related to the capacity of the plants to accumulate stachyose and raffinose than to accumulate sucrose.

The soluble carbohydrates can control events which extend well beyond carbohydrate metabolism, such as regulating vascular differentiation. Sugar-sensing and sensing of the environment appear to be part of a complex regulatory web. In relation to cold, high sugar supply also induces fructan accumulation. Tabaei-Aghdaei et al. (2003) observed similar levels of freezing tolerance and gene expression in barley cell cultures, as occur during cold accilimation in the crown of the whole barley plant. In the cell cultures, cold (6/2°C) did not induce an increase in freezing tolerance or in the expression of detectable levels of *blt 4.9* or *dhn 1* mRNAs when only 1 gL^{-1} sucrose was supplied. When 10 gL^{-1} sucrose was supplied to cells grown in the warm (25°C) condtion increased acclimation to freezing and up-regulation of expression of *blt 4.9* and *dhn 1* mRNAs was observed. Thirty gL^{-1} sucrose induced yet higher levels of freezing tolerance and of *bet 4.9* and *dhn 1* mRNAs in cultures grown in either the cold or the warm environment. The results implicate sugars in the regulation of cold acclimation.

In order to access the contribution of increased osmolality to crop frost hardiness, Hincha et al. (1996) used transgenic tobacco (*Nicotiana tabacum*) plants that accumulate soluble carbohydrates. Leaves from plants of the clone U-ppa-1-10 expressing a bacterial pyrophosphatase gene displayed an increase in frost hardiness of 1.2°C when compared with wild-type control plants. These plants showed a higher capacity to increase their hardiness during exposure to 4°C growth temperature for 10-14 days; frost hardiness increased by 1.1°C in transgenic plants as compared with 0.2°C in wild-type controls. There was no correlation between leaf osmolality and hardiness when leaves from cold acclimated and from non-acclimated wild-type and clones of transformed tobacco were compared.

In the tobacco transgenic plants expressing yeast invertase either in the apoplast or vacuole, hexose accumulation is due to the activity of the enzyme that cleaves the sucrose that is normally exported from the photosynthetically active mesophyll cells. The resulting glucose and fructose are not loaded into the phloem for transport to sink organs such as roots and young leaves. Instead, the monosaccharides are reimported and therefore accumulate in the vacuoles.

When older leaves of tobacco plants expressing the bacterial pyrophosphatase were compared with the leaves of control plants of similar age, leaves from the transgenic plants showed a clear increase in frost hardiness (Fig. 3.1). When such plants were transferred to 4°C growth temperature, their capacity to increase freezing resistance of the leaves strongly increased. When the leaves of different ages were compared from wild-type and transgenic plants, it became obvious that sugar accumulation and the resulting increase in osmolality played at best a minor role in the stress resistance of these leaves. Rather it can be conceived that some of these induced genes are stress related and that their products might increase plant stress resistance (Hincha et al., 1996).

DROUGHT-INDUCED DEHYDRATION

In many higher plants under dehydration stress, carbohydrate metabolism is shifted to favor the conversion of other sugars to sucrose. In all cases increases in sucrose accumulation in response to drought stress is observed, though at varying levels in different plant species, implicating a role for sucrose in the acquisition of desiccation tolerance in these plants. Sucrose is thus thought to function as a typical osmoprotectant, stabilizing cellular membranes and maintaining turgor under drought conditions. As an easily metabolizable reducing sugar, sucrose serves as an immediate energy source upon rehydration. In osmotic stress tolerant cells, presence of solutes (sugars) keeps the membrane surface preferentially hydrated and prevents membrane fusion. Below 0.3 g H_2O g^{-1} dry weight, sugar molecules in tolerant cells replace water in the hydration shell of the membrane, thereby maintaining spacing between the phospholipid molecules. In this chapter, after describing studies on the effects of drought on sugar accumulation in the two important cereals, wheat (Kerepesi and Galiba, 2000) and rice (Yang et al., 2001) and the C_4 grass, *Setaria sphacelata* (da Silva and Arrabaca, 2004), a brief mention of the *in vitro* studies of Wang et al. (2000) on cultured sweet potato cells have been described. After this, more

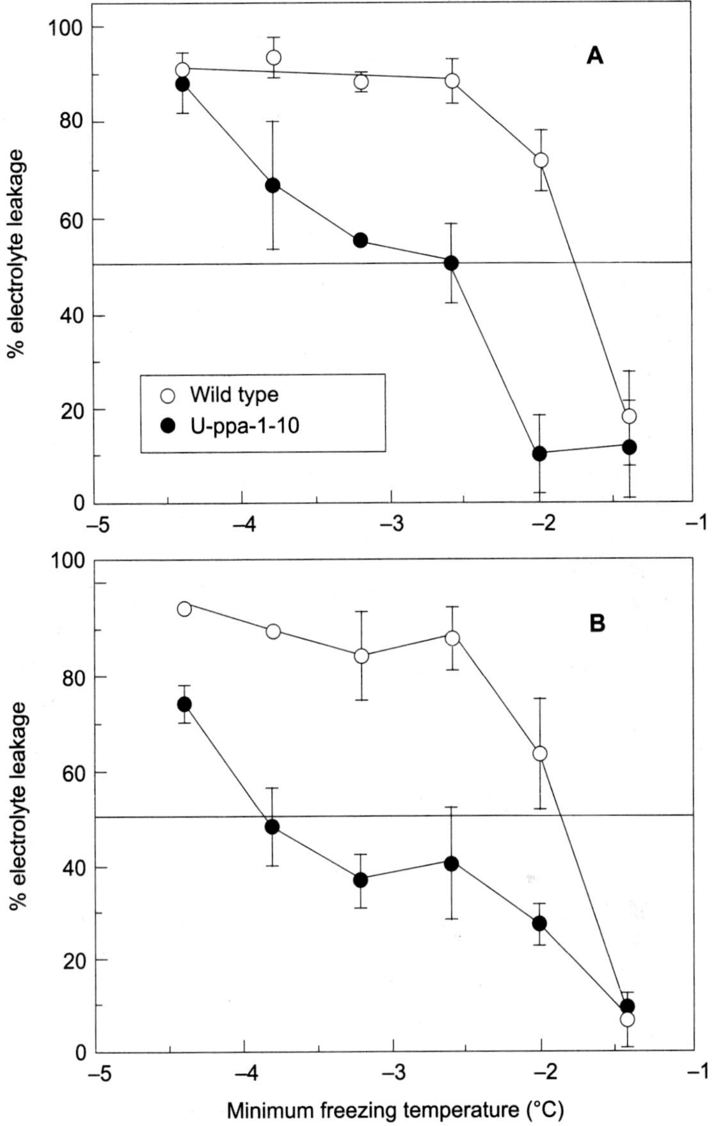

Fig. 3.1 Electrolyte leakage from tobacco leaf discs after a freeze-thaw cycle to different temperatures. Samples were frozen and thawed at 5°C/h and were kept at the minimum temperature for 1 h. Damage was measured by the conductivity method and is expresed as electrolyte leakage relative to the leakage after boiling of the leaf discs. Samples were taken from the leaves of wild-type plants, or of plants expressing a gene for an *E. coli* pyrophosphatase, denoted *U-ppa-1-10.* Plants were either grown under non-acclimating (A) or cold acclimating (B) conditions. The points represent the means and S.D. of three independent experiments with two different plants each. The horizontal lines denote 50% electrolyte leakage. The corresponding temperatures are taken as the LT_{50} of the respective plants. (After Hincha et al., 1996).

emphasis has been given to studies on the orthodox desiccation of seeds during maturation; followed by the reverse changes taking place after seed imbibition and germination, with a view to studying changes taking place in drought tolerance and the soluble sugar levels in seeds during dehydration and rehydration (Hoekstra et al., 1994).

Kerepesi and Galiba (2000) examined the effects of drought and salt stresses on the water soluble carbohydrate content in four varieties (Sakha, Kobomugi, Chinese Spring, and Regina) of wheat seedlings to characterize the involvement of major sugar components in the adaptive processes. Hydroponically grown seedlings of the 4 wheat varieties differing in drought and salt tolerance were exposed to consecutive drought (PEG) and salinity (NaCl) stresses. Total water soluble carbohydrate (glucose, fructose, sucrose and fructans) content of stem were determined. The tolerant genotypes accumulated more soluble carbohydrate than did the sensitive ones (Table 3.1). Both types of stresses (drought and salinity) increased the concentration of reducing sugars and sucrose (Fig. 3.2) and also the fructans. The drought-tolerant genotypes accumulated sucrose to a significantly greater level than did the sensitive ones. However, the changes in fructan content of seedlings after transfer from PEG to NaCl (drought to salinity) containing solutions were genotype-dependent, increasing in salt tolerant and decreasing in salt sensitive genotypes (Fig. 3.3). The initial response of drought stress appears to be an increase in monosaccharides, while the more delayed response was an increase in fructan. On the other hand, fructan content could be a useful indicator of the degree of salt tolerance.

Table 3.1 Water soluble carbohydrate content in stems of four wheat varieties during the consecutive 180 g/kg PEG-induced drought and 200 mM NaCl-induced salt stresses. Data are mean ± SD for nine replicates.(After Kerepesi and Galiba, 2000).

Treatment / wheat varieties	Time in days	Sakha-8	Kobomugi	Chinese Spring	Regina
			⟵ mg/g dry weight ⟶		
Control	2	188.4±20.3	172.2±12.9	164.7±18.3	124.8±11.3
	7	175.7±32.8	168.1±18.9	155.6±20.3	112.9±18.4
	11	168.6±25.6	153.7±17.5	167.6±18.7	124.3±16.9
PEG	2	369.7±24.6	268.4±22.6	270.8±22.5	220.8±25.2
	7	337.1±28.4	612.1±55.2	383.1±36.7	284.6±25.1
NaCl	11	491.8±32.8	247.5±21.7	372.6±36.1	291.4±30.1
Recovery *	11	211.4±24.6	143.7±11.5	197.8±21.2	145.3±15.9

* Recovery = Following the PEG treatment plants were subcultured to Hoagland solution and maintained for four hours.

Yang et al. (2001) studied the activities of starch hydrolytic enzymes and sucrose phosphate synthase in the stems of rice plants subjected to water stress during grain filling. The water stress treatment accelerated reduction of starch in the stems, promoted reallocation of fixed ^{14}C from the stems to grains, shortened the grain filling period, and increased the grain filling rate. More soluble sugars including sucrose were accumulated in stems under water stress than under well-watered treatments. Both α- and β-amylase activities were enhanced by water stress, the former enhanced more than the latter, and were significantly correlated with the concentrations of soluble sugars in the stems. Water stress also increased the sucrose phosphate synthase activity that is responsible for sucrose production. Further, da Silva and Arrabaca (2004) made a comparison between the rapidly and slowly imposed water stress on the contribution of soluble carbohydrates to the osmotic adjustment in the C_4 grass *Setaria sphacelata.* They found an increase in the ratio between free hexoses and sucrose in stressed leaves, but no significant differences were found in the amount of free hexoses, or in the ratio between soluble and insoluble sugars. In long-term stress experiments, a higher amount of soluble sugars and a lower amount of starch were found in stressed leaves. The ratio of free hexoses to sucrose and of soluble to insoluble sugars were also higher in stressed leaves. Contribution of the accumulation of soluble sugars to osmotic adjustment was absent in rapidly stressed leaves and was of minor importance in slowly stressed leaves.

Wang et al. (2000) studied the osmotic stress-induced changes in sucrose metabolism of cultured sweet potato vegetative cells. The stressed cells showed a decrease in hexoses, but a significant increase in sucrose from the start, and the ratio of sucrose to hexoses in the 7-days stressed cells was 12-fold higher than that in the normal. With the onset of 0.6 M sorbitol stress, activity of the cell wall-bound invertase increased strikingly and then declined rapidly at day 7 (Fig. 3.4A). Conversely, the apoplastic capacity remained fairly low in the normal cells, but increased slightly at the 7th day when the external carbon source was almost depleted. Irrespective of the normal and stressed cells, the activity of alkaline invertase was much higher than that of soluble acidic invertase (Fig. 3.4B). The activity of alkaline invertase in stressed cells was significantly higher, and the level remained rather constant (Fig. 3.4C).The results in Figure 3.4D also show that the stressed cells have a higher activity of sucrose synthase, assayed in the direction of sucrose hydrolysis, than that of the normal cells, even though they both showed a downward change in the time course.

The ability to synthesize sucrose was assayed by assaying sucrose synthase and sucrose phosphate synthatase. Both the normal and stressed cells showed a declining tendency for sucrose synthase activity in the

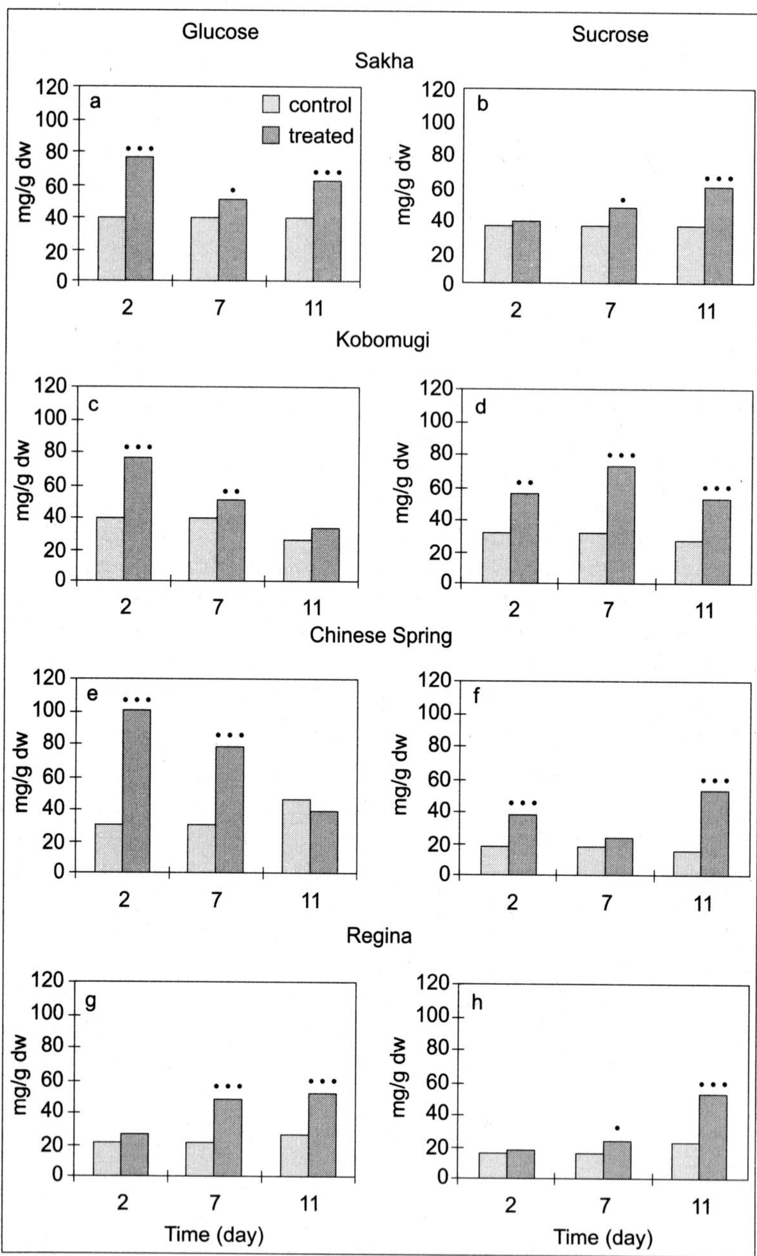

Fig. 3.2 Changes in glucose and sucrose level in stems of four wheat seedlings, grown in 180 g/kg PEG treated culture solution until day 7 and then transferred to equi-osmolar 200 mM NaCl solution until day 11. *, **, and *** indicate significace at P <0.05, 0.01, and 0.001, respectively. (After Kerepesi and Galiba, 2000).

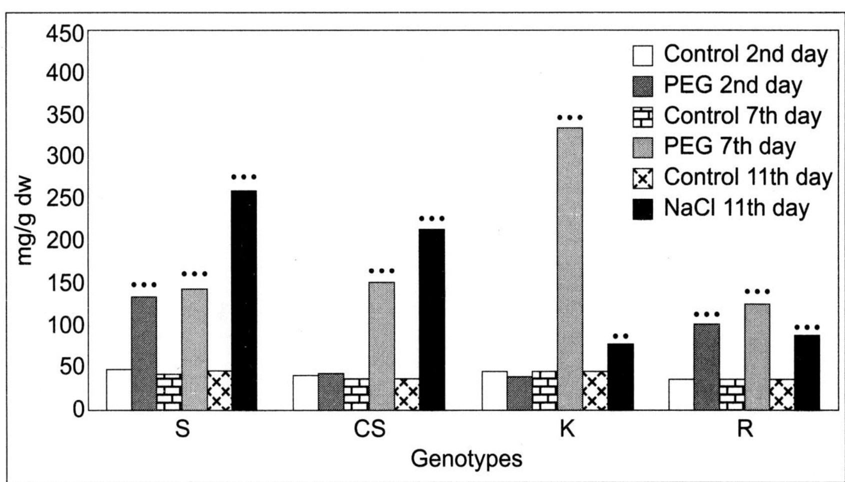

Fig. 3.3 Fructan concentration in the stems of four wheat cultivars exposed to 180 g/kg PEG treated culture until day 7 and then transferred to equi-osmolar 200 mM NaCl solution until day 11. *, **, and *** indicate significance differences between control and stress treatments at P <0.05. 0.01, and 0.001, respectively. Genotypes : S, Sakha-8; K, Kobomugi; CS, Chinese Spring; R, Regina. (After Kerepesi and Galiba, 2000).

direction of sucrose synthesis, but the stressed cells always showed a higher activity (Fig. 3.5A). Unlike sucrose synthase, sucrose phosphate synthatase activity patterns were different between the two types of cells. The sorbitol stress resulted in a substantial increase in sucrose phosphate synthatase activity (Fig. 3.5 B), in good agreement with the very significant increase in sucrose content in the stressed cells (Fig. 3.5 A,B).

An important component of desiccation tolerance is accumulation of a high level of soluble sugars, a characteristics of mature orthodox seeds (Hoekstra et al., 1994). The maturing seeds of soybean accumulate high levels of the raffinose series of oligosaccharides, particularly stachyose in addition to sucrose. The other soluble sugar, trehalose, protects cytosolic components. The hydroxyl constituents of sugars replace the hydration shell around membranes and thus prevent structural damage as water is removed. Sucrose also protects soluble enzymes. Oligosaccharides along with sucrose enhance protection still further by favoring vitrification rather than crystallization. Desiccation of orthodox seeds on the mother plant is a part of their developmental program which allows them to enter a dry quiescent state thereby permitting their storage and survival in various environmental conditions. This terminal phase of seed development, called maturation drying, is also known to ensure the switch from a developmental mode to a germinative mode. The ability of orthodox seeds to withstand severe desiccation generally occurs during the phase of reserve

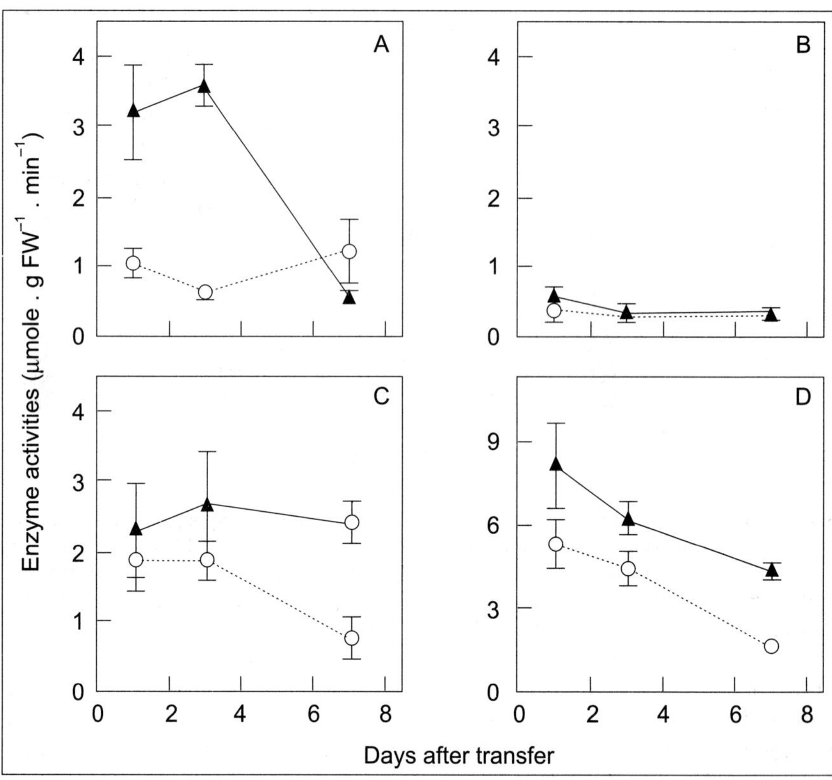

Fig. 3.4 Changes in the activities of enzymes involved in sucrose hydrolysis. The extracts were prepared from sweet potato cells under sorbitol-stressed (▲) and normal (o) conditions. (A) cell wall-bound invertase; (B) acid invertase; (C) alkaline invertase; (D) sucrose synthase assayed in the direction of sucrose cleavage. Each point is average of three independent experiments. (After Wang et al., 2000).

accumulation, but is dependent on the drying rate, which has been shown to affect seed survival after drying.

Numerous cellular and biochemical events appear associated with acquisition of desiccation tolerance of orthodox seeds. They include modifications of ultrastructural characteristics such as vacuolation, synthesis of dehydrins or heat shock proteins, activation of antioxidative defences, accumulation of oligosaccharides, and maintenance of DNA integrity. During slow drying of soybean cv Chippewa 64 seeds, sucrose content increased to 5 times the level present in the axes of seeds held at high relative humidity (128 versus 25 mg/axis, respectively ; Blackman et al., 1992). The stachyose content increased dramatically from barely detectable level upon excision to 483 µg/axis during slow drying but did not increase significantly when seeds were incubated at high relative

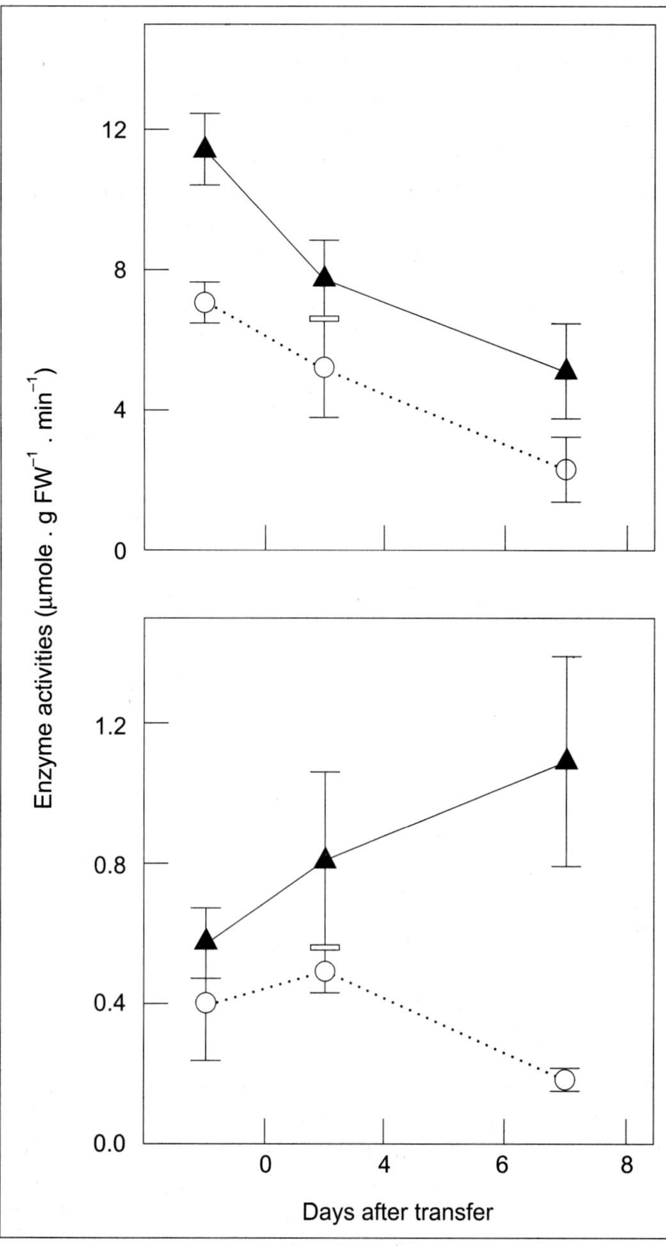

Fig. 3.5 Changes in the activities of enzymes involved in sucrose synthesis. The extracts were prepared from sweet potato cells grown under sorbitol-stressed (▲) and normal (o) conditions. (A) sucrose synthase assayed in the direction of sucrose synthesis; (B) sucrose phosphate synthase. Each point is the average of three independent experiments. (After Wang et al., 2000).

humidity (Fig. 3.6). Galactinol was the only saccharide that accumulated to higher levels in axes from seeds incubated at high relative humidity to axes from seeds that were slowly dried. This suggests that slow drying serves to induce the accumulation of the raffinose series sugars at a point after galactinol biosynthesis . Thus Blackman and associates conclude that stachyose plays an important role in conferring desiccation tolerance. This also suggests that the primary need of the axis for stachyose is not as a storage reserve but as a protectant during desiccation.

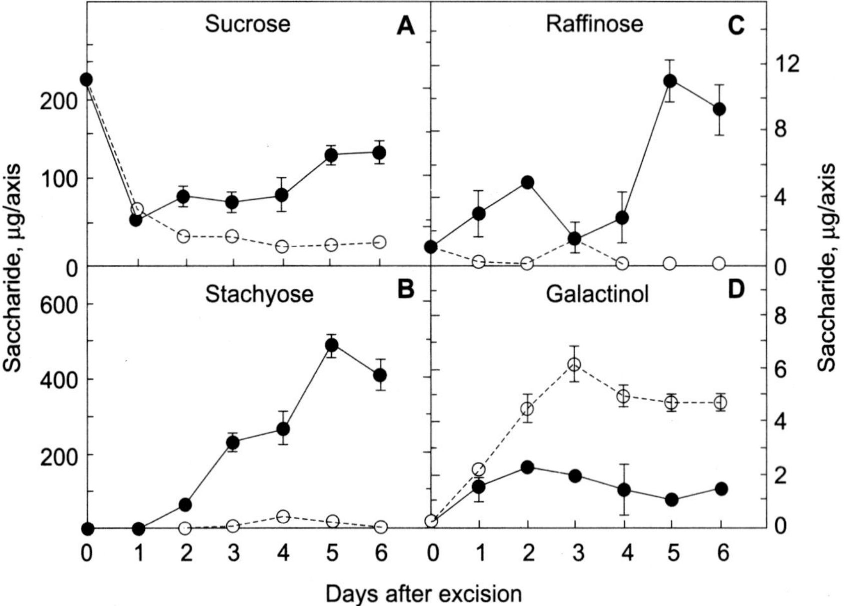

Fig. 3.6 Content of sucrose (A), stachyose (B), raffinose (C), and galactinol (D) in axes of excised developing soybean seeds, undergoing slow drying (●) or high RH control treatment (o). values are the means ± SE of the mean of five samples of five axes each. Note the different scales on the Y axes for each panel. (After Blackman et al., 1992).

In line with the above study, Bailly et al. (2001) collected *Phaseolus* bean seeds throughout their development on the plant and dried at 15°C and 75 percent relative humidity to a final moisture content of about 16 percent to determine whether the onset of tolerance to this drying condition is related to changes in soluble sugars or the activities of the main antioxidant enzymes, viz., superoxide dismutase, catalase, ascorbate peroxidase and glutathione reductase. Measurement of soluble sugars and enzyme activities were made after drying the seeds, and drying tolerance was evaluated by the ability of dried seeds to germinate and to produce normal seedlings. Seeds bacame tolerant to drying at 45 days after anthesis, a time marking physiological maturity. Acquisition of drying tolerance was coincident with an accumulation of raffinose and stachyose. Dried-tolerant seeds were also characterized by a high amount of sucrose and a low content of monosaccharides. The (raffinose + stachyose) : sucrose ratio increased during seed filling, reaching a value close to 1 when all the seeds became tolerant to drying, and maintaining this proportion during the final stages of maturation.

Hoekstra et al. (1994) tried to unravel this problem by studying the reverse changes in soluble sugars in cauliflower seeds during 50 hours of imbibition. Sucrose and stachyose contents decreased , and glucose and fructose accumulated. This occurred in radicles first and subsquently in hypocotyls and cotyledons. Loss of desiccation tolerance in the various seed parts coincided with an increase in glucose and fructose and the complete loss of stachyose, but the sucrose content, the major sugar, was still high. Drying imbibed seeds over silica gel did not evoke resynthesis of stachyose, but did increase sucrose and decrease glucose and fructose contents. Seeds primed in solutions of 30 percent polyethylene glycol (PEG) for 10 days showed a loss of stachyose, while sucrose remained high and glucose and fructose contents were still very low. The redrying of primed seeds did not change the sugar contents, but the primed seeds were still tolerant of desiccation.

Polyethylene glycol (PEG 4000)-induced drought stress increased the water soluble carbohydrate concentrations in 4-wheat, Kobomugi (drought tolerant but salt sensitive), Chinese Spring (moderately drought and salt tolerant), and Regina (both drought and salt sensitive) cultivars. The cultivar Kobomugi which accumulates more water soluble carbohydrates proved more drought tolerant than the cultivar Sakha-8 believed to be more salt and drought tolerant (Kerepesi and Galiba, 2000). However, the rate of increase of water soluble carbohydrates was highest in the most tolerant cultivar Sakha-8 followed by the moderately tolerant cultivar, Chinese Spring and the sensitive cultivars, Regina and Kobomugi. The drought tolerant cultivars accumulated sucrose to a significantly greater level than did the sensitive ones under non-ionic stress conditions.

SALINITY-INDUCED DEHYDRATION

Effects of salt and drought stresses on the water soluble carbohydrate content in wheat seedlings was examined by Kerepesi and Galiba (2000) to determine the involvement of major sugar compounds in the adaptive processes. Four wheat varieties differing in salt and drought tolerance were exposed to PEG and NaCl stresses. Total water soluble carbohydrate, glucose, fructose, sucrose and fructan contents of stems (non-photosynthetic tissue – the amount of distribution of water soluble carbohydrate components in the stem is an indirect indication of translocation and net metabolic activity of sink tissues during stress treatments) were examined. The tolerant genotypes accumulated more soluble carbohydrate than did the sensitive ones (Table 3.1). Both ionic and non-ionic stresses increased the concentration of reducing sugars, sucrose, and fructans. Changes in fructan content of plants after transfer from PEG to NaCl containing solutions is genotype-dependent, increasing in salt tolerant and decreasing in salt sensitive cultivars. The authors propose that the water soluble carbohydrates might be a useful marker for selecting genotypes that are more salt or drought tolerant. However, it may be noted that Nagy and Galiba (1995) have earlier stressed that in wheat varieties, drought and salt tolerance are not necessarily linked.

The fructan content after salt treatment was positively correlated with the degree of salt tolerance (Fig. 3.3). Fructan in the salt resistant (Sakha-8) and moderately tolerant (Chinese Spring) cultivars, increased, while fructan content decreased in the sensitive cultivars (Kobomugi and Regina) following a transfer from PEG to a NaCl medium. Fructan content could be a useful indicator of the degree of salt tolerance since considerably more accumulation of fructan was noted in the salt tolerant genotypes.

SUGARS AND DEHYDRATION-INDUCED ROS

Consistent with the well-established implication of resistance to oxidative stress in tolerance of drought, salinity, and cold stresses, it is relevant to note the accumulation of ROS (reactive oxygen species) owing to increased use of oxygen instead of $NADP^+$ as the photosynthetic electron acceptor. Polle (1996) noted that an increased carbohydrate content in source leaves of transgenic tobacco which expresses yeast invertase in the apoplastic space is associated with enhanced activities of catalase, ascorbate peroxidase and monodehydroascorbate-radical reductase as well as an increased ascorbate/dehydroascorbate ratio compared with source leaves of the wild-type. Thus it appears that an increased availability of reductant in carbohydrate-accumulating pyrophosphatase-expressing plants may increase at least certain aspects of the antioxidative defence capacity of these plants. Then in order to test the effect of increasing dehydration in orthodox seeds on sugar metabolism and ROS tolerance. Bailly et al. (2001) collected *Phaseolus* bean seeds throughout their development on the plant and dried them at 15°C and 75 percent relative humidity to a final moisture content of about 16 percent. They found that acquisition of drying tolerance is related to a reorientation of the enzymatic antioxidant defence system. The drying-tolerant dried seeds displayed high catalase and glutathione reductase activities and low superoxide dismutase and ascorbate peroxidase activities, while the opposite condition was observed in immature dried seeds. These results suggest that oligosaccharide metabolism and enzymatic antioxidant defences may be involved in acquisition of drying tolerance during the bean seed development.

In the same year, Fukushima et al. (2001) investigated the salt tolerance of transgenic tobacco, in which yeast invertase was expressed in the apoplastic (Apo-Inv) spaces. Whereas the photosynthetic activities in wild-type tobacco in light were inhibited under salt stress, the transgenic Apo-Inv tobacco maintained constant photosynthetic activities. The physical appearance of plants under salt stress also indicates that yeast invertase expression in the apoplastic space is beneficial for inducing salt tolerance. Apo-Inv tobacco has a much higher osmotic pressure increase in the cell sap than did the wild-type tobacco under the similar stress condition.

TOXIC ION (ARSENIC) STRESS

Jha and Dube (2004) studied the seedlings of two rice cultivars (Malviya-36 and Pant-12) for effect of increasing levels of arsenic *in situ* on the content of sugars and the activity of several enzymes of starch and sucrose metabolism : α-amylase, β-amylase, starch phosphorylase, acid invertase, sucrose synthase and sucrose phosphate synthase. During a growth period of 10-20 days As_2O_3 at 25 and 50 μM in the growth medium caused an increase in reducing, non-reducing and total soluble sugars. An increased conversion of non-reducing to reducing sugars was observed concomitant with arsenic toxicity. The activities of α-amylase, β-amylase and sucrose phosphate synthase declined, whereas starch phosphorylase, acid invertase and sucrose synthase were elevated. Results indicate that in rice seedlings arsenic toxicity causes perturbations in carbohydrate metabolism leading to the accumulation of soluble sugars by altering enzyme activity. Sucrose synthase possibly plays a positive role in synthesis of sucrose under arsenic toxicity.

HIGH LIGHT-INTENSITY STRESS

Savitch et al. (2000) examined the effects of low temperature stress, cold acclimation and growth at high irradiance in spring wheat cv Katepwa and winter wheat cv Monopol leaves and crowns with respect to sucrose utilization and carbon allocation. In leaves, during either cold stress, cold acclimation or acclimation to high irradiance, the sucrose/starch ratio increased by 5- to 10-fold (Table 3.2) and neutral invertase activity increased by 2- to 2.5-fold (Table 3.3) in both the spring and the winter wheat cultivars. In contrast, Monopol winter wheat, but not Katepwa spring wheat, exhibited a 3-fold increase in leaf sucrose phosphate synthase activity, a 4-fold increase in sucrose : sucrose fructosyl transferase activity

Table 3.2 Effects of low temperature stress, cold or high light acclimation on accumulation (mmol m^{-2}) of carbohydrates in leaves of spring (cv Katepwa) and winter (cv Monopol) wheat. Data represent the mean±SE. n =3. (After Savitch et al., 2000).

Growth conditions	Starch	Sucrose	Glucose	Fructose	←Ratios (w/w)→ Sucrose/ Starch	(Glu + Fru)/ Sucrose
Katepwa						
20°C/250 PFD*	4.34±0.21	0.52±0.06	0.07±0.01	0.07±0.03	0.2	0.14
20-5°C/250 PFD	2.88±0.31	3.65±0.79	0.32±0.06	0.48±0.17	2.4	0.12
5°C/250 PFD	1.87±0.41	3.28±0.23	0.86±0.07	1.09±0.09	3.3	0.31
20°C/800 PFD	3.00±0.17	3.53±0.36	0.67±0.07	0.55±0.03	2.2	0.18
Monopol						
20°C/250 PFD	4.30±0.78	1.35±0.27	0.31±0.09	0.36±0.03	0.6	0.26
20-5°C/250 PFD	2.72±0.47	3.13±0.68	0.46±0.11	0.45±0.13	2.2	0.15
5°C/250 PFD	2.92±0.42	4.59±0.97	1.34±0.12	3.55±0.38	3.0	0.56
20°C/800 PFD	3.30±0.36	5.19±0.31	1.03±0.21	1.04±0.12	3.0	0.21

•PFD = Photon flux density

Table 3.3 Effects of low temperature stress, cold or high light acclimation on the regulation of enzyme activities (mmol m^{-2} s^{-1}) of sucrose and fructan biosynthesis and breakdown in leaves of spring (cv Katepwa) and winter (cv Monopol) wheat. Data represent the mean ± SE. n =3. (After Savitch et al., 2000).

Growth conditions	SPS	SS(s)	SS(d)	Acid invertase	Neutral invertase	SST
Katepwa						
20°C/250 PFD	1.6±0.2	1.2±0.2	ND	0.30±0.02	0.56±0.12	0.22±0.02
20-5°C/250 PFD	1.3±0.2	1.5±0.2	ND	0.31±0.05	0.52±0.02	0.32±0.05
5°C/250 PFD	1.6±0.1	0.8±0.1	1.4±0.1	0.33±0.05	1.13±0.05	0.33±0.04
20°C/800 PFD	2.0±0.1	0.9±0.1	1.4±0.1	0.45±0.05	0.73±0.06	0.39±0.03
Monopol						
20°C/250 PFD	1.6±0.1	1.5±0.2	ND	0.47±0.09	0.73±0.05	0.50±0.06
20-5°C/250 PFD	1.2±0.2	1.8±0.2	ND	0.57±0.13	0.71±0.05	0.80±0.02
5°C/250 PFD	5.0±1.2	2.5±0.2	ND	3.11±0.24	1.83±0.12	1.94±0.19
20°C/800 PFD	2.0±0.2	0.8±0.1	ND	0.52±0.09	1.33±0.12	0.24±0.04

SPS = Sucrose phosphate synthase; SS(s)= Soluble sucrose synthase; SS(d)= Detectable sucrose synthase, SST= Sucrose: sucrose fructosyl transferase; ND = Not detected

and a 6.6-fold increase in acid invertase upon cold acclimation. Although leaves of cold stressed and high-light grown spring and winter wheat showed 2.3- to 7-fold higher sucrose levels than controls, these plants exhibited a limited capacity to adjust either sucrose phosphate synthase or sucrose synthase activity. In addition, the acclimation to high light resulted in a 23–31 percent lower starch abundance and no changes at the level of fructan accumulation in leaves of either winter or spring wheat when compared with controls. However, high light-acclimated winter wheat exhibited a 1.8-fold higher neutral invertase activity and high light-acclimated spring wheat exhibited an induction of detectable sucrose synthase (SS (d)) activity when compared with controls. Crowns of Monopol showed higher fructan accumulation than Katopwa upon cold and high light aclimatiom. The authors suggest that the differential adjustment of CO_2–saturated rates of CO_2 assimilation upon cold acclimation in Monopol winter wheat, as compared with Katepwa spring wheat, is associated with the increased capacity of Monopol for sucrose utilization through the biosynthesis of fructans in the leaves and subsequent export to the crowns. In contrast, the differential adjustment of CO_2 – saturated rates of CO_2 assimilation upon high light acclimation of Monopol appears to be associated with both increased fructan and starch accumulation in the crowns.

HEAT STRESS

Potato genotypes characterized as susceptible and tolerant to heat stress were grown at 19/17°C, and a subset was transferred to 31/29°C by Lafta and Lorenzen (1995) to study the effect of high temperature on plant growth and carbohydrate metabolism. High temperatures reduced growth of tubers more than that of shoots, and increased accumulation of foliar sucrose, but decreased starch accumulation in mature leaves, but did not affect glucose. The sucrose phosphate synthase activity increased significantly in mature leaves of plants subjected to high temperature. The activity of sucrose synthase and ADP-glucose pyrophosphatase was reduced in tubers, but less quickly than the leaf sucrose phosphate synthase activity.

SUCROSE AND STRESS METABOLISM

Photosynthesis converts solar energy to chemical energy which then drives the synthesis of sugars from CO_2 and water. Sugars play multiple roles in all aspects of plant life. The abundance or depletion of sugars or their detrivatives initiate various responses in plants and have profound effects on plant metabolism, growth and development. Sugars provide the main respiratory substrates for the generation of energy and metabolic intermediates that are then used for synthesis of macromolecules and other cell constituents. Ribulose and deoxyribulose sugars form part of the structure of DNA and RNA. Polysaccharides are the major structural elements of plant cell walls, and for proper functioning of many proteins and lipids, linkage to sugar is required. Variations in environmental factors like light intensity and water availability (drought and cold/freezing) lead to a significant decrease in the efficiency of photosynthesis, and thus reduce the supply of sugars to sink organs. Also in germinating seeds under unfavorable environmental conditions, mobilization of carbohydrate reserves in the cotyledons is delayed, which results in depletion of available sugars and a decrease in seedling vigor. Sugar starvation also initiates changes in cellular processes to recycle cellular constituents and dramatically changes the pattern of gene expression.

Dehydration stress (drought, salinity, cold) initially results in reduced photosynthesis and sugar synthesis. During adaptation to the stress, some crop species and varieties are able to increase the compatible solutes (sugars, amino acids, inorganic ions, cyclitols, betaine etc), depending upon intensity of stress and species specificity. Under severe stress, however, sugar level depletes considerably due to the reduced rate of CO_2 assimilation. Sugar starvation generally triggers sequential changes in the following cellular events : (a) arrest of cell growth, (b) rapid consumption of cellular carbohydrate content and decrease in respiration rate, (c) degradation of lipids and proteins, (d) increase in accumulation of Pi, phosphorylcholine and free amino acids, and (e) decline in glycolytic enzymatic activities (Yu, 1999). The changes in metabolism are involved in adaptation response of plant cells to sugar starvation so as to survive, at least for some time, in absence of sugar. There is a decrease in enzymatic activities related to sugar metabolism and respiration; nitrate reduction and assimilation; and protein synthesis (Fig. 3.7). Decreases in these enzymatic activities presumably protect cells against nutrient stress by switching off biosynthesis (i.e. growth) to conserve energy. At the same time, an increase in enzymatic activities related to catabolism of fatty acids, amino acids, and proteins occurs. Such a change can constitute protein and lipid catabolism for sugar catabolism to sustain respiration and metabolic processes (Yu, 1999). The findings of Devaux et al. (2003) also support this above view of Dr. Yu. Devaux and co-workers found that in sugar starved roots of tomato, starch, proteins and amino acids (aspergine and glutamine) decreased but ammonium ions increased up to 4 days and then decline started (Fig. 3.8).

At the gene level, sugar favors the expression of enzymes in connection with biosynthesis, utilization, and storage of reserves (including starch, lipid, and proteins), and represses the expression of enzymes involved in photosynthesis and reserves mobilization (Koch, 1996). Generally, gene expression repressed by sugar is up-regulated by sugar starvation, whereas that enhanced by sugar is down-regulated. The alteration of gene expression by sugar starvation results in induction of synthesis of pre-existing or new proteins and repression of normally expressed proteins.

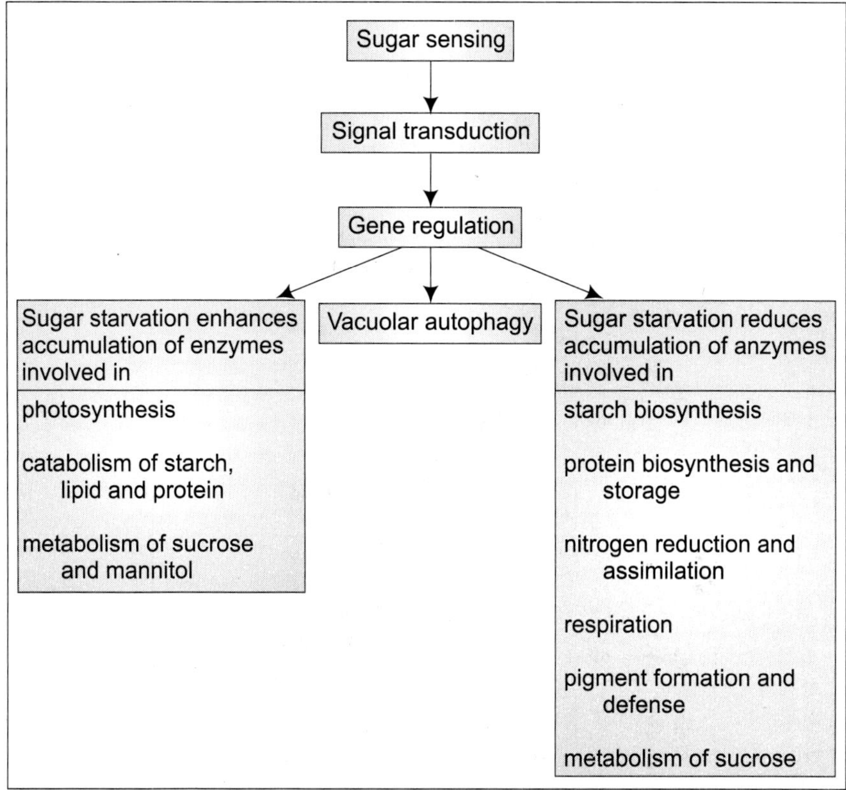

Fig. 3.7 Events in cellular responses to sugar starvation in plants (After Yu et al., 1999).

At the initiation of rice seed germination, active metabolism and a rise in the respiration rate cause rapid sugar depletion in the embryo, which then triggers the expression of α-amylase gene and degradation of starch in this tissue (Yu et al., 1996). Sugar depletion is the primary factor in initiating the synthesis of phytohormone gibberellin in the embryo, since sugar reduces the quantity of gibberellin in this tissue. Sugar repression of α-amylase gene expression involves control of transcription and mRNA stability (Lu et al., 1998). A sugar response complex in the promoter region of a sucrose-deprivation-induced α-amylase gene, α-*Amy 3,* has been identified. Sugar repression of mannitol use allows large amounts of mannitol to be stored as a reserve carbohydrate and osmoprotectant, while sugars are preferentially used in central metabolism. However, when energy and carbon demand are high (e.g. in meristems), sugar pools rapidly deplete and become growth limiting. Subsequent depression of mannitol dehydrogenase makes stored mannitol available as a carbon and energy

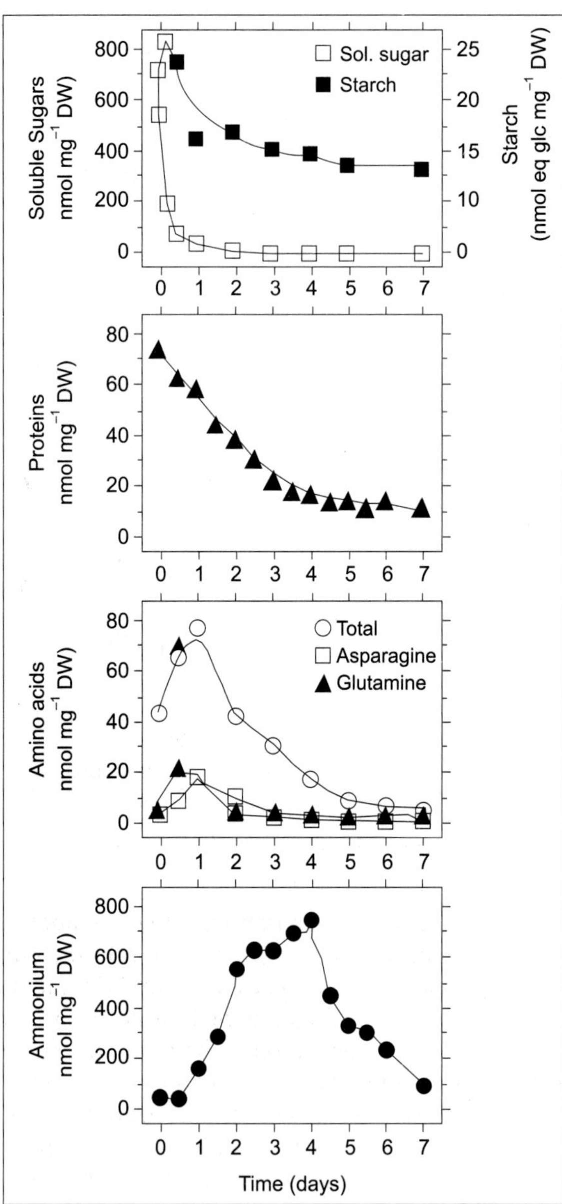

Fig. 3.8 Effects of carbohydrate depletion on the metabolite content of excised tomato roots. Changes in total soluble sugar content (sucrose, glucose, fructose), starch content, protein content, amino acid content (asparagine, glutamine, total), and NH_4^+ content were measured in excised tomato roots from 4-week-old plants. After excision, the roots were cultured . The data represent the means (±SD) of three assays from three independent experiments. (After Devaux et al., 2003).

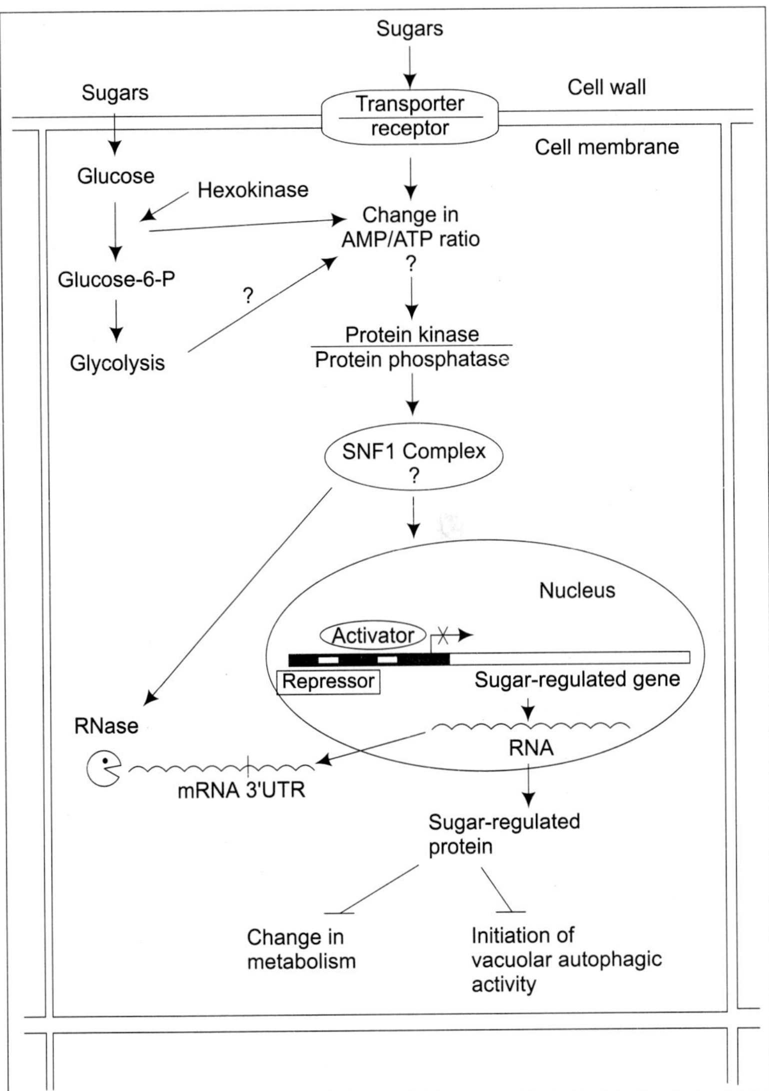

Fig. 3.9 Hypothetical model of genetic and cellular responses of plant cells to sugar, including a sugar signal transduction pathway and a mechanism of gene regulation. Three elements function as sugar sensors: a hexokinase, a sugar transporter, and a change in the AMP/ATP ratio. Protein phosphatase and protein kinase are involved in the signal transduction pathway. In some cases, the SNF1 complex may be a component of the signal transduction pathway. The expression level of sugar-regulated genes is determined by the control of promoter transcriptional activity and mRNA stability. In the presence of sugar, the expression of sugar-starvation-induced genes is suppressed and there is no change in metabolism or vacuolar autophagic activity. Under sugar starvation, an opposite action of these events most likely occurs (After Yu, 1999).

source for continued growth. Under salt stress, repression of mannitol dehydrogenase expression by both sugar and salt, results in an even larger accumulation of mannitol (Williamson et al., 1998).

Nuclear proteins from rice suspension cells that bind to the motif *TATCCA* element in a sequence specific and sugar-dependent manner have also been identified (Lu et al., 1998). The requirement for a *NSF1*–related protein kinase-1 (*Snrk1*) in Suc-activated expression of a Suc synthase gene was demonstrated in potato by an antisence RNA approach (Purcell et al., 1998)

Thus, based on the available information, a model of sugar sensing, signal transduction, and mechanisms of gene regulation in plant cells is shown in Figure 3.9. An understanding of how plants respond to sugar starvation and regulate the mobilization of stored carbohydrates can help us to design crops with higher stress tolerance capacity and is thus of biotechnological importance.

REFERENCES

Bailly, C.; C. Andigier; F. Ladome; et al. 2001. Changes in oligosaccharide content and antioxidant enzyme activities in developing bean seeds as related to acquisition of drying tolerance and seed quality. J. Exp. Bot. 52 (357): 701-08.

Blackman, S.A.; R.L. Obendorf and A.C. Leopold, 1992. Maturation proteins and sugars in desiccation tolerance of developing soybean seeds. Plant Physiol. 100: 225-30.

Bruni, F.B. and A.C. Leopold, 1991. Glass transitions in soybean seed. Relevance to anhydrous biology. Plant Physiol. 96: 660- 63.

Buitink, J.; M.M.A.E. Laessens; M.A. Hernmings and F.A. Hoekstra, 1998. Influence of water content and temperature on molecular mobility and intracellular glasses in seeds and pollen. Plant Physiol. 118: 531-41.

Castonguay, Y.; P. Nadeau; P. Lechasseur and L. Choninard, 1995. Differential accumulation of carbohydrates in alfalfa cultivars of contrasting winter hardiness. Crop Sci. 35: 509-16.

Chen, Y. and J.S. Buris, 1990. Role of carbohydrates in desiccation tolerance and membrane behavior in maturing maize seed. Crop Sci. 30: 971-75.

da Silva, J.M. and M.C. Arrabaca, 2004. Contributions of soluble carbohydrates to the osmotic adjustment in the C_4 grass *Setaria sphacelata* : a comparison between rapidly and slowly imposed water stress. J. Plant Physiol. 161: 551-55.

Devaux, C.; P. Baldet; J. Joubes; et al. 2003. Physiological, biochemical and molecular analysis of sugar starvation responses in tomato roots. J. Exp. Bot. 54(385): 1143-51.

Dionne, J.; Y. Castonguay; P. Nadeau and Y. Desjardins, 2001. Freezing tolerance and carbohydrate changes during cold acclimation of green type annual bluegrass (*Poa annua* L.) ecotype. Crop Sci. 41: 443-51.

Fukushima, E.; Y. Arata; T. Endo; et al. 2001. Improved salt tolerance of transgenic tobacco expressing apoplastic yeast-derived invertase. Plant Cell Physiol. 42: 245-49.

Guy, C.L.; J.L.A. Huber and S.C. Huber, 1992. Sucrose phosphate synthase and sucrose accumulation at low temperature. Plant Physiol. 100 : 502-08.

Hincha, D.K.; U. Sonnewald; L. Willmitzer and J.M. Schmitt, 1996. The role of sugar accumulation in leaf frost hardiness—Investigations with transgenic tobacco expressing a bacterial pyrophosphatase or a yeast invertase gene. J. Plant Physiol. 147: 604-10.

Hoekstra, F.A. and J. Buitink, 2001. Mechanisms of plant desiccation tolerance. Trends Plant Sci. 8: 431-38.

Hoekstra, F.A.; L.M. Crowe and J.H. Crowe, 1989. Differential desiccation sensitivity of crown and *Pennisetum* pollen linked to their sucrose content. Plant Cell Environ. 12: 83-91.

Hoeksra, F.A.; A.M. Haigh; F.A.A. Tetteroo and T. van Roekel, 1994. Changes in soluble sugars in relation to desiccation tolerance in cauliflower seeds. Seed Sci. Res. 4: 143-47.

Jha, A.B. and R.S. Dube, 2004. Carbohydrate metabolism in growing rice seedlings under arsenic toxicity. J. Plant Physiol. 161: 867-72.

Kerepesi, I. and G. Galiba, 2000. Osmotic and salt stress-induced alteration in soluble carbohydrate content in wheat seedlings. Crop Sci. 40: 482-87.

Koch, K.E. 1996. Carbohydrate-modulated gene expression in plants. Annu. Rev. Plant Physiol. Plant Mol. Biol. 47: 509-40.

Krapp, A. and M. Stitt, 1994. Influence of high-carbohydrate content on the activity of plastidic and cytosolic isoenzyme pairs in photosynthetic tissues. Plant Cell Environ. 17: 861-66.

Lafata, A.M. and J.H. Lorenzen, 1995. Effect of high temperature on plant growth and carbohydrate metabolism in potato. Plant Physiol. 109: 637-43.

Leopold, A.C.; W.Q. Sun and L. BernalLugo, 1994. The glassy state in seeds : analysis and function. Seed Sci. Res. 4: 267-74.

Lu, C.A.; E.K. Lin and S.M. Yu, 1998. Sugar response sequence in the promoter of a rice α-amylase gene reserve as a transcriptional enhancer. J. Biol. Chem. 273: 10120-10131.

Matsuura-Endo, C.; A. Kobayashi; T. Noda ; et al. 2004. Changes in sugar content and activity of vacuolar acid invertase during low temperature storage of potato tubers from six Japanese cultivars. J. Plant Res. 117: 131-37.

Menendez, C.M.; E. Ritter; R. Schafer-Prege ; et al. 2002. Cold sweetening in diploid potato : mapping quantitative trait loci and candidate genes. Genetics 162: 1423-34.

Nagy, Z. and G. Galiba, 1995. Drought and salt tolerance are not necessarily linked : A study on wheat varieties differing in drought resistance under consecuitive water and salinity stresses. J. Plant Physiol. 143: 168-74.

Polle, A. 1996. Developmental changes of antioxidative systems in tobacco leaves as affected by limited sucrose export in transgenic plants expressing yeast-invertase in apoplastic space. Planta 198: 253-62.

Purcell, P.C.; A.M. Smith and N.G.. Halford, 1998. Antisense expression of a sucrose-nonfermenting-l-related protein kinase sequence in potato results in decreased expression of sucrose synthase in tubers and loss of sucrose-inducibility of sucrose synthase transcripts in leaves. Plant J. 14: 195-202.

Savitch, LV.; T. Harney and N.P.A. Huner, 2000. Sucrose metabolism in spring and winter wheat in response to high irradiance, cold stress and cold acclimation. Physiol. Plant. 108: 270-78.

Tabaei-Aghdaei, S.R.; R.S. Pearce and P. Harrison, 2003. Sugars regulate cold-inducd gene expression and freezing tolerance in barley cell cultures. J. Expt. Bot. 54 (387): 1565-75.

Wang, H-L.; P. Du Lee; W.L. Chen; et al. 2000. Osmotic stress-induced changes of sucrose metabolism in cultured sweet potato cells. J. EXP. Bot. 51 (353): 1991-99.

Williamson, J.D.; W.W. Guo and D.M. Pharr, 1998. Cloning and characterization of a genomic clone (Accession No. AF 067082) encoding mannitol dehydrogenase from celery (*Apium graveolens*). Plant Physiol. 118: 329.

Yang, J.; J. Zhang; Z. Wang and Q. Zhu, 2001. Activities of hydrolytic enzymes and sucrose phosphate synthase in the stems of rice subjected to water stress during grain filling. J. Exp. Bot. 52 (364): 2169-79.

Yu, S.M. 1999. Cellular and genetic responses of plants to sugar starvation. Plant Physiol. 121: 687-93.

Yu, S..; Y.C. Lee; S.C. Fang; et al. 1996. Sugars act as signal molecules and osmotica to regulate the expression of α-amylase genes and metabolic activities in germinating cereal grains. Plant Mol. Biol. 30: 1277-89.

Zabotina, O.A.; D.A. Ayupova; I.A. Larskaya; et al., 1998. Physiologically active oligosaccharides accumulating in the roots of winter wheat during adaptation to low temperature. Russian J. Plant Physiol. 45: 221-26.

SUGGESTED READINGS

Bailly, C.; C. Audigier; F. Ladonne; et al. 2001. Changes in oligosaccharide content and antioxidant enzyme activities in developing bean seeds as related to acquisition of drying tolerance and seed quality. J. Exp. Bot. 52 (357) : 701-08.

Hincha, D.K.; U. Sonnewald; L. Willmitzer and J.M. Schmitt, 1996. The role of sugar accumulation in leaf frost hardiness – Investigations with transgenic tobacco expressing a bacterial pyrophosphatase or a yeast invertase gene. J. Plant Physiol. 147: 604-10.

Kerepesi, I. and G. Galiba. 2000. Osmotic and salt-induced alteration in soluble carbohydrate content in wheat seedlings. Crop Sci. 40: 482-87.

Koch, K.E. 1996. Carbohydrate-modulated gene expression in plants. Annu. Rev. Plant Physiol. Plant Mol. Biol. 47: 509-40.

4

COMPLEX SUGARS — CYCLITOLS

Introduction

The complex sugars like raffinose family oligosaccharides* (RFOs) are of almost ubiqutous occurrence in crop seeds (the maturation process of seeds mimics the activities taking place during dehydration, and germination mimics the activities taking place during rehydration). RFOs accumulate during seed development and disappear rapidly during germination. The biosynthesis of raffinose, the first member of the series, proceeds by addition of a galactosyl unit to sucrose. Galactinol, a galactosyl derivative of *myo*-inositol, acts as a galactosyl donor. It is synthesized from UDP- D-galactose and *myo*-inositol. Stachyose, verbascose and ajugose, the next higher RFOs, are either synthesized by galactinol-dependent galactosyltransferases or by transfer of galactosyl units between two RFO molecules. In seeds, the metabolism of methylated inositols, such as D-ononitol and D-pinitol, is linked with the RFO pathway. In contast to *myo*-inositol, these cyclitols** are galactosylated by transfer of galactosyl residues from galactinol and not from UDP- D-galactose. However, the resulting galactosyl cyclitols can replace galactinol as galactosyl donors for the biosynthesis of stachyose. These branches of the RFO pathway are active in the seeds of a range of crop species, especially in legumes (Peterbauer and Richter, 2001). They studied the accumulation pattern of RFOs in maturing seeds of two pea lines with contrasting RFO composition. Seeds of the line SD1 accumulated stachyose as the predominant RFO, whereas verbascose, the next higher homolog of stachyose, was almost absent. However, in the seeds of line RRRbRb, a high level of verbascose

---

* The raffinose family oligosaccharides are sucrose galactosides of sugar, 5 and 6 rings based on two molecules, pyran and furan.

** The nomenclature of "cyclitols" provides the 1973 recommendations of the International Union of Pure and Applied Chemistry (IUPAC) Commission on the Nomenclature of Organic Chemistry (CNOC) and the IUPAC-IUB Commission on Biochemical Nomenclature (CBN). 1. Cyclitols with only hydroxyl or substituted hydroxyl groups (inositol and other cyclitols), and 2. Cyclitols with groups other than hydroxyl or substituted hydroxyl (inositol detrivatives and other cyclitols).

was accumulated alongside with stachyose. Further, Peterbauer et al. (2002) characterized a multifunctional stachyose synthase from developing pea seeds. The protein, a member of family 36 of galactoside hydrolases, catalyses the synthesis of stachyose, the tetrasaccharide of the raffinose series, by galactosyl transferase from galactosyl to raffinose.

The enzyme, galactosyl-ononitol synthase catalyses reversible galactosyl transfer from galactinol to D-ononitol yielding galactosylononitol and *myo*-inositol. Earlier, Peterbauer et al. (1998) demonstrated that galactosylononitol can substitute for galactinol in the stachyose synthase catalysed synthesis of stachyose. Galactosylononitol and galactosylononitol synthase activity were initially detected half-way through *Vigna umbellata* seed development. In mature seeds, stachyose was the predominant sugar (3.6 mμmol/seed), followed by sucrose (1.5 mμmol/seed) and galactosylononitol (1.2 mμmol/seed). Stachyose is the major soluble carbohydrate in seeds of a number of important crop species. High levels of stachyose mRNA are transiently accumulated midway through seed development, and the enzyme is also present in mature adzuki bean (*Vigna angularis*) seeds and during germination (Peterbauer et al., 1999).

Pinitol (1D-3O- methyl-chiro-inosital) is the major cyclitol present in developing soybean seed tissues. As the seed tissue progresses towards maturity, there are strong correlations between the decrease of *myo*-inositol and starch, and the increasing raffinose saccharides (Kuo et al., 1997). Pinitol is correlated negatively with raffinose saccharides in both seed coats and axes, but this correlation is insignificant in cotyledons. Pinitol concentration in all tissues fluctuates more than other components during seed development, and it decreases sharply as axes turns yellow, and rapidly accumulates raffinose saccharides. These observations suggest that pinitol may represent a transient component of the carbon pool in soybean seed tissues during development toward physiological maturity.

Accumulation of high levels of certain oligosaccharides, particularly stachyose, are necessary for *lea* proteins (maturation proteins in orthodox seeds) to be effective for providing desiccation tolerance. The galactosides are stored together with raffinose series oligosaccharides. The cyclitol galactosides serve as carbon storage material that promote desiccation tolerance. For instance, in young seeds of *Vigna* species, sucrose is the predominant sugar, besides glucose, fructose and cyclitols *myo*-inositol and D-ononitol. Accumulation of cyclitol galactosides starts in the late stages of seed maturation, and the same is true for raffinose and stachyose. Galactinol is the galactosyl donor for the biosynthesis of raffinose series oligosaccharides and cyclitol galactosides. Parallel to the accumulation of α-galactosides, the amounts of reducing sugars, sucrose and free cyclitols start to deplete. Stachyose is the major storage carbohydrate in mature seeds, whereas galactosyl ononitol is stored at relatively low amounts. In leaves and pod walls neither cyclitol galactosides nor raffinose series oligosaccharides are synthesized.

During germination of *Vigna* seeds, however, α-galactosides are hydrolyzed relatively fast and sucrose becomes the predominant sugar in the seedling. The breakdown of cyclitol galactosides and raffinose series oligosaccharides start right at the beginning of imbibition and goes on throughout the phase of germination. Synthesis of D-ononitol is known to function as compatible solutes for the seedling.

Further, *myo*-inositol is the central component of several biochemical pathways. It is a part of at least four cycles in which it is either cycled or shunted to end products with slow turnover : (a) inositol phosphates are essential to signaling in plants; inositol hexaphosphate provides for phosphate storage, (b) inositol containing lipids are the components of membranes, (c) galactinol synthesized from UDP-galactose and inositol is the basic substrate for the raffinose series of sugars. The sugars have been implicated in stress tolerance and possibly also in carbohydrate transport, and (d) inositol may be conjugated to auxins, preventing biological activity and allowing long-distance transport within the plant. In these cases, enzyme cycles exist for the regeneration of inositol without significant net *de novo* synthesis (Nelson et al., 1998). There are two pathways : (a) D-glucuronate is synthesized directly from inositol and is used in the production of cell wall components, glycoproteins, gums, and mucilage, and (b) monomethylation or dimethylation of inositol. Some of the derivatives of this pathway may not be metabolized further. Inositol itself and these methylated derivatives increase in some plant cell types in association with high external NaCl concentrations and dehydration. *De novo* synthesis of inositol proceeds from glucose-6-phosphate through inositol-1-phosphate in two steps catalyzed by inositol-1-phosphate synthetase and inositol monophosphatase. Inositol-1-phosphatase synthetase is tightly controlled : both inhibition by inositol and activation by choline lead to efficient partitioning of substrate from the main flux of carbon. The promoter elements and two component regulators for feedback inhibition of synthase activity by inositol and enhancement of transcription by choline have been identified. Repression of transcription during periods that do not require inositol and activation during periods of active growth represent fine-tuning of this essential pathway, which also coordinates enzymes that synthesize products downstream of inositol (Nelson et al., 1998).

Mannitol, a six carbon non-cyclic sugar alcohol, is the most abundant polyol (cyclitol) in nature which plays various roles in plants including carbon storage, osmoregulation, free radical scavenging, and serving as a compatible solute. Mannitol is a phloem translocatable photoassimilate constituting to almost 50 percent of the total assimilate in some crop species, the remainder largely being sucrose. Mannitol is synthesized in cytosol of photosynthetic leaf cells by the NADPH-dependent reduction of mannose-6-phosphate to mannitol-6-phosphate, constituting a reduction of carbon-1. The consequent removal of phosphorus from mannitol-6-phosphate by a phosphatase produces mannitol (Fig. 4.1). For unravelling mannitol catabolism, Fellman and Loescher (1987) supplied radioactively labelled sucrose to celery excised leaves in which it was readily converted to CO_2, whereas the conversion of mannitol to CO_2 is largely restricted to actively growing sinks. Petioles of celery contain a substantial storage pool of mannitol, with ~80 percent of the mannitol being in vacuoles of the petiole storage parenchyma cells and the remainder in the cytosol. The activity of the mannitol synthetic enzyme, mannose-6-phosphate reductase, is extremely low in petiole parenchyma tissue, supporting the hypothesis that mannitol is accumulated from the phloem translocation

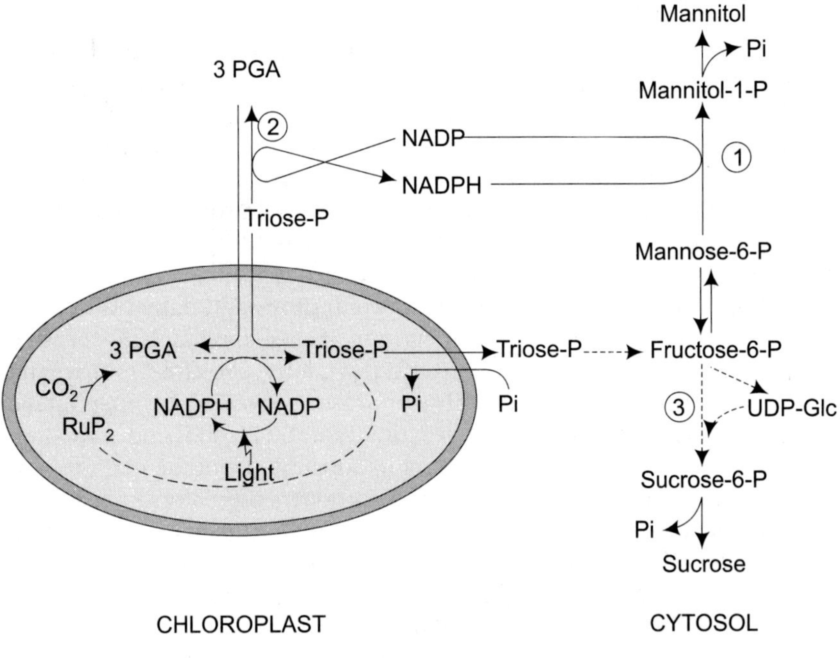

Fig. 4.1 Simplified scheme showing the biosynthesis of mannitol from photosynthetically fixed CO_2 (After Pharr et al., 1995).

stream (Pharr et al., 1995). Mannitol can perform the function of an osmolyte as their water-like OH groups may mimic the structure of water and maintain an artificial sphere of "hydration" around the macromolecules (they protect large molecules during dehydration and rehydration, acting as stabilizers against rapid osmotic changes and oxidative stress (Tarszynski et al., 1993). This compatible solute may also function as a scavenger of membrane-damaging hydroxyl radicals (ROS) generated during stress, thereby preventing peroxidation of lipids and resulting cell damage.

Malate also accumulates preferentially in vacuole whereas mannitol is found in both vacuole and cytosol. These two compounds are also distributed within the extracellular compartments and the apoplastic space of leaves. Mannitol, which may be loaded into the sieve elements of the phloem, is translocated in long-distance transport from the apoplast of the leaves to other parts of the plant.

Sheveleva et al. (1998) analyzed transgenic tobacco expressing *Stpd l,* a cDNA encoding sorbitol-6-phosphate dehydrogenase from apple, under the control of a cauliflower mosaic virus ^{35}S promoter. In 125 independent transformants variable amounts of sorbitol ranging from 0.2 to 130 µmol g^{-1} fr wt was found. Plants that accumulated up to 2 to 3 µmol g^{-1} fr wt sorbitol had slower growth as sorbitol amounts increased. Plants accumulating sorbitol at 3 to 5 µmol g^{-1} fr wt occasionally showed regions in which chlorophyll was partially lost, but at higher sorbitol amounts young leaves of all plants lost chlorophyll in irregular spots that developed into necrotic lesions. The adverse effect of sorbitol could be caused by a disturbance of the glucose-6-phosphate pool by sorbitol-6-phosphate dehydrogenase, which might then affect either UDP-glucose or glucose amounts, leading to altered sugar sensing in plants. *Stpd l* expression in the transgenics led to sorbitol accumulation, whereas sorbitol was not detectable in the *asr* 1 progenitor line. The sorbitol-producing transgenic plants differed from mannitol-producing plants. Plants with sorbitol concentrations less than approximately 2 µmol g^{-1} fr wt had normal growth patterns, whereas plants with more than approximately 3 µmol g^{-1} fr wt sorbitol showed stunted growth. The higher the sorbitol concentration, the more numerous the lesions and the larger the areas of necrotic tissues.

Like mannitol- or ononitol-accumulating plants, the sorbitol-accumulating plants tolerate both salt stress and drought better than the wild-type. Depending on sorbitol amounts in the plants, absolute photosynthesis rates of line S5C in the absence of stress could be less than 50 percent of those of control plants. However, rates of photosynthesis declined less in S5C than in SR 1 controls when the plants were stressed by addition of 150 mM NaCl. Sorbitol in the cytosol might interfere with the export of carbohydrates from the chloroplasts, and this could be responsible for starch increases in these leaves. The perturbations in carbohydrate levels can be affected in three different ways : (a) sorbitol-6-phosphate dehydrogenase could affect the rate of sucrose biosynthesis by competing for glucose-6-phosphate needed for the activation of sucrose-phosphate synthase, (b) hexose phosphates do not serve as substrates for sucrose biosynthesis but are rather directed toward glycolysis and respiration, and (c) UDP-glucose as well as *myo*-inositol are essential precursors for cell wall biosynthesis. A causal relationship is also suggested by the fact that under salt stress and during drought, *myo*-inositol increases in wild-type tobacco. This is also observed in sorbitol producing plants, which show less lesion formation when stressed by the addition of 150 mM NaCl (Sheveleva et al., 1998). Further, supplementing *myo*-inositol in seedlings and young plants prevent lesion formation. Thus hyper accumulation of sorbitol, which interferes with inositol biosynthesis, seems to lead to osmotic imbalance, possibly acting as a signal affecting carbohydrate allocation and transport.

The sugar alcohols, *myo*-inositol and D-pinitol function as osmoprotectants during environmental stresses. Inositol and its isomers are methylated to form a variety of products, such as D-ononitol and D-pinitol. These cyclitols (cyclic polyols) resemble similar products associated with plants that grow naturally in salt and drought conditions (Swapan, 2002). Accumulation of D-pinitol in response to drought stress is observed in *Vigna umbellata, Cajanus cajan* and *Glycine max.* Likewise, D-pinitol levels are positively correlated with salt tolerance of plants. Ononitol accumulating (up to 10 μmol g^{-1} fr wt) plants also display a normal phenotype. When these plants are exposed to drought stress and high salinity, they accumulate even higher amounts of ononitol (up to 35 μmol g^{-1} fr wt) as a result of a stress-induced increase in *myo*-inositol, which then serves as a substitute for additional ononitol production (Sheveleva et al., 1997). It is noteworthy that the growth of plants remains normal during this stress-inducible accumulation of ononitol. They transferred a cDNA encoding *myo*-inositol O-methyl transferase (IMT1) into *Nicotiana tabacum* cultivar SR1. During drought and salt stress, transformants (I5A) accumulated the methylated inositol D-ononitol in amounts exceeding 35 mμmol g^{-1} fr wt. In I5A, photosynthetic CO_2 fixation was inhibited less during salt stress and drought, and the plants recovered faster than wild-type. One day after rewatering drought-stressed plants, I5A photosynthesis had recovered 75 percent versus 57 percent recovery with cultivar SR1 plants. After 2.5 weeks of 250 mM NaCl in hydroponic solution, I5A fixed 4.9 ± 1.4 mμmol CO_2 m^{-2} s^{-1}, whereas SR1 fixed 2.5 ± 0.6 mμmol CO_2 m^{-2} s^{-1}. *myo*-Inositol, the substrate for IMT1, increases in tobacco under stress. Preconditioning of I5A plants in 50 mM NaCl increased D-ononitol amounts and resulted in increased protection when the plants were stressed subsequently with 150 mM NaCl. Proline, sucrose, fructose, and glucose showed substantial diurnal fluctuations in amounts, but D-ononitol did not. Plant transformation resulting in stress-inducible, stable solute accumulation appears to provide better protection under drought and salt-stress conditions than strategies using osmotic adjustment by metabolites that are constitutively present.

After salinity stress, the amount of L-*myo*-inositol 1-phosphate synthase enhances in leaves but depresses in roots (Nelson et al., 1998). The absence of *myo*-inositol synthesis in roots is compensated by inositol/ononitol transport in the phloem. Mobilization of photosynthate through *myo*-inositol translocation links root metabolism to photosynthesis. By inducing expression of *myo*-inositol O-methyltransferase and increasing *myo*-inositol synthesis, the anabolic end products accumulate, fascilitating sodium sequestration and protecting photosynthesis. *myo* –Inositol, D-pinitol and D-ononitol are also capable of scavenging hydroxyl radicals (Smirnoff, 1998).

COLD STRESS

Cold treatment (14 day at 10/3°C day/night) of *Ajuga repens* increased the total non-structural carbohydrate content of leaves by a factor of about 10, mainly because of increase in raffinose family oligosaccharides (Bachmann et al., 1994). The degree of polymerization of the RFO increases sequentially. A novel, galactinol-independent galactosyltransferase enzyme is found, forming from two molecules of RFO. RFOs are the main carbohydrates translocated in the phloem, with stachyose being by far the most dominant form. The RFO synthesis could be detected even before the commencement of export, suggesting the existence of a nonphloem-linked RFO pool, even in very young leaves. Taken together, it seems that *Ajuga* leaves contain two pools of RFO metabolism, a pronounced long-term storage pool in the mesophyll, possibly also involved in frost resistance, and a transport pool in the phloem. Further, Bachmann and Keller (1995) investigated the inter- and intra-cellular compartmentation of anabolic RFO metabolism by comparing whole-leaf tissue with mesophyll protoplasts and vacuoles. Mesophyll appears to be the primary site of RFO formation upon *in vitro* $^{14}CO_2$ photosynthesis. The components of RFO storage pool-galactinol synthase, stachyose synthase, *myo*-inositol, galactinol, and sucrose are extracellular (cytosolic), whereas galactinol-independent galactosyltransferase and higher RFO oligomers (with degree of polymerization 4) are vacuolar. Raffinose is found in both locations and serves as a cryoprotectant.

Induction of galactinol synthase (*GaS*) transcripts in crowns of winter hardy alfalfa cultivars occurs within 8 hours of exposure to 2°C, and is intensified by exposing plants to -2°C for two weeks (Cunnigham et al., 2003). Galactinol synthase transcripts increase in November in crown and root tissues of winter hardy alfalfa plants (Fig. 4.2). This increase is accompanied by large increases in root RFO concentrations between October and December (Fig. 4.3). A close positive association between RFO accumulation in roots in December and genetic differences in winter survival is observed in these alfalfa populations (4 cultivars with each having low and high fall dormancy selections). Understanding the mechanisms regulating *GaS* gene expression and subsequent RFO accumulation in roots and crowns provides an opportunity to genetically improve alfalfa winter hardiness.

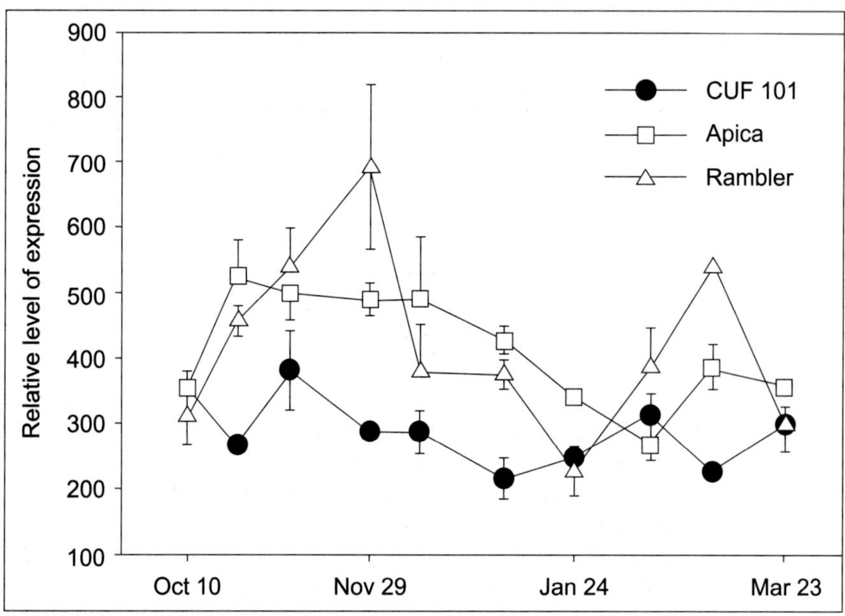

Fig. 4.2 Relative level of expression of galactynol synthase gene (*msa*CIF) in crowns of the cold-tolerant cv Apica and Rambler and the cold sensitive cv CUF 101. Plants were acclimated to natural temperature conditions in an unheated greenhouse. Means ± SEM (n=3). (After Cunnigham et al., 2003).

SALINITY STRESS

Gilbert et al. (1997) studied the effects of salinity (60 mM NaCl : 12 mM $CaCl_2$) on RFO metabolism in the variegated leaves of coleus (*Coleus blumel*). During the initial 5-10 days of exposure to salinity, mature source leaves showed strongly diminished rates of photosynthesis, which gradually recovered to close to the control rates by the end of the experiment. In

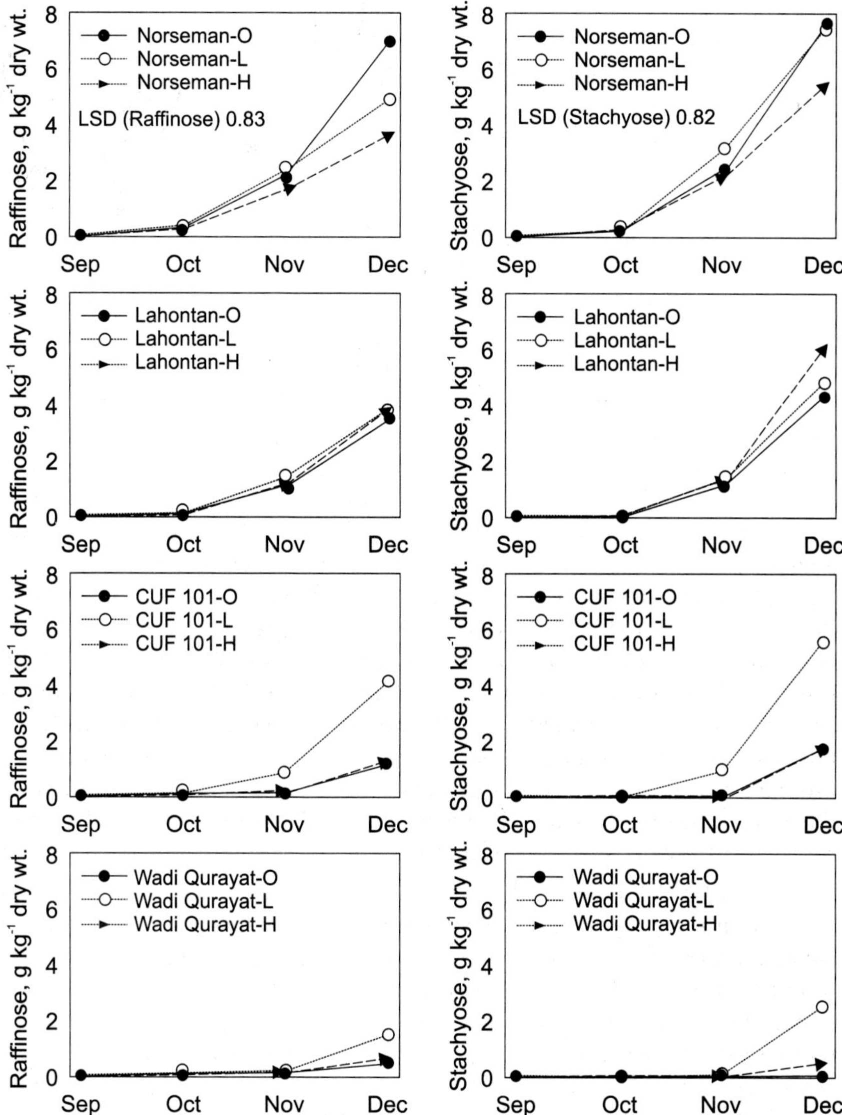

Fig. 4.3 Raffinose and stachyose accumulation in taproots of cold acclimating alfalfa cultivars differing in fall dormancy (FD) and winter hardiness. Norseman (FD = 1) is winter hardy, Lahontan (FD = 7) has intermediate winter hardiness,while CUF 101 (FD =9) and Wadi Qurayat (FD =11) are not winter hardy. Responses of the parent cultivar (O) and populations from this cultivar selected for greater (L) and less (H) fall dormancy are shown. Data were averaged over two year. The least significant difference (LSD) at the 5% level of probability is provided in the top panel of each column. (After Cunnigham et al., 2003).

green leaf tissues, starch levels showed the same transient decline and recovery pattern. Low starch levels were accompanied by the appearance of several novel carbohydrates, including high molecular weight RFOs with a degree of polymerization of 5 to 8, and an O-methylated inositol. New enzyme activities, including galactan : galactan galactosyltransferase, for the synthesis of high degree polymerization RFOs and *myo*-inositol 6-O-methyltransferase or O-methylation of *myo*-inositol, were induced by salinity stress. Phloem-sap analysis showed that in the stressed condition, substantially more sucrose than RFO was exported, as was the O-methylated inositol. In white sink tissues these phloem sugars are used to synthesize high degree polymerization RFOs but not O-methylated inositols. In sink tissues galactan : galactan galactosyltransferase but not *myo*-inositol 6-O-methyltransferase, is induced by salinity stress.

Tobacco, a non-mannitol producing plant, was metabolically engineered to produce mannitol, by introducing a single gene from *Escherichia coli* under the control of ^{35}S cauliflower mosaic virus promoter (Tarczynski et al., 1992). The gene *mtlD* encodes mannitol-1-P dehydrogenase, an enzyme that catalyzes the reversible reduction of fructose-6-P to mannitol-1-P and uses NADH as the electron donor and NAD as electron acceptor. Plants transformed with *E. coli mtlD* gene expressed the mannitol-1-P dehydrogenase enzyme and accumulated mannitol throughout the plant. Further, Tarczynski et al. (1993) demonstrated that the mannitol accumulating, transformed tobacco exhibited greater salt tolerance than the non-transformed tobacco plants (mannitol concentrations exceeded 6 μmol g^{-1} fr wt in leaves and in the roots of some transformants). Growth of the non-transformed tobacco, in the presence of 250 mM NaCl in hydroponic nutrient culture, was strongly reduced compared to that of the transformed mannitol-producing tobacco.

The non-stressed mannitol-producing tobacco plants were 20-25 percent smaller than the non-stressed, non-transformed (wild-type) plants subjected to both salinity and drought stress (Karakas et al., 1997). However, salt stress reduced dry weight in wild-type plants by 44 percent, but did not reduce the dry weight of transgenic plants. Transgenic plants adjusted osmotically by 0.57 MPa, whereas the wild-type plants did not adjust osmotically in response to salt stress. However, calculations of the solute contribution to osmotic adjustment showed that mannitol contributed only 0.003-0.004 MPa to the 0.2 MPa difference in full turgor osmotic potential between salt-stressed transgenic and wild-type plants. Assuming cytoplasmic location for mannitol and that the cytoplasm contributed 5 percent of the total water volume, mannitol accounted for only 30-40 percent of the change in osmotic potential of the cytoplasm. Also, inositol, a naturally occurring cyclitol in tobacco, accumulated in response to salt stress in both transgenic and wild-type plants, and was 3-fold more abundant than mannitol in transgenic plants. In some plant species, inositol provides substrate for the production and accumulation of the compatible solutes ononitol and pinitol. Inositol methylation is catalyzed by *myo*-inositol-O-methyltransferase. Production of ononitol is epimerized to pinitol, which accumulates to high amounts during salt stress. Both ononitol and pinitol most likely lower the cytoplasmic osmotic potential and balance sodium accumulation in the plant vacuole. Thus inositol and ononitol constitute the major phloem carbohydrates that seem to have at least two functions in roots : inositol is essential for continued root growth, and the amounts of inositol and ononitol are correlated with the concentration of sodium translocated to leaves through the xylem. After salinity stress, the amount of inositol 1-phosphate synthase (INPS) is enhanced in leaves but repressed in roots (Nelson et al., 1998). *myo*-Inositol-O-methyltransferase (IMP) is induced in all cell types. The absence of *myo*-inositol synthesis in roots is compensated by inositol/ononitol transport in the phloem. Mobilization of photosynthate through *myo*-inositol translocation links root metabolism to photosynthesis. This model integrates the transcriptional control of a specialized metabolic pathway with physiological reactions in different tissues. The tissue specific differential regulation of INPS, which leads to a gradient of *myo*-inositol synthesis, supports the root growth and sodium uptake. By inducing expression of IMT and increasing *myo*-inositol synthesis, metabolic end products accumulate, facilitating sodium sequestration and protecting photosynthesis.

Sodium accumulates to high concentrations in vacuoles, and cyclitols (*myo*-inositol, D-ononitol, and D-pinitol) accumulate in the cytosol. The cyclitol synthesis is regulated by NaCl and involves induction and repression of gene expression. Nelson et al.(1999) analyzed the effects of exogenous application of *myo*-inositol and found that the NaCl-inducible *myo*-inositol 1-phosphate synthase is repressed in roots, and sodium uptake from root to shoot increases without stunting growth. Sodium uptake and transport through the xylem was coupled to a 10-fold increase of *myo*-inositol and ononitol in the xylem. The results indicate an interdependence of sodium uptake and alterations in the distribution of *myo*-inositol. Thus *myo*-inositol might serve not only as a substrate for the production of compatible solutes but also as a leaf-to-root signal and promotes sodium uptake.

Ononitol-accumulating (up to 10 µmol g^{-1} fr wt) plants display a normal phenotype (Vernon et al., 1993 ; Sheveleva et al., 1997). When these transgenic tobacco plants were exposed to drought stress and high salinity, they accumulated higher amounts of ononitol (up to 35 µmol g^{-1} fr wt) as a result of stress-induced increase of *myo*-inositol, which then serves as the substrate for additional ononitol production (Sheveleva et al., 1997). Growth of the plants was normal during this stress-induced accumulation.

Deguchi et al. (2004) introduced a cDNA encoding sorbitol-6-phosphate dehydrogenase, which is the key enzyme in sorbitol biosynthesis into the Japanese persimmon (*Diospyros kaki*) to increase salt tolerance. The resultant plants exhibited salt-tolerance with dwarfing phenotypes. The lines PS1 and PS6 accumulated high and moderate levels of sorbitol, respectively. The average length of shoots was significantly shorter as compared with the wild-type in line PS1, while no such decrease was observed in line PS6. Although the *myo*-inositol content was decreased in PS1 plants, the decrease was much smaller than that observed in transgenic tobacco that accumulates sorbitol.

DROUGHT STRESS

Under drought conditions, the stomatal conductance and leaf photosynthetic rates are reduced. In green leaf tissues, drought stress also greatly decreases the diurnal light-period levels of the RFOs stachyose and raffinose, as well as those of other nonstructural carbohydrates (glactinol, sucrose, hexoses, and starch). However, drought has little effect on soluble carbohydrate content of white, nonphotosynthetic leaf tissues of the varigated coleus (*Coleus bluemel*) leaf; relative water turgor reduced from 80 to 60 percent for 21 days (Pattanagul and Madore, 1999). In green leaf tissues, galactinol synthase activity is depressed by drought stress. An accumulation of O-methyl-inositol is also observed, which is consistent with the induction of *myo*-inositol 6-O-methyltransferase activity seen in the stressed green tissues. In source tissues, RFO metabolism is apparently reduced by drought stress through a combined effect of decreased photosynthesis and reduced galactinol synthase activity. Moreover, a further reduction in RFO biosynthesis may be due to a switch in carbon partitioning to O-methyl-inositol biosynthesis, creating competition for *myo*-inositol, a metabolite shared by both biochemical pathways (Pattanagul and Madore, 1999).

Karakas et al. (1997) transformed tobacco plants with a mannitol-1-phosphate dehydrogenase gene resulting in mannitol accumulation. Drought stress reduces leaf relative water content, leaf expansion, and dry weight of transgenic, mannitol-accumulating and wild-type tobacco plants. However, water potential was not significantly reduced by drought stress in transgenic or wild-type plants, despite an increase in nonstructural carbohydrates and mannitol in droughted plants. Karakas et al. concluded that: (a) mannitol is a relatively minor osmolyte in transgenic tobacco, but may indirectly enhance osmotic adjustment and salt tolerance, (b) mannitol cannot be substituted by inositol in this role, (c) slower growth of the transgenic plants, without the presence of mannitol *per se*, may be the cause of greater salt tolerance, and (d) mannitol accumulation is enhanced by drought stress but it does not affect the osmotic potential or drought tolerance.

NUTRIENT STRESS

In contrast to wild-type tobacco, transgenic tobacco plants containing sorbitol exhibit a marked increase in within plant boron mobility and a resultant increase in plant growth and yield when grown with limited or interrupted soil boron supply (Brown et al., 1999). Boron is phloem mobile in transgenic tobacco but is immobile in the control lines. These results demonstrate that the transgenic enhancement of within-plant nutrient mobility is a viable approach to improve plant tolerance to nutrient stress. In the same year, Bellaloui et al. (1999) transformed tobacco line S11 with sorbitol-6-phosphate dehydrogenase; the production of sorbitol was accompanied by an increase in the concentration of boron in plant tissues and an increased uptake of boron compared with either tobacco line A4 or the wild-type tobacco line SR1 (zero sorbitol producer). These results demonstrate that the concentration of the boron-complexing sugar alcohol in the plant tissue has a significant effect on boron uptake and distribution in plants (boron is mobile in sorbitol producing tobacco line S11). This suggests that selection or transgenic regeneration of cultivars with an increased sugar alcohol content can result in increased boron uptake, with no apparent negative effect on short-term growth.

THE STRESS OF ROS

Besides its function in osmotic adjustment, mannitol improves tolerance to stress through scavenging hydroxyl radical and stabilization of macromolecular structures (Shen et al., 1997 a; b). Reactive oxygen species (ROS) in general react aggressively with biological molecules and can cause lipid peroxidation, breakdown of macromolecules and damage to nucleic acids. Under limited water availability, osmolytes can form hydrogen bonds with macromolecules and thus prevent formation of intramolecular H-bonds that would otherwise irreversibly change their three-dimentional structure (Crowe et al., 1992). Unlike osmotic adjustment, hydroxyl radical scavenging and other protective functions require only small amounts of mannitol, and it is likely that the improved performance of transgenic wheat is a result of the hydroxyl radical scavenging and/or improved stability of macromolecular structures.

Protection offered by mannitol at low levels (5 or less than 100 mM) may not be based on osmotic mechanism, rather it might act through several different mechanisms, viz.: (a) stabilization of the structure of proteins under stress, (b) might act as radical scavengers, or (c) might serve as low molecular weight chaperons (Shen et al., 1997 a). Mannitol is known as a hydroxyl radical scavenger. Efficient scavenging of active oxygen species is essential for chloroplasts to maintain photosynthesis because even low amounts of H_2O_2 inhibits the activity of dark reaction enzymes. Under normal conditions, ROS are scavenged by radient enzyme systems and non-ezymatic antioxidants. The enzyme systems include superoxide dismutases, which catalyze the reaction from superoxide to H_2O_2, and ascorbate peroxidases, which are responsible for the conversion of H_2O_2 to water. Ascorbate can be regenerated by the ascorbate-glutathione cycle (Allen, 1995). Overexpression of superoxide dismutase, ascorbate peroxidase, and glutathione reductase in transgenic plants has been shown to lead to an increased resistance to oxidative stress. The nonenzyme oxidants include α-tocopherol, ascorbate, glutathione, and carotenoids.

To investigate the potential role of mannitol in oxidative stress protection, Shen et al. (1997 a) targetted a bacterial mannitol-1-phosphate dehydrogenase gene to tobacco chloroplasts by the addition of an amino-terminal transit peptide. The transgenic tobacco lines accumulated mannitol at concentrations ranging from 2.5 to 7.0 μmol g^{-1} fr wt. Line BS1-31 accumulated approximately 100 mM mannitol in chloroplasts and was identical to the wild-type in phenotype and photosynthetic performance. Presence of mannitol in chloroplasts resulted in an increased resistance to the methyl viologen-induced oxidative stress, as seen by the increased retention of chlorophyll in transgenic leaf tissue following methyl viologen treatment. In presence of methyl viologen, isolated mesophyll cells of BS-1-31 exhibited higher CO_2 fixation as compared to the wild-type. When the hydroxyl radical probe dimethyl sulfoxide was introduced into cells, the initial formation rate of methane sulfinic acid was significantly lower in cells containing mannitol in the chloroplast compartment than in wild-type cells, indicating an increased hydroxyl radical scavenging capacity in BS-1-31 tobacco. Thus, Shen et al. (1997 a) suggest that the chloroplast location of mannitol can supplement endogenous radical-scavenging mechanisms and reduce oxidative damage of cells by hydroxyl radicals.

ROS attack amino acid residues in proteins, especially tryptophan, phenylalanine, methionine, and cystine, and form carbonyl derivatives. Also, ROS promote intra- and inter-molecular cross-linking, such as -S-S- bonding and protein fragmentation. Such modifications mark proteins for degradation. The damage of membranes by ROS has been documented by increased formation of malondialdehyde, a product of lipid peroxidation, during oxidative stress. Superoxide is mainly produced from photoreduction of oxygen via the reduction of oxygen by PS I. Most H_2O_2 in chloroplasts is then produced through disproportionation of superoxide by superoxide dismutases (SODs). Additionally, hydroxyl radicals are produced through the interaction of H_2O_2 and superoxide or directly from H_2O_2 in presence of transition metals, such as Fe^{2+} and Cu^{2+}. ROS affect both the photosynthetic electron transport machinery and Calvin cycle enzymes.

Increased concentrations of both H_2O_2 and free Fe have been observed in conditions of drought or cold stress. Free Fe and Cu increased 1.5- and 2.5-fold, respectively, under water stress and reached a concentration as high as 16 µM in the chloroplast stroma (Moran et al., 1994). As discussed earlier, presence of mannitol in chloroplasts can protect plants against oxidative damage by hydroxyl radicals (Shen et al., 1997 a). Shen et al. (1997 b) further observed that the activity of phosphoribulosekinase is reduced by 65 percent in illuminated thylakoids producing hydroxyl radicals. Mannitol (125 mM) and sodium formate (15 mM), both hydroxyl radical scavengers, and catalase (3000 units per ml) prevent loss of phosphoribulosekinase activity. Shen et al. suggest that the stress-protective role of mannitol may be to shield susceptible thiol-regulated enzymes like phosphoribulokinase plus thioredoxin, ferridoxin, and glutathione from inactivation by hydroxyl radicals in plants. Thus the accumulation of mannitol may have dual functions : facilitating osmotic adjustment and supporting redox control (Shen et al., 1999).

IMPROVEMENT WORK

Raffinose family oligosaccharides (RFOs) have been implicated in mitigating the effects of abiotic stresses on plants. In seeds, proposed roles for RFOs include protecting cellular integrity during desiccation and/or imbibition; extending longevity in the dehydrated state, and providing substrates for energy generation during germination. A gene encoding galactinol synthase (*Gols)*, the first committed enzyme in the biosynthesis of RFOs was cloned from tomato cv Moneymaker seeds (Downie et al., 2003). The gene (*LeGols-1)* mRNA accumulates in developing tomato seeds concomitant with maximum dry weight deposition and the acquisition of desiccation tolerance. *LeGols-1* mRNA is present in mature, desiccated seeds but declines within 8 hours of imbibition in wild-type seeds. *LeGols-1* mRNA accumulation is induced by dehydration but not by cold in germinating seeds, whereas both stresses induce *LeGols-1* mRNA accumulation in seedling leaves. The extent of RFO accumulation is controlled by the initial substrates, *myo*-inositol and sucrose, rather than by *Goels* activity alone in tomato seeds and seedlings (Downie et al., 2003). Thus, Downie and coworkers were able to improve tomatoes genetically for salt and drought tolerance.

As the high levels of cyclitols correlate with tolerance to osmotic stress in a number of crop species; gene encoding cyclitol biosynthesis enzyme from a halophyte *Mesembryanthemum crystallinum* was introduced into tobacco (Vernon et al., 1993). The gene, *Imt 1,* encodes a *myo*-inositol

O-methyl transferase that in *M. crystallinum* catalyses the first step in stress-induced accumulation of the cyclitol pinitol. Tobacco transformed with the *Imt 1* cDNA under the control of the *CaMV* ^{35}S promoter appears phenotypically normal and exhibits *IMT1* enzyme activity. Transformants accumulate a carbohydrate product not detectable in non-transformed control plants. This product is identified as ononitol. Ononitol is the major carbohydrate constituent of the leaf tissue of plants expressing *Imt 1* gene, accumulating up to 25 percent of the level of sucrose in transformed seedlings. Identification of ononitol as the *IMT1* product and the specific accumulation of this compound in transformed tobacco supports a role for ononitol as a stable intermediate in pinitol biosynthesis, and indicates lack of epimerization activity in tobacco responsible for the conversion of ononitol to pinitol. Thus the transgenic system might serve as a useful model to test the ability of cyclitols such as ononitol to confer tolerance to environmental stress in glycophytic crop plants. Encouraged with this observation, Sheveleva et al. (1997) transferred a cDNA encoding *myo*-inositol O-methyltransferase (IMT1) into tobacco cv SR1. During drought and salt stress, the transformants (I5A) accumulated D-ononitol in amounts exceeding 35 mµmol g^{-1} fr wt. In I5A, photosynthetic CO_2 fixation was less inhibited during salt stress and/or drought stress, and the plants recovered faster than the wild-type. One day after rewatering drought-stressed plants, I5A photosynthesis recovered 75 percent vs 57 percent recovery with cv SR1 plants. After 2.5 weeks of 250 mM NaCl in hydroponic solution, I5A fixed 4.9 ± 1.4 mµmol CO_2 m^{-2} sec^{-1}, whereas SR1 fixed 2.5 ± 0.6 mµmol CO_2 m^{-2} sec^{-1} . *myo-* Inositol , the substrate for IMT1, increases in tobacco under stress. Preconditioning of I5A plants in 50 mM NaCl increased D-ononitol amounts and resulted in increased protection when the plants were stressed subsequently with 150 mM NaCl (Sheveleva et al., 1997).

Later, Abebe et al. (2003) transformed wheat with the *mtlD* gene of *E. coli.* Tolerance to water stress and salinity was evaluated using calli and T_2 plants transformed with (+ *mtlD*) or without (-*mtlD*) *mtlD.* The T_2 plants were stressed by withholding water or by adding 150 mM NaCl to the nutrient medium. In *-mtlD* plants, fresh weight, dry weight, plant height, and flag leaf length were reduced by 70, 56, 40 and 45 percent compared with 40, 8, 18 and 29 percent, respectively, in +*mtlD* plants (Table 4.1). Salt stress reduced shoot fresh weight, dry weight, plant height by 77, 73, 25 and 36 percent in –*mtlD* plants, respectievly, compared with 50, 30, 12 and 20 percent in +*mtlD* plants. However, the amount of mannitol accumulated in the callus and mature 5th leaf (1.7 – 3.7 µmol g^{-1} fr wt in the callus and 0.6 – 2.0 µmol g^{-1} fr wt in the leaf) was too small to protect against stress through osmotic adjustment. Abebe and associates thus conclude that the improved growth performance of mannitol accumulating calli and mature leaves is due to other stress-protective functions of mannitol, although this study cannot rule out possible osmotic effects in growing regions of the plant. In this way Abebe and coworkers were able to improve the wheat variety for improved salt and drought stress.

Table 4.1 Shoot weight, root weight, plant height, length of the flag leaf, length of root, and number of tillers of transgenic wheat plants exposed to NaCl (After Abebe et al., 2003).

		Shoot					Root		
Stress level	Plant type	Fresh wt (g)	Dry wt (g)	Plant height (cm)	Flag leaf length (cm)	No. of tillers	Fresh wt(g)	Dry wt(g)	Length (cm)
Unstressed	*–mtlD*	28.3c	4.8c	50.0b	34.7c	6.0c	17.8c	1.1c	63b
	+mtlD	24.2c	3.7c	51.0b	33.7c	7.0d	16.9c	1.1c	61b
Stressed	*-mtlD*	6.4a	1.3a	37.7a	22.7a	2.0a	3.5a	0.2a	43a
	+mtlD	12.2b	2.6b	45.0b	27.0b	3.0b	8.5b	0.5b	47a
	$LSD_{0.05}$	5.7	1.1	6.6	2.3	0.5	4.2	0.2	12.1

T_2 plants (P1-13-1, -mtlD) were grown hydroponically, and stress was imposed by raising the salt concentration of the nutrient medium to 150 mM NaCl. Measurements were taken after 30 days of stress. Data are means of three replications; means followed by the same letter in a column are not significantly different at $P < 0.05$ as determined by Fisher's protected LSD test.

Thus addition of a gene to synthesize mannitol should satisfy the critics of genetically modified (GM) crops, as the gene occurs naturally in many food crops and is also routinely used as an additive in several processed foods. Wheat engineered to accumulate mannitol in leaf tissues significantly improved productivity under stress from water deficit or salinity (Professor A.C. Guenzi of Oklahoma State University—Personal communication). A. C. Guenzi and T. Abebe have also improved stress tolerance by introducing into wheat a chimeric, or hybrid gene derived from maize and two common bacteria.

REFERENCES

Abebe, T.; A.C. Guenzi; B. Martin and J.C. Cushman, 2003. Tolerance to mannitol-accumulating transgenic wheat to water stress and salinity. Plant Physiol. 131: 1748-55.

Allen, R. D. 1995. Dissection of oxidative stress tolerance using transgenic plants. Plant Physiol. 107: 1049-54.

Bachmann, M. and F. Keller, 1995. Metabolism of the raffinose family oligosaccharides in leaves of *Ajuga reptans* L. Inter- and intra-cellular compartmentation. Plant Physiol. 109: 991-98.

Bachmann, M.; P. Matile and F. Keller, 1994. Metabolism of raffinose family oligosaccharides in leaves of *Ajuga repens* L. Cold acclimation, translocation, and sink to source transition: Discovery of chain elongation enzyme. Plant Physiol. 105 : 1335-45.

Bellaloui, N.; P.H. Brown and A.M. Dandekar, 1999. Manipulation of *in vitro* sorbitol production alters boron uptake and transport in tobacco. Plant Physiol. 119: 735-42.

Brown, P.H.; N. Bellaloui and A.M. Dandekar, 1999. Transgenically enhanced sorbitol synthesis facilitates phloem boron transport and increases tolerance of tobacco to boron deficiency. Plant Physiol. 119: 17-20.

Crowe, J.H.; F.A. Hoekstra and L.M. Crowe, 1992. Anhydrobiosis. Annu. Rev. Physiol. 54: 579-99.

Cunnigham, S.M.; P. Nadeau; Y. Castonguay; et al. 2003. Raffinose and stachyose accumulation, galactinol synthase expression, and winter injury of contrasting alfalfa germplasms. Crop Sci. 43: 562-70.

Deguchi, M.; Y. Koshita; M. Gao; et al. 2004. Engineered sorbitol accumulation induces dwarfism in Japanese persimmon. Plant Physiol. 161: 1177-84.

Downie, B.; S. Gurusinghe; P. Dahal; et al. 2003. Expression of galactynol synthase gene in tomato seeds is up-regulated before maturation desiccation and again after imbibition whenever radicle protrusion is prevented. Plant Physiol. 131: 1347-59.

Fellman, J.K. and W.H. Loescher, 1987. Comparative studies of sucrose and mannitol utilization in celery (*Apium graveolens*). Plant Physiol. 69: 609-16.

Gilbet, G.A.; C. Wilson and M.A. Madore, 1997. Root-zone salinity after raffinose oligosaccharide metabolism and transport in coleus. Plant Physiol. 115: 1267-76.

Karakas, B.; P. Ozias-Atkins; C. Stushnoff; et al. 1997. Salinity and drought tolerance of mannitol accumulating transgenic tobacco. Plant Cell Environ. 20: 609-16.

Kuo, T.M.; C.A. Lowell and T.C. Nelson, 1997. Occurrence of pinitol in developing soybean seed tissues. Phytochemistry 45: 29-35.

Moran, J.F.; M. Becana; I. Iturbe-Ormaetxe; et al. 1994. Drought induces oxidative stress in pea plants. Planta 194: 346-52.

Nelson, D.E.; G. Rammesmayer and H.J. Bohnert, 1998. Regulation of cell specific inositol metabolism and transport in plant salinity tolerance. Plant Cell 10: 753-64.

Nelson, D.E.; M. Koukoumanos and H.J. Bohnert, 1999. *myo*-Inositol-dependent sodium uptake in ice plant. Plant Physiol. 119: 165-72.

Pattanagul, W. and M.A. Madore, 1999. Water deficit effects of raffinose family oligosaccharide in coleus. Plant Physiol. 121: 987-93.

Peterbauer, T. and A. Richter, 2001. Biochemistry and physiology of raffinose family oligosaccharides and galactosyl cyclitols in seeds. Seed Sci. Res. 11: 185-87.

Peterbauer, T.; M. Puschenreiter and A. Richter, 1998. Metabolism of galactosyl-ononitol in seeds of *Vigna umbellata.* Plant Cell Physiol. 39: 334-41.

Peterbauer, T.; J. Mucha; U. Mayer; et al. 1999. Stachyose synthesis in seeds of adzuki bean (*Vigna angularis*) : molecular cloning and functional expression of stachyose synthase. Plant J. 20: 509-18.

Peterbauer, T.; L.B. Lahuta; A. Blöchl; et al. 2001. Analysis of the raffinose family oligosaccharide pathway in pea seeds with constrasting carbohydrate composition. Plant Physiol. 127: 1764-72.

Peterbauer, T.; J. Mucha and A. Richter, 2002. Chain elongation of raffinose in pea seeds. Isolation, characterization, and molecular cloning of multifunctional energy catalyzing the synthesis of stachyose and verbascose. J. Biol. Chem. 277: 194-200.

Pharr, D.M.; J.M.H. Stoop; J.D. Williamson; et al. 1995. The dual role of mannitol as osmoprotectant and photoassimilate in celery. HortScience 30: 1182-88.

Shen, B.; R.G. Jensen and H.G. Bohnert, 1997 a. Increased resistance to oxidative stress in transgenic plants by targetting mannitol biosynthesis to chloroplasts. Plant Physiol. 113: 1177-83.

Shen, B.; R.G. Jensen and H.J. Bohnert, 1997 b. Mannitol protects against oxidation by hydroxyl radicals. Plant Physiol. 115: 527-32.

Shen, B.; S. Hohmann; R.G. Jensen and H.G. Bohnert, 1999. Roles of sugar alcohols in osmotic stress adaptation. Replacement of glycerol by mannitol and sorbitol in yeasts. Plant Physiol. 121: 45-52.

Sheveleva, E.V.; W. Chmara ; H.J. Bohnert and R.G. Jensen, 1997. Increased salt and drought tolerance by D-ononitol production in transgenic *Nicotiana tabacum* L. Plant Physiol. 115: 1211-19.

Sheveleva, E.V.; S. Marquez; W. Chmara; et al. 1998. Sorbitol-6-phosphate dehydrogenase expression in transgenic tobacco. High amounts of sorbitol lead to necrotic lesions. Plant Physiol. 117: 831-39.

Smirnoff, N. 1998. Plant resistance to environmental stress. Curr. Opinion in Biotechnol. 9: 214-19.

Swapan, K.D. 2002. Plant developments in transgenics for abiotic stress tolerance in rice. JIRCAS Working Report 43-53.

Tarczynski, M.C.; R.G. Jensen and H.J. Bohnert, 1992. Expression of a bacterial *mtlD* gene in transgenic tobacco leads to production and accumulation of mannitol. Proc. Nat. Acad. Sci. 89: 2600-04.

Tarczynski, M.C.; R.G. Jensen and H.J. Bohnert, 1993. Stress protection of transgenic tobacco by production of the osmolyte mannitol. Science 259: 508-10.

Vernon, D.M.; M.C. Tarczynski; R.G. Jensen and H.J. Bohnert, 1993. Cyclitol production in transgenic tobacco. Plant J. 4: 199-205.

SUGGESTED READINGS

Peterbauer, T. and A. Richter, 2001. Biochemistry and physiology of raffinose family oligosaccharides and galactosyl cyclitols in seeds. Seed Sci. Res. 11 (3): 185-99.

Sheveleva, E.V.; W. Chmara; H.J. Bohnert and R.G. Jensen, 1997. Increased salt and drought tolerance by D-ononitol production in transgenic *Nicotiana tabacum* L. Plant Physiol. 115: 1211-19.

Shen, B.; R.G. Jensen and H.J. Bohnert, 1997. Increased resistance to oxidative stress in transgenic plants by targetting mannitol biosynthesis in chloroplasts. Plant Physiol. 113: 1177-83.

5

PROLINE

Introduction

In defence of the abiotic-osmotic stresses (drought, salinity, cold) plants have developed mechanisms of accumulating certain compatible osmolytes (compatible osmolytes do not interfere with normal biochemical reactions and act as osmoprotectants during osmotic stress which are species and cultivar specific) which provide a certain degree of protection from the stress-environment that they are exposed to. Proline is one such osmolyte which has high solubility (14 kg kg^{-1} water) and no perturbing effect on macromolecules. Proline protects membranes and proteins against the adverse effects of high concentrations of inorganic ions and temperature extremes. Proline also functions as a protein compatible hydrotrope and as a hydroxyl radical scavenger, and represents over 80 percent of the free amino acid pool (Rhodes and Handa, 1989). The concentration of proline in the cytoplasm could exceed 200 mM and, therefore, may contribute substantially to cytoplasmic osmotic adjustment. Proline synthesized in the cytoplasm alleviates cytoplasmic acidosis, and maintains $NADP^+$/ NADPH ratios at values compatible with metabolism (Hare and Cress, 1997). Rapid catabolism of proline upon relief of stress provides reducing equivalents that support mitochondrial oxidative phosphorylation and the generation of ATP for recovery from stress and repair of stress-induced damage.

Proline biosynthesis in stressed plants

Proline is synthesized in two ways which are regulated at the level of both enzyme activity and gene expression. Increased proline synthesis during osmotic stress involves a loss of feedback regulation of the first step(s) in the pathway leading to the formation of Δ^1- pyrroline-5-carboxylate (P5C), identifying this reaction(s) as the rate-limiting step in proline biosynthesis. Proline can be synthesized either from glutamate or ornithine (Fig. 5.1). The pathway from glutamate is the primary route for synthesis of proline under conditions of osmotic stress and nitrogen limitation, while the pathway from ornithine predominates at high levels of available nitrogen (Delauney et al., 1993). The other important factor that controls levels of proline in plants is the degradation or metabolism of proline itself. l-Proline is oxidized to P5C in plant mitochondria by proline dehydrogenase (oxidase), and P5C is converted to l-glutamate by P5C dehydrogenase. Such oxidation of proline is inhibited during the accumulation of proline under water stress and is activated in rehydrated plants. Some other plants synthesize proline from ornithine, a conversion which proceeds via two routes, both involving transamination of ornithine followed by cyclization and reduction. If the α-amino group of ornithine is transaminated, the product would be α-keto-δ-amino valerate which cyclizes to Δ′-pyrroline-2-carboxylate (P2C) and is then reduced to proline (Fig. 5.1). Alternatively, transamination of the δ-amino group yields glutamic-δ-semialdehyde (GAS) which is converted to proline via P5C. Much of the available evidence suggests that the latter pathway is the one which actually occurs but there are conflicting data from *in vitro* labeling experiments implicating the possible route via P2C in plants. However, characterization of a moth bean (*Vigna aconitifolia*) ornithine aminotransferase cDNA clone suggests that the encoded enzyme is a δ-amino-transferase (Delauney et al., 1993) confirming the formation of P5C.

Metabolism of proline is inhibited when proline accumulates during dehydration and it is activated when rehydration occurs. Under dehydration conditions, when expression of the gene for P5C synthase is strongly induced, expression of the gene for proline dehydrogenase is inhibited. By contrast, under rehydration conditions, when expression of the gene for proline dehydrogenase is strongly induced, expression of the gene for P5C synthase is inhibited (Yoshida et al., 1997). Thus, P5C synthase (P5CS), which acts during the biosynthesis of proline and proline dehydrogenase, appears to be a rate-limiting factor under water stress. Further, water stress also triggers production of ABA, which in turn, induces various genes. As the gene for P5C synthase is also induced by exogenous ABA application, it seems likely that this gene is also one of the many ABA-inducible genes helping in water deprivation tolerance.

Proline accumulation is associated with the ability of proline to act as an osmolyte, as a protective agent for cytoplasmic enzymes, as a reservior of nitrogen and carbon sources for post stress growth, or even as a stabilizer of the machinery for protein synthesis. Proline acts as a cytoplasmic osmoticum counteracting the effect of accumulated salt in the vacuole. Proline in addition to acting as an osmoprotectant, also serves as a sink for energy to regulate redox potentials, as a hydroxy radical scavenger, as a solute that protects macromolecules against denaturation, and as a means of reducing acidity in the cell. Accumulation of proline is due primarily to *de novo* synthesis, although a reduced rate of catabolism has also been

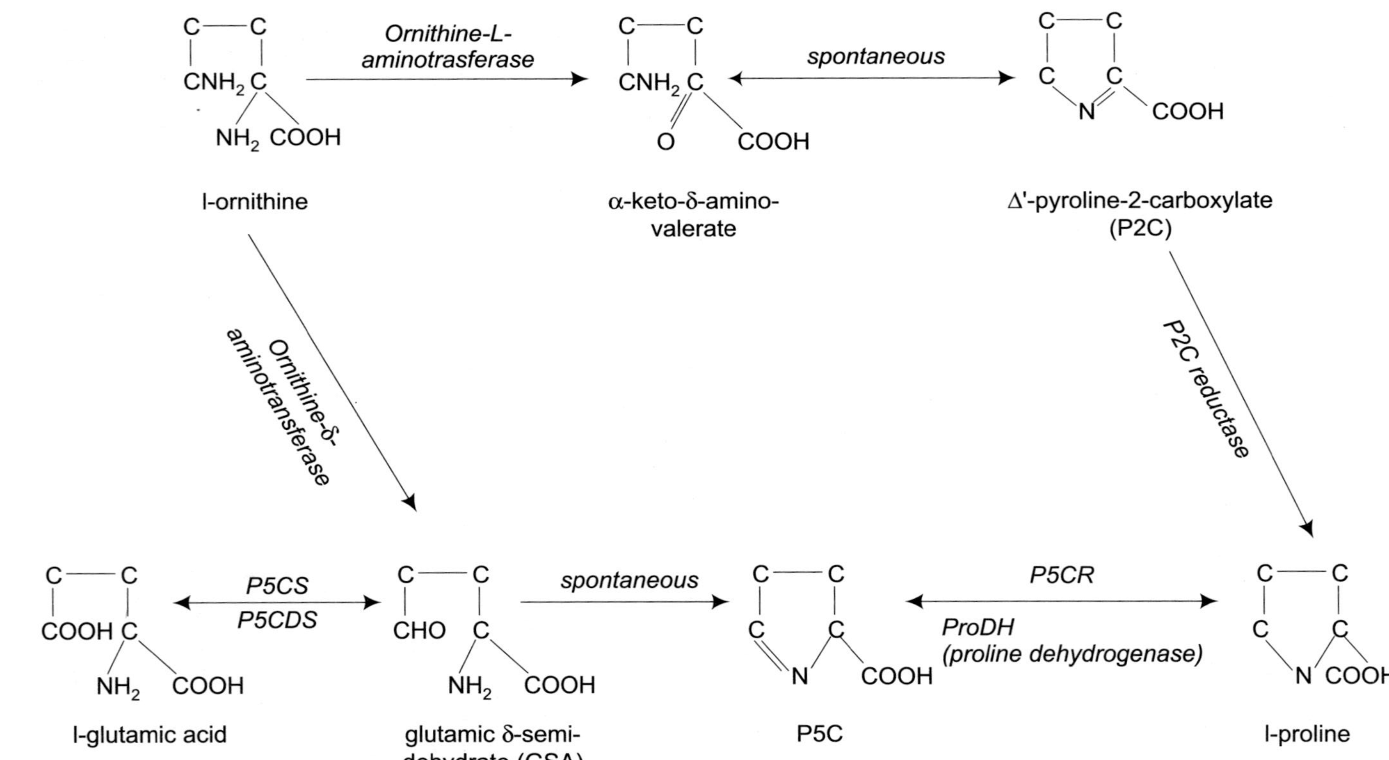

Fig. 5.1 Pathways of proline biosynthesis in plants (After Deauney and Verma, 1993, and Yoshida et al., 1997).

observed. Since proline accumulation also occurs under the stresses of heavy metals (e.g. lead nitrate with as low as –0.001 to –0.06 M compared to -0.125 of mineral growth medium), mineral deficiency (e.g. Fe, K, Mg), and UV radiation, which do not result in hyperosmotic, surely the induced proline accumulation has some important role other than osmoregulation (Sharmila, P. – personal communication). With expression of moth bean P5C synthase in transgenic plants, 80 percent higher free proline accumulates as against the wild-type, and the plants grow more rapidly in toxic cadmium concentrations (100 μM), and bind 4-fold more cadmium (Siripornadulsil et al., 2002). It appears that the free proline levels are correlated with reduced glutathione (GSH) redox state and malondialdehyde levels in heavy metal treated plants. Free proline most likely acts as an antioxidant in Cd-stressed cells.

Osmotic stress and the discrepancies in the absorption and use of photosynthetically active radiation (PAR) by the chlorophyllous organs have destructive consequences. Since the primary capture of photon energy is insensitive to stress, plants under adverse conditions are not able to utilize the whole radient energy received for carbon assimilation. If the generation of $NADP^+$ is limited under conditions of continued photon absorption, redox imbalance is likely to result in photoinhibition and enhanced use of CO_2 instead of $NADP^+$ as the electron accepter in photosynthesis. Thus, in addition to any biophysical protective effects of free proline, which may be only of secondary importance, a stress induced increase in the transfer of reducing equivalents into proline by P5C synthase and P5C reductase (P5CS) may be a protective mechanism whereby species ameliorate shifts in cellular redox potential which accompany all biotic and abiotic stresses, including those that do not cause cellular dehydration (Hare and Cress, 1997). This may assist in counteracting the photoinhibitory damage under adverse conditions. Furthermore, a cycle of proline synthesis and degradation might enable sensitive regulation of cellular redox potential in the cytosol or plastid as well as adenylate charge in the mitochondria.

The other benefit of the replacement of NADP supply by proline synthesis may be to support redox cycle which is important in the antioxidant defence mechanism. There are also other evidences indicating an important role of proline synthesis in potentiating the pentose phosphate pathway activity. A signal derived from proline biosynthetic and catabolic pathways may control gene expression in response to osmotic stress (Hare and Cress, 1997). Since the prompt accumulation of transcripts encoding P5C synthase and proline dehydrogenase within one hour of the imposition of severe stress precedes proline accumulation by several hours, Hare and Cress (1997) are of the opinion that the elevated levels of proline accumulated after stress may simply be symptomatic of a metabolic adjustment of benefit, only in the early stages of accumulation. Upon relief from the stress, the degenerated proline may be providing carbon, nitrogen and energy for recovery.

Iyer and Caplan (1998) found that intermediates in proline biosynthesis and catabolism such as glutamine and P5C can increase the expression of several osmotically regulated genes in rice including *salT* and *dhn4.* P5C- and 3,4-dehydroproline (an analog of P5C)-treated plants consumed less oxygen, had reduced NADPH levels, had increased NADH levels, and accumulated many osmolytes associated with osmotically stressed rice. These results indicate that osmotically induced increases in the concentrations of one or more intermediates in proline metabolism might influence some of the characteristic responses to osmotic stress.

Localization and transport

Accumulation of compatible solutes occurs preferentially in the cytosol under water stress. By the accumulation of compatible solutes like proline in the sytosol, osmotic potential inside the cell is lowered. Total amount of proline in the cells increases and the amount of proline in the vacuole decreases, when the cells are exposed to water stress. Thus the vacuoles play an important role in the accumulation and transportation of proline during the dehydration stress. Synthesis and transport of proline in maize seedlings at low water potentials was studied by Raymond and Smirnoff (2002). They observed that the maize seedling-primary roots accumulate most proline at low water potentials. Most of the proline is synthesized in the endosperm. They also observed that the rate of proline utilization (oxidation and protein synthesis) exceeds the rate of biosynthesis by 5-fold at high and low water potentials. Proline constitutes 10 percent of the amino acids released from the endosperm. ^{14}C-proline is transported from the scutellum to other parts of the seedling and reaches the highest concentration in the root tips. Though lesser proline is transported at lower water potentials, because of the lower rate of protein synthesis and oxidation, more proline accumulates in the root tips. Despite the low biosynthesis capacity of the roots, the extent of proline accumulation in relation to water potential is precisely controlled by the transport and utilization rate. The increased transport of proline to the root tip serves as the source of low water potential-induced proline accumulation (Verslues and Sharp, 1999).

A proline transporter gene, *ProT2,* is strongly induced by water and salt stress in *Arabidopsis thaliana* (Rantsch et al., 1996). Homologous proline transporter genes have been identified in tomato : *LeProT1* is strongly expressed in mature and germinating pollen, and may encode a general transporter for compatible solutes (Schwacke et al., 1999). *LeProT1* transports proline and GABA with low affinity and glycinebetaine with high affinity (Schioacke et al., 1999).

STRESS-INDUCED PROLINE ACCUMULATION AND STRESS MITIGATION

Salinity stress

Increasing NaCl concentrations (100 – 200 mol/m^3) increased the leaf proline content of the four potato species (*Solanum tuberosum, S. andigena, S. curtilobum* and *S. juzepczuckii*) (Martinez et al., 1996). However, the frost-resistant genotypes *S. juzepczuckii* and *S. curtilobum* accumulated more proline than the less frost-resistant *S. andigena,* and the non-resistant *S. tuberosum* (Fig. 5.2). In *S. juzepezuckii* proline content increased ~ 18-fold as compared to the control. The higher proline content in *S. juzipczuckii* and *S. curtilobum* resulted in higher shoot height and higher fresh weight as compared to *S. andigena* and *S. tuberosum* (Fig. 5.3). A close relationship was observed between proline content, frost tolerance and also salt tolerance in the four potato species. Thus Martinez et al. suggest that proline can be used as a biochemical marker for salt tolerance in potato and possibly other crop species.

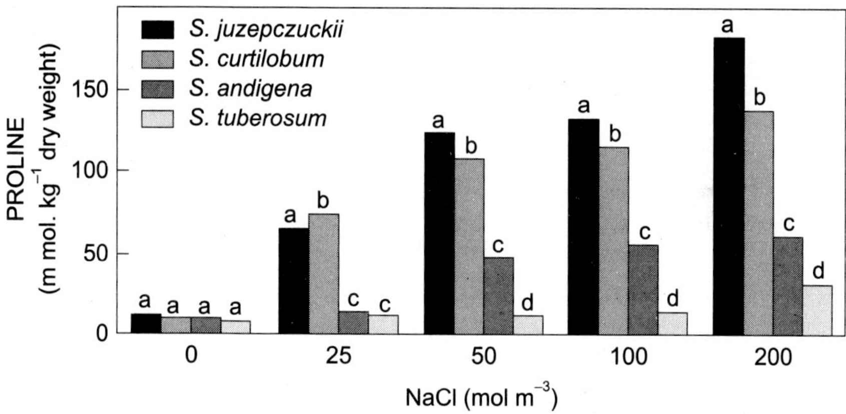

Fig. 5.2 Effect of NaCl levels on the proline content of leaves of four *Solanum* species. Different letters denote significant difference between species at P = < 0.01. (After Martinez et al., 1996).

A possible correlation between proline content and salt tolerance has been found in a wide range of crop species, viz. *Solanum melongena, Nicotiana sylvestris, Cicer arietinum* and *Lycopersicon esculentum* cells

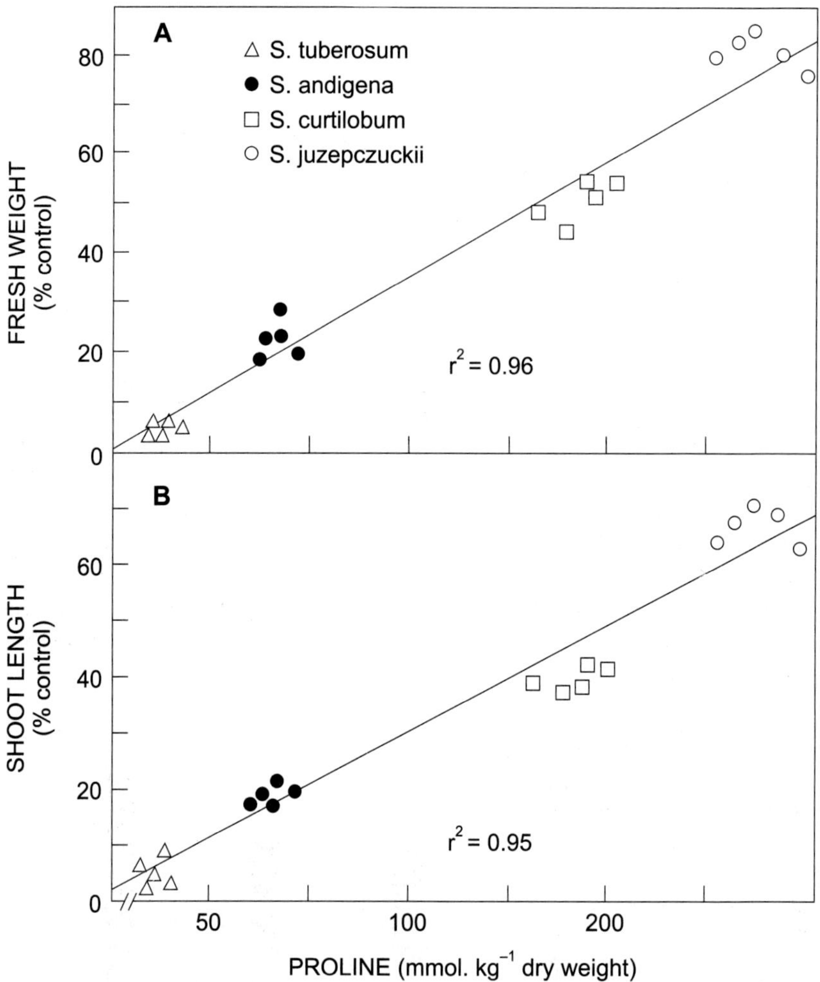

Fig. 5.3 Relations between proline content and fresh weight (A) and shoot length (B) of plantlets of four *Solanum* species under 200 mol/m³ NaCl stress. (After Martinez et al., 1996).

adapted to water stress. Possibly the Na and Cl ions that accumulate intracellularly in salt-adapted cells are compartmentalized to a greater extent in the vacuole and the organic osmolytes, e.g. proline, serve as compatible solutes in the cytoplasm, counteracting the effect of salt accumulated in the vacuole. Proline has a protective effect on the enzyme exposed to high ionic concentration in the cells and also serves as a reservoir of carbon and nitrogen for utilization during salinity stress. Roy et al. (1992) reported that the activity of Δ^1- pyrroline-5-carboxylate reductase (P5CR) was increased twice under 100 mol/m^3 NaCl in salt resistant cultivars of rice. Simultaneously the activity of l-proline dehydrogenase (participating in proline catabolism) is significantly decreased.

Overexpression of proline enhances biomass production and flower development under salt-stress conditions (Kavi Kishore et al., 1995). They observed significant differences between control and P5CS-transgenic tobacco plants grown under salt- stress conditions. When the transgenic plants overexpressed the moth bean P5CS, a bifunctional enzyme able to catalyze conversion of glutamate to P5C, which is then reduced to proline. The transgenic plants synthesized 10- to 18-fold more proline than the control plants. The roots on transgenic plants were 40 percent longer and had 2-fold greater biomass. Rascio et al. (1994) observed that accumulation of Cl^- , K^+ , and proline was concurrent with an increase in the "binding strength" of water in wheat leaves. Primary function of the accumulation of these solutes may be the regulation of intracellular water activity (La Rosa et al., 1991). Both proline and glycinebetaine under water-stressed conditions may be able to induce the formation of strong H-bonded water around the protein, preserving the native state of the cell biopolymers (Rascio et al., 1994). These osmolytes have been suggested to directly influence protein solvation by protecting from dehydration-induced thermodynamic perturbations in proteins. Also the externally added proline has been demonstrated to protect the structural and functional integrity of enzymes *in vitro.*

Drought stress

Proline is one of the most common osmolytes in water-stressed plants. Accumulation of proline in dehydrated plants is increased both by activation of the biosynthesis of proline and by inactivation of the degradation process. l-Proline is synthesized from l-glutamic acid via P5C by two enzymes P5CS and P5CR. l-Proline is metabolized to l-glutamate via P5C by two enzymes : proline dehydrogenase (oxygenase) and P5C dehydrogenase. Such a metabolism of proline is inhibited when proline accumulates during dehydration and it is activated when rehydration occurs. Under dehydration conditions, when expression of the gene for P5CS is strongly induced, expression of the gene for proline dehydrogenase is inhibited. By contrast, under rehydration conditions, when expression of the gene for proline dehydrogenase is strongly induced, expression of the gene for P5CS is inhibited. Thus, P5CS which acts during the biosynthesis of proline, and proline dehydrogenase which acts during metabolism of proline, appear as rate-limiting factors under water stress. Therefore, it is suggested that the levels of proline are regulated at the level of transcriptional, the genes of these two enzymes during dehydration and rehydration (Yoshida et al.,

1997). Moreover, it has been demonstrated that proline acts as an osmoprotectant and that overproduction of proline results in increased tolerance to osmotic stress of transgenic plants. Genetically engineered crop plants that overproduce proline might acquire dehydration tolerance.

Many plants have the ability to adapt to water stress at the cellular level and that proline is involved in tolerance to osmotic stress acting as a compatible osmolyte. Some of the observations that substantiate this are cited : Itai and Paleg (1982) demonstrated that proline ameliorates dehydration-induced perturbation in proteins and the exogenously supplied proline confers some osmotic tolerance to barley and cucumber plants and cultured cells. Handa et al. (1986) and Rhodes et al. (1986) reported that tomato cells cultured under water stress rapidly accumulated about 300 times more proline than the non-water-stressed cells, and they adapted to water stress. The transgenic tobacco plants that expressed a cDNA for moth bean P5CS under the control of ^{35}S promoter of cauliflower mosaic virus produced a high level of the enzyme and subsequently accumulated 10- to 18-fold more proline than the control plants (Kavi Kishore et al., 1995). Kavi Kishore and associates also reported that overproduction of proline enhanced root biomass production and the development of flower in transgenic plants exposed to drought. In the apical millimeter of maize roots, proline represents a major solute, reaching concentrations of 120 mM in roots growing at a water potential of –1.6 MPa (Voetberg and Sharp, 1991). The accumulated proline accounts for a significant fraction (~ 50 %) of the osmotic adjustment in this region. Proline accumulation in maize root apical meristems in response to water deficits involves increased proline deposition to the growing region, and appears to regulate abscisic acid. Thus it seems likely that proline-overproducing crop plants obtained by genetic engineering might acquire osmotolerance. It is further substantiated by the findings of Igarashi et al. (1997) who isolated a cDNA for P5CS from rice plants and compared the level of mRNA for P5CS and the level of proline accumulation in a salt-sensitive rice cv IR28 and a salt-tolerant cv Dee-gee-woo-gen under high salinity conditions. The transcript of the rice gene for P5CS appeared within 10 hours after the start of salt treatment, and the level of the transcript increased for up to 48 hours in both the rice cultivars, but the level of the transcript in the salt-tolerant cultivar was higher than in the salt-sensitive cultivar. The level of proline that accumulated in the salt-tolerant cultivar was also higher than in the salt-sensitive cultivar. These observations suggest that expression of the gene for P5CS and the accumulation of proline might be correlated with salt tolerance in rice.

The expression of genes for P5CS and P5CR was analyzed under dehydration conditions in moth bean (Verbruggen et al., 1993). Also the transgenic tobacco plants expressing moth bean P5CS produced a large amount of the enzyme and subsequently accumulated 10-fold more proline than did control plants (Kavi Kishore et al., 1995). Dehydration causes osmotic stress and leads subsequently to elevated levels of proline in plant cells as a result of *de novo* biosynthesis, but cannot induce the expression of the gene for proline dehydrogenase because of repression by osmotic stress. When the plants are rehydrated, expression of this gene becomes inducible by proline because of the removal of osmotic stress and absence of repression (Kiyosue et al., 1996). The gene for proline dehydrogenase is induced by rehydration but is repressed by dehydration. Moreover, its expression is induced by proline and repressed by osmotic stress. Expression of the gene for proline dehydrogenase is inducible by an elevated level of the intracellular proline during redydration, but it is repressed by osmotic stress under dehydration conditions. In dehydrated plants, accumulation of proline occurs as a result of both the activation of its biosynthesis and the inactivation of its degradation, whereas a decrease in the accumulation of proline occurs as a result of both the inactivation of biosynthesis and the activation of degradation in rehydrated plants (Yoshida et al., 1997).

COLD AND FREEZING STRESS

Proline acts as a cryoprotectant in maize (Duncan and Widholm, 1987), tobacco (Bornman and Jensson, 1980) and wheat leaves (Dörffling et al., 1990). In grapevine, proline concentration was affected by cold treatment, but the changes were less in dormant buds (stage 1) than in swollen (stage 3) or greentip (stage 5) buds (Fig. 5.4). At stage 1, no change was observed in proline concentration when buds were exposed to –14°C or lower temperatures. The temperature effect was large at stage 3. Proline concentration increased by exposure to –14°C but decreased by cooling to –24°C. At stage 5, proline concentration increased at –6°C, but was unaffected by exposure to –24°C (Barka and Audran, 1997).

IMPROVEMENT

Van Swaaij et al. (1986) selected hydroxyproline (an analog of P5C)-resistant potato callus which overproduced proline. Proline accumulation and increased frost tolerance were exhibited in the leaves of plants regenerated from this callus. Selection for hydroxyproline-resistant mutants of barley and winter wheat has succeeded in identifying lines that accumulate greater quantities of proline than the wild-types (Kueh and Bright, 1981; Dörffling et al., 1993). In winter wheat the hydroxyproline-resistant lines are significantly more frost tolerant than the wild-type (Dörffling et al., 1993). Salt tolerant and PEG-resistant mutants of *Nicotiana*

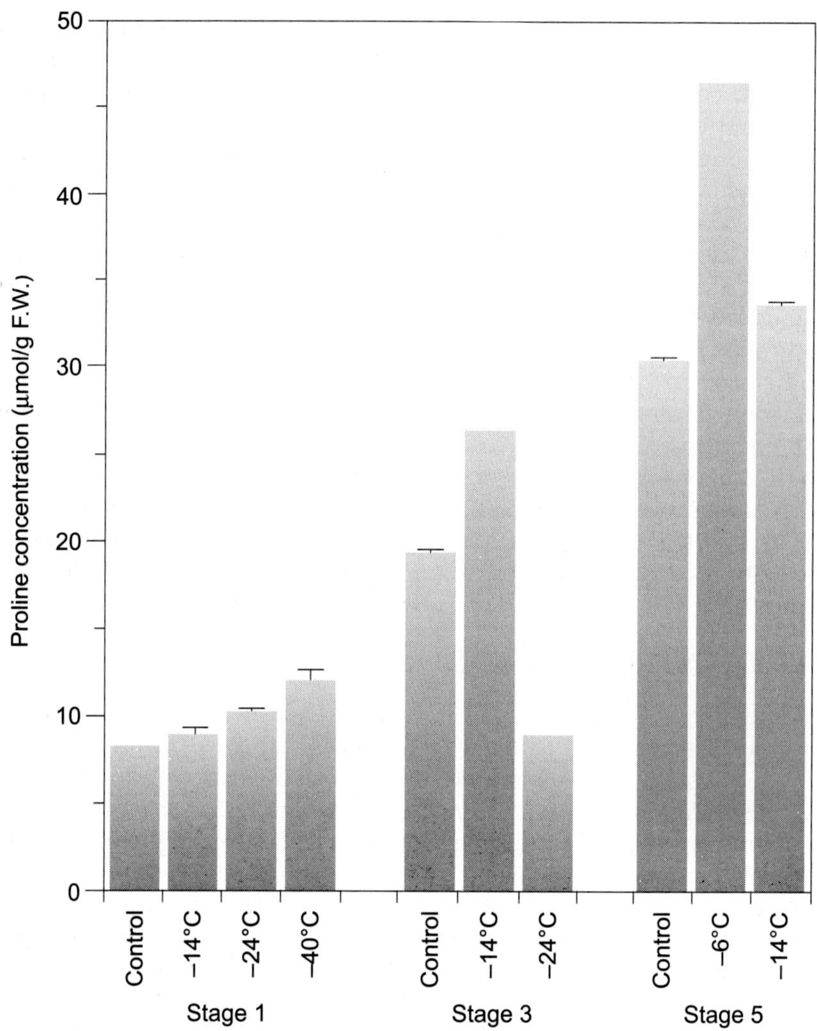

Fig. 5.4 Proline concentration of vine buds at different phenological stages after exposure to different low temperatures. (After Barka and Audran, 1997).

plumbaginifolia have been derived from protoplast culture and appear to have enhanced proline accumulation in comparison to wild-type (Sumaryati et al., 1992). In the same year, Galiba et al. (1992) found that drought sensitive wheat cv Capplle Desprez chromosomes associated with osmotic (0.7 M mannitol) stress-induced free amino acid are 5A and 5D. Further, Deane et al. (1995) worked with cauliflower in which *in vitro* mutagenesis and hydroxyproline-resistant selections were made to obtain proline-accumulating shoots as a strategy for improving frost tolerance of this crop. The concentration of 3 mM hydroxyproline was considered suitable for selecting resistant shoots. However, continuous subculturing in the absence of selection led to segregation of plants into proline accumulating and nonaccumulating types.

Intermediates of proline metabolism and catabolism such as glutamine and P5C can increase the expression of several osmotically regulated genes in rice, including *salT* and *dhn4*. The osmotically induced increases in the concentrations of one or more intermediates of proline metabolism can influence some of the characteristic responses to osmotic stress (Iyer and Caplan, 1998). P5CS is the rate limiting enzyme in proline accumulation in plants and is subjected to feedback inhibition by proline. Overexpression of P5CS results in greater accumulation of proline in transgenic plants. Using site-directed mutagenesis, a mutant form of P5CS enzyme (P5CS F 129A) was created (Hong et al., 2002). The transgenic plants expressing P5CS F 129A accumulated more proline over the plants expressing the wild-type P5CS. The elevated proline effectively reduces free radical levels in response to osmotic stress, as measured by malondialdehyde production, and significantly improves the ability of the transgenic seedlings to grow in media containing up to 200 mM NaCl (Hong et al., 2002). It has been suggested that the feedback regulation of P5CS is lost in plants under stress conditions. Hong et al. (2002) compared proline levels in transgenic tobacco (*Nicotiana tabaccum*) plants expressing a wild-type form of *Vigna aconitifolia* P5CS and a mutated form of the enzyme (*P5CS F 129A*) whose feedback inhibition by proline was removed by site-directed mutagenesis. Transgenic plants expressing *P5CS F 129A* accumulated about 2-fold more proline than the plants expressing *Vigna aconitifolia* wild-type P5CS. This difference was further increased in plants treated with 200 mM NaCl. These results demonstrate that the feedback regulation of P5CS plays a role in controlling the level of proline in plants under both normal and stress conditions. The elevated proline also reduced free radical levels in response to osmotic stress, as measured by malondialdehyde production, and significantly improved the ability of the transgenic seedlings to grow in medium containing up to 200 mM NaCl. These findings shed light on the regulation of proline biosynthesis in plants and the role of proline in reducing oxidative stress induced by osmotic stress, in addition to its accepted role as an osmolyte.

Yonamine et al. (2004) reported that the transgenic tobacco cells overexpressing *NtHAL 3a* gene from *Arabidopsis thaliana* have increased intracellular ratio of proline. They further reported that *NtHAL 3* proteins are involved in the coenzyme A biosynthetic pathway in tobacco cells. Overexpression of *NtHAL 3a* which improved salt, osmotic and lithium tolerance in cultured tobacco cells also showed increased intracellular ratio of proline.

Drought stress also triggers production of endogenous ABA which in turn induces transcription of several genes which protect cells from dehydration. Changes in free proline level and endogenous ABA content in leaves of several crop species show a correlation between the biotypes having elevated levels of proline and ABA, and the enhanced drought (PEG) tolerance (Zheng and Li, 2000). Zheng and Li suggest that the proline and ABA level measurements under PEG 6000 (6%) treatment can be used for evaluating drought tolerance-ability among the soybean cultivars (Table 5.1). However, such a correlation was not found between proline and ABA levels, and salt tolerance, probably because 0.5% NaCl used was not enough (Table 5.2).

Table 5.1 Effects of 6 percent PEG 6000 (drought) treatment on the contents of proline (µg/g dry wt.) and ABA (µg/g dry wt) in 7 drought tolerant and 4 drought sensitive soybean cultivars (Modified from Zheng and Li, 2000).

Cultivars	←Proline level→			←ABA level→		
	0 h stress	36 h stress	Times of control	0 h stress	26 h stress	Times of control
T1 *	116.2	311.5	2.7	8.13	12.76	1.6
T2	141.0	274.9	2.0	11.81	28.83	2.4
T3	187.8	320.0	1.7	16.45	25.48	1.5
T4	105.3	346.5	3.3	12.66	18.66	1.5
T5	95.0	441.2	4.6	4.59	26.57	5.6
T6	133.2	375.3	2.8	10.95	92.54	8.5
T7	125.1	286.0	2.7	7.90	28.74	3.6
In range of	9.5-187.8	274.9-441.2	1.7-4.6	4.59-16.46	12.76-92.54	1.5-8.5
S1 **	113.4	212.7	1.9	22.73	16.62	0.7
S2	107.9	145.6	1.3	10.00	11.99	1.1
S3	143.2	258.9	1.8	11.49	14.05	1.2
S4	106.0	180.0	1.7	14.35	8.79	0.6
In range of	106.0-143.2	145.6-258.9	1.3-1.9	10.00-22.73	8.79-16.62	0.6-1.2

* Cultivars = Seven drought tolerant (T) and four drought sensitive (S) cultivars selected on the basis of grain yield, pod set, plant height and dry weight, and test weight (100 seed weight) under moisture stress at a high temperature.

** Treatment = Soybean seeds were germinated and grown in vermiculite for 3-4 weeks. Roots of the soybean seedlings were tested in Hoagland solution containing 6% PEG (6000) for 36 – 48 hours.

Table 5.2 Effects of 0.5% NaCl treatment** on the contents of proline (μg/g dry wt) and ABA (μg/g dry weight) in 7 drought tolerant and 4 drought sensitive soybean cultivars (Modified from Zheng and Li, 2000).

Cultivars *	← Proline contents after→			←ABA contents after→		
	0 h stress	36 h stress	Times of control	0 h stress	36 h stress	Times of control
T1	116.2	175.6	1.5	8.13	14.7	1.8
T2	141.0	226.0	1.6	11.81	13.1	1.1
T2	187.8	233.4	1.2	16.45	24.9	1.5
T4	105.3	181.6	1.7	12.66	4.47	0.4
T5	95.0	123.3	1.3	4.59	5.00	1.1
T6	133.2	203.8	1.5	10.95	12.69	1.2
T7	125.1	181.8	1.5	7.90	8.64	1.1
In range of	95.0-187.8	123.3-233.4	1.2-1.7	4.59-16.46	4.47-24.9	0.4-1.8
S1	113.4	179.9	1.6	22.73	13.81	0.6
S2	107.9	150.0	1.4	10.00	26.29	2.6
S3	143.2	154.9	1.1	11.49	11.64	1.0
S4	106.0	154.7	1.5	14.35	16.06	1.1
In range of	106.0-143.2	154.7-179.9	1.1-1.6	10.00-22.73	11.64-26.29	0.6-1.2

* Treatment = Soybean seeds were germinated and grown in vermiculite for 3-4 weeks . Roots of soybean seedlings were treated in Hoagland solution containing 0.5% NaCl for 36 hours.

** Cultivars = 7 drought tolerant (T) and 4 drought sensitive (S) cultivars selected on the basis of grain yield, pod set, plant height and dry weight, and test weight (100 seed weight) under moisture stress at a high temperature.

REFERENCES

Barka, E.A. and J.C. Audran, 1997. Response of Champenoise grapevine to low temperatures: Changes of shoot and bud proline concentrations in response to low temperatures and correlations with freezing tolerance. J. Hort. Sci. 72: 577-82.

Bornman, C.H. and E. Jansson, 1980. *Nicotiana tabacum* callus studies : ABA increases resistance to cold damage. Physiol. Plant. 48: 491-93.

Deane, C.R.; M.P. Fuller and P.J. Dix, 1995. Selection of hydroxyproline-resistant proline-accumulating mutants of cauliflower (*Brassica oleracea* var. *botrytis*) Euphytica 85: 329-34.

Delanuey, A.J.; C.A. Hu; P.B. Kavi Kishore and D.P.S. Verma, 1993. Cloning of orthithine δ-aminotransferase cDNA from *Vigna aconitifolia* by trans-complimentation in *Escherichia coli* and regulation of proline biosynthesis. J. Biol. Chem. 268: 18673-78.

Dörffling, K.; S. Schulenburg; G. Lesselich and H. Dörffling, 1990. Abscisic acid and proline levels in cold-hardened winter wheat leaves in relation to variety-specific differences in freezing resistance. J. Agron. Crop Sci. 165 : 230-39.

Dörffling, K.; H. Dörflling and G. Lesselich, 1993. *In vitro* selection and regeneration of hydroxyproline-resistant lines of winter wheat with increased proline content and increased frost tolerance. J. Plant Physiol. 142: 222-25.

Duncan, D.R. and J.M. Widholm, 1987. Proline accumulation and its implication in cold tolerance of regenerable maize callus. Plant Physiol. 83: 703-08.

Galiba, G.; L. Simonsarkadi; G. Kocsy; A. Salgo and J. Sutka, 1992. Possible chromosomal location of genes determining the osmoregulation of wheat. Theo. Appl. Genet. 85: 415-18.

Handa, S.; A.K. Handa; P.M. Hasegawa and R.A. Bressan, 1986. Proline accumulation and the adaptation of cultured plant cells to water stress. Plant Physiol. 80: 938-45.

Hare, P.D. and W.A. Cress, 1997. Metabolic implications of stress-induced proline accumulation in plants. Plant Growth Regul. 21: 79-102.

Hong, Z.; K. Lakkineni; Z. Zhang and D.P.S. Verma, 2002. Removal of feedback inhibition of Δ^1–pyrroline-5-carboxylate synthase results in increased proline accumulation and protection of plants from osmotic stress. Plant Physiol. 122: 1129-36.

Igarashi, Y.; Y. Yoshida; Y. Sandra; et al. 1997. Characterization of the gene for Δ^1-pyrroline-5-carboxylase synthetase and correlation between the expression of the gene and salt tolerance in *Oryza sativa* L. Plant Mol. Biol. 33: 857-65.

Itai, C. and L.G. Paleg, 1982. Response of the water stress on *Hordeum dictichum* L. and *Cucumis sativus* to proline betaine. Plant Sci. Lett. 25: 329-35.

Iyer, S. and A. Caplan, 1998. Products of proline catabolism can induce osmotically regulated genes in rice. Plant Physiol. 116: 203-11.

Kavi Kishore, P.B.; Z. Hong; G.H. Miao; et al. 1985. Overexpression of D^1-pyrroline-5 carboxylate synthetase increases proline production and confers osmotolerance in transgenic plants. Plant Physiol. 108: 1387-94.

Kiyosue, T.; Y. Yoshiba; K. Yamaguchi-Shinozaki and K. Shinozaki, 1996. Nuclear gene, encoding mitochondrial proline dehydrogenase, an enzyme involved in proline metabolism, is upregulated by proline but down regulated by dehydration in *Arabidopsis.* Plant Cell 8: 1323-35.

Kueh, J.S.H. and S.W.J. Bright, 1981. Proline accumulation in a barley mutant resistant to trans-4-hydroxy-l-proline. Planta 153 : 166-71.

La Rosa, P.C.; D. Rhodes; J.C. Rhodes; et al. 1991. Elevated accumulation of proline in NaCl adapted tobacco cells is not due to altered Δ^1-pyrroline-5-carboxylate reductase. Plant Physiol. 96: 245-50.

Martinez, C.A.; M. Maestri and E.G. Lani, 1996. *In vitro* salt tolerance and proline accumulation in Andean potato (*Solanum* spp) differing in forst resistance. Plant Sci. 116: 177-84.

Rascio, A.; C. Plantani; G. Sealfati; et al. 1994. The accumulation of solutes and water binding strength in durum wheat. Physiol. Plant. 90: 715-21.

Raymond, M.J. and N. Smirnoff, 2002. Proline metabolism and transport in maize seedlings at low water potential. Ann. Bot. 89: 813-23.

Rantsch, D.; B. Hirner; E. Schmeizer and W.B. Frommer, 1996. Salt stress-induced proline transporters and salt stress-repressed broad specificity; amino acid permeases identified by suspension of a yeast-amino acid permease-targeting mutant. Plant Cell 8: 1437-46.

Rhodes, D. and S. Handa, 1989. Amino acid metabolism in relation to osmotic adjustment in plant cells. pp 41-62. In: Biochemical and Biophysical Mechanisms Associated with Environmental Stress Tolerance. (ed) J.H. Cherry. Berlin, Springer-Verlag.

Rhodes, D.; S. Handa and R.A. Bressan, 1986. Metabolic changes associated with adaptation of plant cells to water stress. Plant Physiol. 82: 890-903.

Roy, D.; A. Bhunia; N. Basu and S.K. Banerjee, 1992. Effect of NaCl-salinity on metabolism of proline in salt-sensitive and salt-resistant cultivars of rice. Biol. Plant. 34: 159-62.

Schioacke, R.; S. Grallath; K.E. Breitkreuz; et al. 1999. *LeProT1,* a transporter for proline, glycinebetaine, and d-aminobutyric acid in tomato pollen. Plant Cell 11: 377-91.

Siripornadulsil, S.; S. Traina; D.P.S. Verma and R.T. Sayre, 2002. Molecular mechanisms of proline mediated tolerance of toxic heavy metals in transgenic microalgae. Plant Cell 14: 2837-47.

Sumaryati, S.; I. Negrutiu and M. Jacobs, 1992. Characterization and regeneration of salt- and water-stress mutants from protoplast culture of *Nicotiana plumbaginifolia* (Viviani). Theo. Appl. Genet. 83: 613-19.

Van Swaaij, A.C.; E. Jacobsen; J.A.K.W. Kiel and W.J. Feenstra, 1986. Selection, characterization and regeneration of hydroxyproline-resistant cell lines of *Solanum tuberosum*. Tolerance of NaCl and freezing stress. Physiol. Plant. 68: 359-66.

Verbruggen, N.; R. Villarrole and M. van Montagu, 1993. Osmoregulation of a pyrroline- 5-carboxylate reductase gene in *Arabidopsis thaliana.* Plant Physiol. 103: 771-81.

Verslues, P.E. and R.E. Sharp, 1999. Proline accumulation in maize (*Zea mays* L.) primary roots at low water potentials. II. Metabolic source of increased proline deposition in the elongation zone. Plant Physiol. 119 : 1349-60.

Voetberg, G.S. and R.E. Sharp, 1991. Growth of the maize primary root at low water potentials. III. Role of increased proline deposition in osmotic adjustment. Plant Physiol. 96: 1125-30.

Yonamine, I.; K. Yoshida; K. Kido; et al. 2004. Overexpression of *NtHAL3,* gene confers increased levels of proline biosynthesis and the enhancement of salt tolerance in cultured tobacco cells. J. Exp. Bot. 55 (396): 387-95.

Yoshida, Y.; T. Kiyosue; K. Nakashima; et al. 1997. Regulation of levels of proline as an osmolyte in plants under water stress. Plant Cell Physiol. 38: 1095-1102.

Zheng, Y.Z. and T. Li, 2000. Changes of proline levels and abscisic acid content in tolerant/ sensitive cultivars of soybean under osmotic condition. Soybean Genetics Newsl. jhp@ivy.nenu.edu.cn.

SUGGESTED READING

Delauney, A.J. and D.P.S. Verma, 1993. Proline biosynthesis and osmoregulation in plants. Mini review. Plant J. 4 (2): 215-23.

Kavi Kishore,s P.B.; Z. Hong; G.H. Miao; et al. 1995. Overexpression of Δ^1-pyrroline-5-carboxylate synthetase increases proline production and confers osmotolerance in transgenic plants. Plant Physiol. 108: 1387-94.

Yoshiba, Y; T. Kiyosue; K. Nakashima; et al., 1997. Regulation of levels of proline as an osmolyte in plants under water stress. Mini review. Plant Cell Physiol. 38 (10): 1095-1102.

6

GLYCINEBETAINE

Introduction

Betaine (glycinebetaine, N,N,N-trimethylglycine), a compatible solute, accumulates in plants in response to water-deficit caused by drought, salinity, cold/freezing or high temperature stress. Even at high concentrations, glycinebetaine (GB) does not interfere with cytoplasmic functions and stabilizes the structure and function of macromolecules. It is a dipolar but electrically neutral molecule at physiological pH. GB accumulation is induced under stress conditions and its level correlates with the degree of enhancement in stress tolerance. Also, the exogenous application (to leaves or roots) of GB to several crop (both accumulater and nonaccumulater) species has resulted in improved growth and survival under several abiotic stress conditions. GB stabilizes the quaternary structure of enzymes and complex proteins, and membrane integrity under reduced moisture conditions, high salt concentrations and extremes of temperature.

Glycinenetaine has been shown to protect enzymes and membranes from cold (Krall et al., 1989), freezing (Zhao et al., 1992), heat (Jolivert et al., 1982) and salinity (Jolivert et al., 1983). Glycinebetaine also stabilizes PSII protein-pigment complex when stressed with NaCl salinity (Papageorgiou and Murata, 1995). Thus genetic engineering of the biosynthesis of GB from choline (choline has a similar structure to GB except that the carboxyl function of glycine is replaced by an alcohol function) has been the focus of considerable attention as a potential strategy for increasing stress tolerance in stress sensitive plants that are incapable of synthesizing the compatible/protective (GB and some other osmoregulators discussed in this volume have a protective rather than osmotic effect *in vivo* and should not be considered exclusively as an osmoregulator) solute (Sakamoto and Murata, 2000). The osmotic protection offered by GB has been questioned because of the osmotically ineffective concentration of the accumulated GB. As assuming a cell water content of 85 percent, a vacuolar volume of 50 percent of the cell, and compartmentation of GB; the amount of GB accumulated in the shoots of treated plants can at best produce a freezing point depression of only 0.2°C. This is considerably less than the apparent protection produced, and thus an alternative mechanism has to be looked into; probably the maintenance of membrane integrity and arresting denaturation of macromolecules. To test this possibility, Zhao et al. (1992) sprayed *Medicago sativa* seedlings with 0.2 M GB to run-off. When these plants were subjected to subfreezing temperatures, GB reduced the loss of ions from the shoot tissues, as measured by electrical conductivity of the bathing solution. It was particularly apparent at –6°C, which correlated with tissue survival as 67 percent of crowns survived 1 hour at –6°C if treated with GB whereas no control crowns survived this treatment. Glycinebetaine also enhanced and stabilized the oxygen-evolving activity of the PSII protein complexes of higher plants (Papageorgiou and Murata, 1995).

The amounts of GB accumulated suggest that the enhancement of stress tolerance can hardly be attributed to any action of betaine that involves osmotic adjustment to the external environment. The major role of GB might be to protect membranes and macromolecules from the damaging effects of stress. It is also possible that GB might be compartmentalized within cells at certain sites; concentration of GB might be high enough to confer substantial protection against stress even when the overall level of accumulation is low. Such a possibility is supported by the observations that GB concentrated exclusively in the cytoplasm, and not in the vacuoles of leaves of salt-grown halophytes, and the levels of GB as low as 5 mM g^{-1} fr wt can protect some natural accumulaters from the damaging effects of stress (Arakawa et al., 1990; Ishitani et al., 1993). Glycinebetaine might also play a role in promoting transcription and replication under high salt concentrations (Rajendrakumar et al., 1997). Alia et al. (1999) suggest that GB might also accelerate *de novo* protein synthesis during recovery of plants from stress.

BIOSYNTHESIS IN STRESSED PLANTS

Unlike proline, GB is not degraded rapidly upon relief from stress. Glycinebetaine is synthesized as a result of two-step oxidation of choline via betaine aldehyde, a toxic intermediate (Fig. 6.1). The relevant enzymes are choline monooxygenase (CMO), a ferridoxin-dependent soluble protein, and betaine aldehyde dehydrogenase (BADH), a soluble NAD^+-dependent enzyme. These enzymes are found mostly in chloroplast stroma and their activities, as well as levels of betaine increase in response to stress. As reduced ferridoxin (Fd) is generated by photosynthetic electron transport, CMO activity *in vivo* is strongly light dependent. Both drought and salinity induce CMO expression. Betaine synthesis places a substantial demand on pools of its precursors. Despite an ability to down-regulate choline synthsis, GB deficiency is associated with accumulation of serine and a significant expansion of the free choline pool (Yang et al., 1995). The second step of GB synthesis is catalyzed by BADH (Fig. 6.1) and NAD-dependent dehydrogenase that has been characterized and cloned (McNeil et al., 1999). BADH is a dimer of identical 54 kDa subunits which has been efficiently targeted to chloroplasts in transgenic tobacco (Rathinasabapathi et al., 1994). In Graminae, however, BADH which is induced by osmotic stress is peroxisomal (Nakamura et al., 1997).

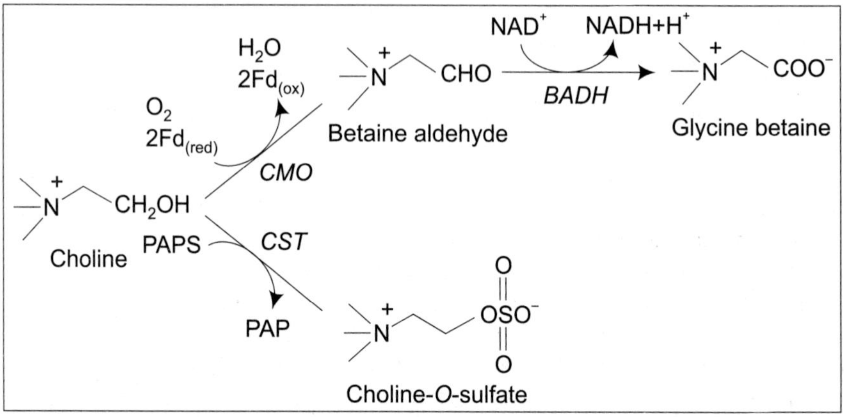

Fig. 6.1 Biosynthetic pathways of glycinebetaine in Chenopodiaceae and of choline-*O*-sulfate in Plumbaginacease. The product of the CMO reaction is the hydrate from the betaine aldehyde. (CST= Choline sulfotransferase; PAP= 3′-Phoshoadenosine-5′-phosphate; PAPS= 3′-Phosphoadenosine-5′-phosphosulfate) (From Mc Neil et al., 1999).

In C_3 plants subjected to stress, the endogenous catalase activity during day may be inadequate for destruction of H_2O_2 generated during photorespiration, and BADH may be more abundant in peroxisomes than in chloroplasts. The peroxisomal GB synthesis might compete with the photorespiratory enzyme serine:glyoxylate aminotransferase for serine (Hare et al., 1998). Ideally, hypotheses that suggest an important link between osmolyte accumulation and photorespiration should account for the apparently poor distinction between the patterns of osmolyte accumulation in C_3 plants and in those displaying C_4 photosynthesis or Crassulacean acid metabolism (CAM). The C_4 species which constitute about 10 percent of the higher plants, mechanisms that concentrate CO_2 at the site of carboxylation, result in reduced photorespiratory rates.

GB-INDUCED MECHANISM OF STRESS TOLERANCE

In certain plant species GB accumulates to the extent of 400 µmol g^{-1} dry wt and osmoprotection occurs, but in most non-GB-accumulating species, GB acccumulating transgenes have resulted in some stress tolerance at as low accumulated-GB as 5 µmol g^{-1} fr wt. At this concentration of GB, no osmoprotection can be expected. Thus the GB has protective rather than osmotic effect. It has been noted that GB protects PSII complex from photoinduced inactivation in transgenic cells. The GB protection of the photosynthetic machinery against photoinduced damage can be attributed to acceleration of recovery of the PSII complex from such a damage (Holmström et al., 2000). The accumulated GB might be preventing the destructive reorganization of the membrane lipids, e.g. in seeds allowed to imbibe water at extreme temperatures.

The fact that GB accumulating transgenic plants exhibit tolerance to several types of abiotic stresses suggests that GB might contribute to the maintenance of cellular functions of fundamental importance under stress conditions. There is some evidence for the involvement of GB in the protection of transcriptional and translational machinery under stress conditions. Rajendrakumar et al. (1997) reported that GB decreases the melting temperature *in vitro* of double-stranded DNA. Allard et al. (1998) further observed that exogenous application of GB to wheat seedlings induced the expression of cold-inducible genes, suggesting that GB has the ability to enhance transcription *in vivo* of genes that are involved in stress tolerance. Further, Bourot et al. (2000) suggested that like a chaperonin, GB may stablize the transcriptional and translational machinery for the efficient expression of genes under stress conditions. Also under high-intensity light, GB might accelerate protein synthesis *de novo* , as Alia et al. (1999) noted that the treatment with lincomycin, an inhibitor of protein synthesis in chloroplasts, dismisses the protective effect of GB.

Yang et al. (1995) developed a series of near-isogenic GB-containing and –deficient F_8 pairs of maize lines. The pairs of lines differ for alternative alleles of a single locus; the wild-type allele conferring GB accumulation is designated *Bet1* and the mutant (recessive) allele is designated *bet1*. The near-isogenic lines were used to investigate whether GB deficiency affects the pool size of the GB precursor, choline, using a new method for GB and choline determination : stable isotope dilution plasma desorption mass spectroscopy. Glycinebetaine deficiency in maize was associated with a significant expansion of the free choline pool, but the difference in choline pool size was not equal to the difference in GB pool size, suggesting that choline must down-regulate its own synthesis. Consistent with this, GB deficiency was also associated with the accumulation of the choline precursor, serine. A randomly amplified polymorphic DNA marker was identified that detects the *bet1* allele.

Yang and Lu (2005) investigated the effects of exogenous application of GB (10 mM) on growth, leaf water content, water use efficiency, photosynthetic gas exchange, and PSII photochemistry in maize plants subjected to salt stress (50 and 100 mM NaCl). In salt-stressed plants, GB application improved growth, leaf water content, net photosynthesis, and the apparent quantum yield of photosynthesis. Glycinebetaine application also increased stomatal conductance, leaf evaporation rate, and water use efficiency. These results suggest that photosynthesis was improved by GB application in salt-stressed plants and such an improvement was associated with an improvement in stomatal conductance and the actual PSII efficiency.

Subcellular localization of GB

The subcellular localization of GB and the site of its biosynthesis seem to be crucial for the efficient GB-protection of the plants under stress. Accumulation of GB in chloroplasts has been demonstrated convincingly in natural accumulaters of GB (Mc Neil et al., 1999), and such accumulation also seems to occur in transgenic plants when the appropriate enzyme is targeted to the chloroplasts (Sakamoto et al., 2000). Overexpression of choline oxidase in chloroplasts has a more significant impact on the stress tolerance of transgenic plants than the overproduction in the cytosol (Huang et al., 2000). Glycinebetaine protects the physiological and metabolic activities of chloroplasts, particularly the photosynthetic machinery by stablizing the quaternary structures and functions of protein complexes such as the PSII complex (Sakamoto and Murata, 2001). Targeting the enzyme for the biosynthesis of GB to mitochondria and the cytosol also enhances tolerance of the photosynthetic machinery to salt and chilling stresses (Holmström et al., 2000; Takabe et al., 1998).

STRESS-INDUCED GB ACCUMULATION AND STRESS MITIGATION

Drought stress

Choline monooxygenase (CMO) catalyses the committing step in the synthesis of GB, accumulated by many plants in response to drought and salinity. When water was withheld, leaf relative water content declined to 59 percent and the levels of CMO mRNA, protein, and enzyme activity rose 3- to 5-fold; rewatering reversed these changes. An elite maize inbred line DH 4866 was transformed with the *betA* gene from *Escherichia coli* encoding choline dehydrogenase, a key enzyme in the biosynthesis of GB from choline (Quan et al., 2004). The transgenic maize plants accumulated higher levels of GB and were more tolerant to drought stress than the wild-type plants (nontransgenic) at germination and the young seedling stages. Most importantly, the grain yield of transgenic plants was significantly higher than that of wild-type plants after drought stress. The enhanced GB accumulation in transgenic maize provides greater protection of the integrity of the cell membrane and greater activity of enzymes compared with wild-type plants in conditions of drought stress.

In drought stressed spring barley genotypes, Ladyman et al. (1983) and Grumet et al. (1985) observed genotypic variability in GB accumulation, and showed that the ability to accumulate GB is a nucleus-encoded, predominantly additive trait with relatively high heritability in the narrow sense.

Salinity stress

By introducing the bacterial *betA* gene, encoding choline dehydrogenase, into tobacco both salt and choline-resistant phenotype was developed (Lilius et al., 1996). As measured by dry weight, there was an 80 percent increase in salt tolerance between the transgenic and wild-type plants at 300 mM NaCl. Further, accumulation of GB in transgenic tobacco expressing the two *E. coli* genes *betA* and *betB* was demonstrated (Holmström et al., 2000). They showed that the transgenic line expressing only *betA* accumulated GB, though the accumulation increased by 2 – 3 times in lines producing both *betA* and *betB* genes. However, the root tissues of both GB accumulating and GB-nonaccumulating sorghum genotypes contain low GB levels under both control and salinized conditions. This observation is contrary to that observed in wheat (Krishnamurthy and Bhagwat, 1990). These transgenic lines showed increased salt tolerance. The increase in salt tolerance of the transgenic GB producing tobacco lines is probably related to the increase in protection of the photosynthetic apparatus, as the production of GB resulted in enhanced tolerance to photoinhibition at low temperature and improved the capacity of PSII to recover from photoinhibition caused by salt stress and high intensity light (Holmström et al., 2000). Protection by GB could involve stabilization of particular protein complexes and membranes. Relatively low concentrations of GB can improve stress tolerance in photosynthesizing organisms, possibly by protecting protein complexes involved in the photosynthetic processes.

After gradual salinization (NaCl : $CaCl_2$ = 5.7 : 1, mol/mol), CMO mRNA, protein, and enzyme levels in sugarbeet leaves increased 3- to 7-fold at 400 mM salt, and returned to uninduced levels when the salt was removed (Russell et al., 1998). Beet roots also expressed CMO, most strongly when salinized. Salt-inducible CMO mRNA, protein, and enzyme activity were readily detected in leaves of *Amaranthus caudatus* L. The transgenic tobacco expressing the modified *betA* gene enhanced biomass production under salt stress and the tolerance of the photosynthetic machinery to photoinhibition under salt and chilling stresses. In this transgenic tobacco that was transformed to synthesize betainealdehyde dehydrogenase (BADH), the resultant progeny produced both choline dehydrogenase (CDH) and BADH, and accumulated nearly 2-fold higher levels of GB than the transgenic plants with CDH alone. However, the coexistence of CDH and BADH did not further improve stress tolerance (Holmström et al., 2000).

Glycinebetaine concentration usually exceeds that of choline and may even form a large percentage of the total plant nitrogen. Both choline and GB concentrations increase with increasing levels of NaCl salinity in both roots and shoots of wheat cv Sonalika (Fig. 6.2 , Krishnamurthy and Bhagwat, 1990). Even though choline and GB are synthesized from the same precursor ethanolamine, choline content of shoots and roots always exceed GB at a wide range of salinity treatment (Fig. 6.3). Percent increase in accumulation of these two compounds in both shoot and root are linearly correlated with salinity of the growth medium.

Sakamoto et al. (1998) transformed a japonica type of rice with two chimeric constructs in which the *codA* gene was under control of the ^{35}S promoter and which included an intron that originated in rice to enhance gene expression. One construct encoded choline oxidase with a signal for targeting to chloroplasts, whereas the other lacked such a signal sequence, allowing choline oxidase to remain in the cytosol. Level of the subcellular GB in the leaves was 1µmol g^{-1} fr wt with the first construct and 3 to 5 µmol g^{-1} fr wt with the second. The targeting of choline oxidase to the chloroplasts provided more effective protection of the PSII complex under high-salt and low-temperature conditions, suggesting that the subcellular localization of the synthesis of GB might be of crucial importance in the protection of the photosynthetic machinery against high-salt and cold stress conditions (Sakamoto and Murata, 2001). The transgenic plants that accumulated GB survived better than the control plants during recovery from salt and drought stress. Furthermore, the photosynthetic machinery was more tolerant in transgenic plants than in wild-type plants to salt stress.

Rice plants are sensitive to various environmental stresses and they do not synthesize GB. The relative simplicity of systems for the transformation of rice and the refinement of systems for the expression of transgenes have enabled to endow this important staple cereal crop with the ability to produce GB (Sakamoto et al., 1998). However, in sorghum genotypes salinity stress markedly reduced GB accumulation and the GB nonaccumulating genotypes of maize did not produce GB even under salinity stress. But in GB accumulating maize genotypes, GB accumulation increased by 4- to 5-fold in young expanding leaves in response to 150 mM NaCl (Rhodes et al. 1989). This accumulation of GB in GB-accumulating maize genotypes occurred in proportion to a decline in leaf relative water content (RWC) induced by salinity stress.

Cold stress

Nomura et al. (1995) reported that barley plants accumulate high levels of GB in response not only to water and salt stress, but also to cold stress.

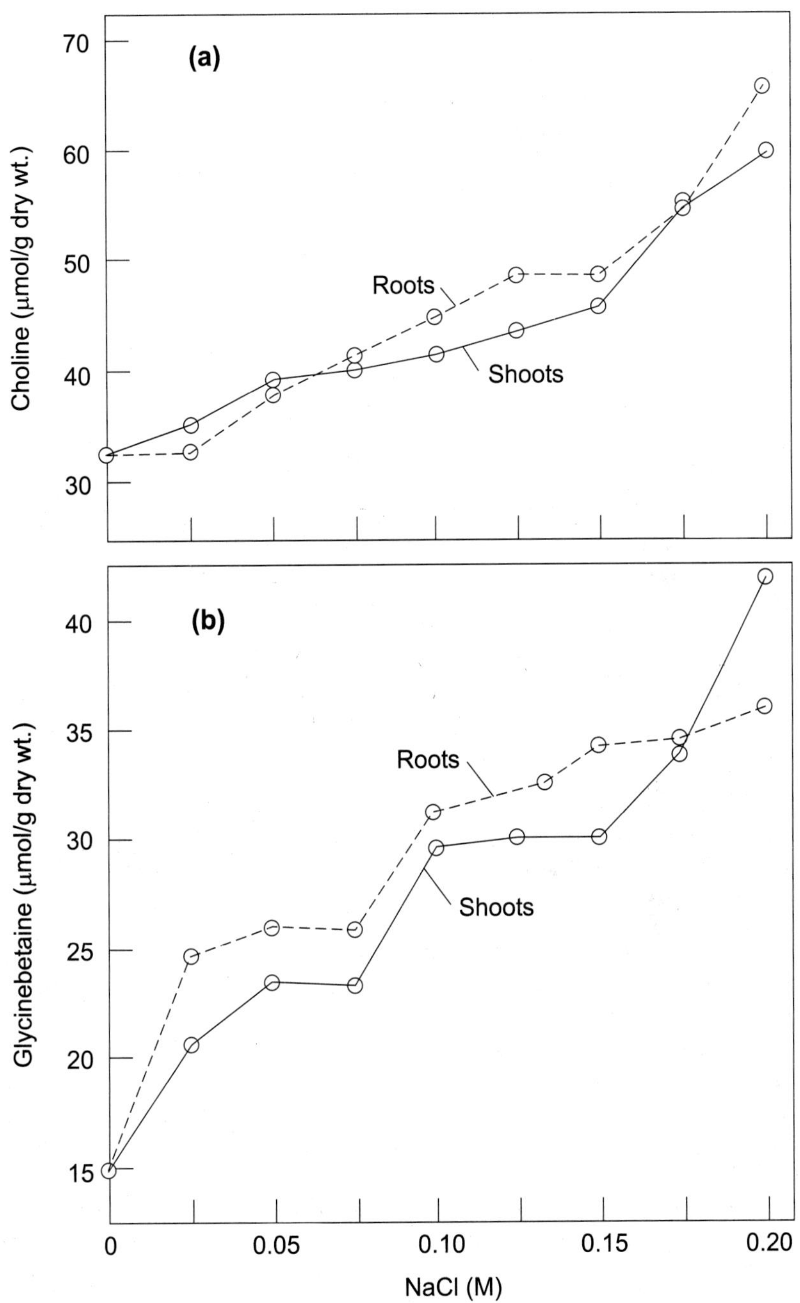

Fig. 6.2 Choline (a) and glycinebetaine (b) accumulation in shoots and roots of wheat at different salinity levels of growth medium (After Krishnamurthy and Bhagwat, 1990).

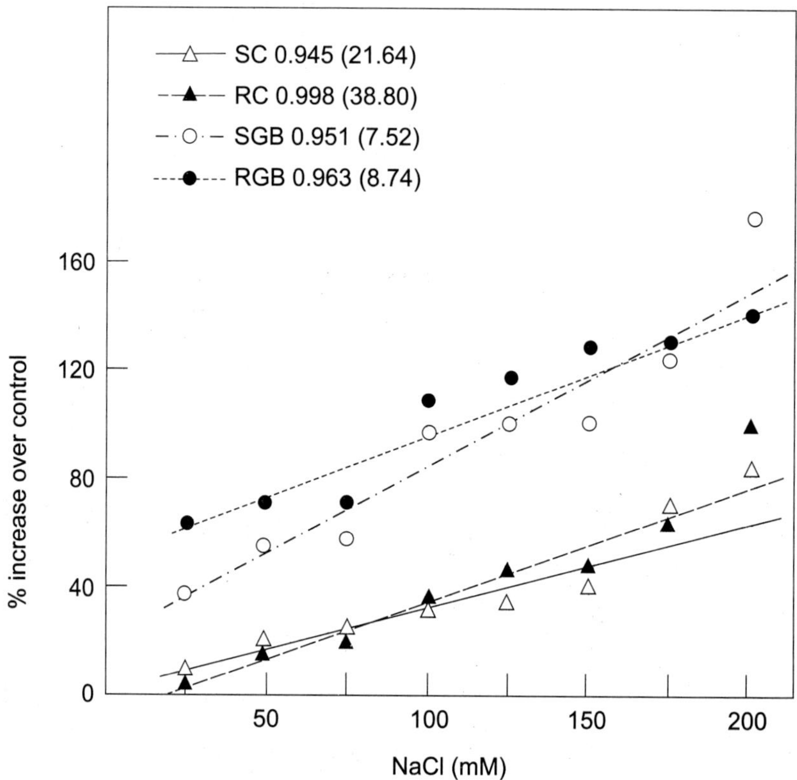

Fig. 6.3 Linear relationship between accumulation of choline and glycinebetaine in shoots and roots of wheat seedlings and salinity of growth medium. (SC= Shoot choline; RC= Root choline; SGB= Shoot glycinebetaine; RGB= Root glycinebetaine). Figures in parentheses are *t* values. (After Krishnamurthy and Bhagwat, 1990).

Such an accumulation of GB during acclimation to cold is associated with freezing tolerance in leaves of barley plants. The levels of GB accumulated during cold acclimation might be associated with earliness or lateness of the maturity of cultivars, *viz.* late cultivars accumulate more GB than the early cultivars. Also, Kishitani et al. (1994) using two winter barley cultivars and the two corresponding sets of near-isogenic lines, with each line differing only in a single 'spring' gene, namely, sh_2^{I}, sh_2^{II} and sh_3, reported that GB accumulates in response to cold stress and that the accumulation of GB during cold acclimation is associated to some extent with freezing tolerance in leaves of barley plants. Glycinebetaine accumulated to 5-times the basal level over the course of 3 weeks at low temperature (5°C) in the winter-type cultivar and in a spring-*sh* line having the *sh* gene for the spring-type growth habit, but the level was only doubled in the spring-sh_3 line, which carried the sh_3 gene for the spring-type growth habit.

In wheat seedlings, Naidu et al. (1991) reported GB accumulation in response to cold stress. Later, Allard et al. (1998) found a correlation between GB accumulation and freezing tolerance during cold acclimation of two wheat (Glenlea and Fredrick) cultivars. The exogenous GB application resulted in a large increase in total osmolarity mostly due to betaine accumulation. The betaine accumulating plants were more tolerant to freezing stress. Betaine treatment (250 mM) during cold acclimation increased freezing tolerance in an additive manner since the LT_{50} reached –14°C (Glenlea) and –22°C (Fredrick) compared to –8°C (Glenlea) and –16°C (Fredrick) for plants that are cold acclimated in the absence of betaine. These results show that betaine treatment can improve freezing tolerance by more than 5°C in both nonacclimated and cold-acclimated wheat plants. The betaine treatment resulted in induction of a subset of low temperature responsive genes, that are also induced by salinity or drought stresses. In addition to these genetic responses, betaine treatment was also able to improve the tolerance of photoinhibition of PSII and the steady-state yield of electron transport over PSII in a manner that mimicked cold-acclimated plants. These data also suggest that betaine improves freezing tolerance by eliciting some of the genetic and physiological responses associated with cold acclimation.

Rajashekar et al. (1999) observed that GB levels in the leaves of strawberry plants increased nearly 2-fold after 4 weeks of cold acclimation during which cold tolerance of leaves increased from –5.8 to –17°C (Fig. 6.4). Exogenous application of ABA (100 μM) to plants triggered GB accumulation and also increased the cold tolerance of leaves in both unhardened and cold-hardened plants. Similar to ABA, exogenous GB application was also effective in inducing cold tolerance in unhardened and cold-hardened plants. Two mM GB application to unhardened plants increased cold tolerance to 2-fold within 72 hours of application (Table 6.1). In addition, it also improved freezing survival and growth of the whole plant. Rajashekar and coworkers suggested that GB is involved in inducing cold tolerance both in response to exogenous ABA and during natural cold acclimation.

Heat stress

Both membrane integrity and photochemical activity of PSII are markedly impaired by thermal stress in GB-deficient maize lines. Yang et al. (1996) tested pairs of homozygous near-isogenic GB-containing(*Bet 1/Bet 1*) and GB-deficient (*bet 1/bet 1*) F_8 lines of maize for differences in their responses to high temperature. When whole plants, excised whole leaves or leaf segments were subjected to temperature above 45°C, *Bet1/ Bet1* lines exhibited less membrane injury (electrolyte leakage) than their near-isogenic *bet1/ bet1* sister lines. An average difference of 2°C was observed between

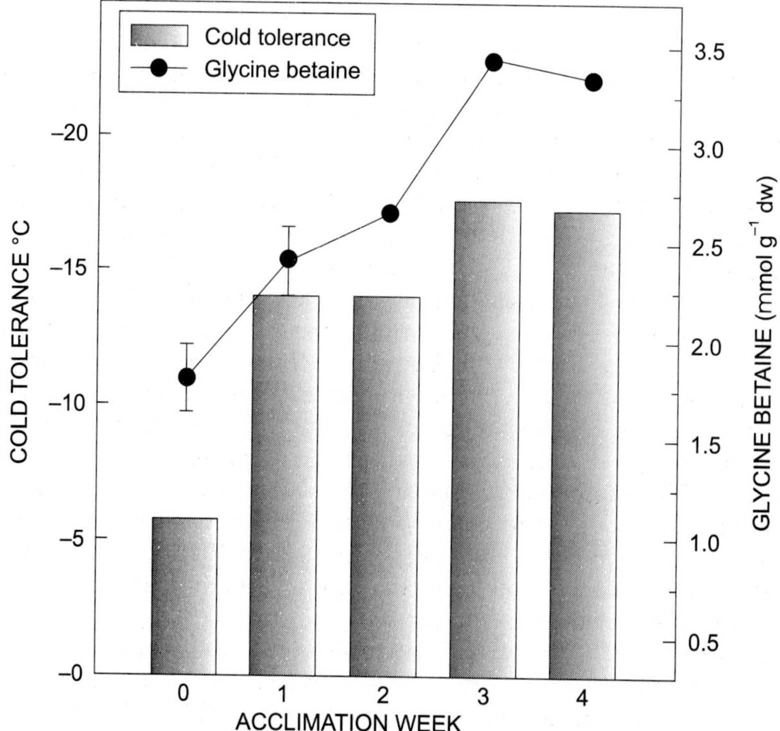

Fig. 6.4 Changes in leaf cold tolerance and endogenous glycinebetaine levels of strawberry leaves during cold acclimation. Strawberry plants were cold acclimated at 4/2°C (day/night) with 10 hour photoperiod. The data for glycinebetaine levels are means (n=4) with SE. $LSD_{0.05}$ for cold tolerance was 0.377. (After Rajashekar et al., 1999).

Table 6.1 Freezing survival of strawberry plants treated with glycinebetaine[a] (After Rajashekar et al., 1999).

Temperature (°C)	Percentage survival	
	Control	Glycinebetaine
0 to –6	95	100[b]
-8	50	80[b]
-10	5[b,c]	30[b,c]
-12	0	20[b,c]
-15	0	0
-17	0	0

a = Plants were treated with 2 mM glycinebetaine as a foliar spray. Potted plants were cooled at 2°C/h to various test temperatures and were grown in a growth chamber after thawing at 4°C. Data on survival of shoots were collected after 2 days of the freezing test while those on regrowth were collected over a 2-4 week period.

b = Plants produced flowers after 2 weeks of freezing.

c = Regrowth of new shoots from the crown was observed in freeze-injured plants after 2 weeks of the freezing test.

Bet1/Bet1 and *bet1/bet1* lines in the critical temperature threshold which triggered catastrophic membrane damage *in vitro. Bet1/Bet1* lines also exhibited greater thermostability of PSII function. The authors suggest that a single gene conferring GB accumulation confers protection against heat destabilization of leaf plasma membranes as well as the photochemical reactions of PSII *in vitro.* Thus GB has a protective role against heat destabilization of plasma membranes and thylakoid membranes in maize. The high temperature stability of PSII is of potential significance to maize crop productivity. An increase in heat tolerance of perhaps only a few degrees might influence survival because heat-induced inactivation of photosynthesis occurs within a narrow range of temperature (Havaux, 1992).

Williams et al. (1992) found that GB led to increases in threshold temperature for PSII-mediated electron transport of the heat-stressed chloroplast membranes. They attributed the protective effect of GB to its capacity to stabilize the lipid-lipid and protein-protein interactions within the thylakoid membranes. Similarly, PSII function was protected against freezing stress in spinach thylakoid preparations by addition of GB, and GB protected PSII from NaCl concentrations that inactivated oxygen evolution or caused dissociation of several extrinsic proteins from the PSII complex.

Yang et al. (2005) stablished genetically engineered tobacco with the ability to synthesize GB by introducing the BADH gene for betaine aldehyde dehydrogenase from spinach. Genetic engineering enabled the plants to accumulate GB mainly in chloroplasts and resulted in enhanced tolerance to high temperature stress during growth of young seedlings. Moreover, CO_2 assimilation of transgenic plants was significantly more tolerant to high temperatures than that of wild-type plants. The results suggest that under high temperature stress, GB maintains the activation of Rubisco by preventing the sequestration of Rubisco activase to the thylakoid membranes from the soluble stroma fractions and thus enhances the tolerance of CO_2 assimilation to high temperature stress. The authors suggest that engineering of the biosynthesis of GB by transformation with the BADH gene might be an effective method for enhancing high temperature tolerance of plants.

Oxidative stress

Parvanova et al. (2004) studied the reaction of the oxidative component of freezing in several tobacco lines, transformed with genes coding for enzymes involved in the synthesis of osmoprotectants (proline, fructan or GB) along with their wild-type. The levels of some oxidative stress markers (leakage of electrolytes, hydrogen peroxide and malondialdehyde) as well as the activity of antioxidative enzymes catalase and guaiacol peroxidase were followed at acclimation, 12- and 24-hour freezing and at recovery. Freezing for 24 hours resulted in severe damage for the wild-type. A corresponding increase of electrolyte leakage, hydrogen peroxide and malondialdehyde contents, a rise of peroxidase activity and inhibition of catalase activity occurred in the non-transformants. Similar, but significantly lower trend of the same parameters has been found for the transgenic lines. Moreover, the oxidative markers returned to their normal levels when the transformants were able to recover from freezing. It could be speculated that transfer of genes, coding for accumulation of osmoprotectants, is related to reduced intensity of freezing-induced oxidative processes.

IMPROVEMENT

Glycinebetaine is the most effective osmolyte which naturally occurs in crops like spinach, sugarbeet, cotton, etc. but others like tomato, tobacco, potato, rice, legumes, etc. lack the ability to synthesize GB. Engineering GB production in these crops will increase yield and/or survival under drought, salinity and low or high temperature limitation conditions in lands presently considered marginal.

Variability

For any improvement work, trait variability has to be studied followed by selection, conventional breeding and genetic engineering. Thus the GB-non-accumulating genotypes of maize and sorghum were identified, although in sorghum, lines that do not accumulate GB are rare. The relative concentrations of GB vary both among and within species. The levels of GB found in sorghum are as much as 10-fold higher than those observed in maize; and within maize GB-accumulating lines, a wide range of relative levels exists. Glycinebetaine accumulation appears to be caused by recessive alleles of a single locus in maize that results in an inability to convert choline to betaine aldehyde, the first committed step in the synthesis of GB. At least 13 diverse GB-non-accumulating maize genotypes are allelic with respect to this locus. Near-isogenic maize lines that accumulate GB are more salt tolerant than their GB-non-accumulating sister lines (Saneoka et al., 1995) further supporting the suggestion that GB accumulation is an important stress tolerance mechanism in maize. In maize, recessive alleles of a single locus are responsible for the phenotype of GB non-accumulation (Yang et al., 2003).

In an analysis of 240 sorghum genotypes, total quaternary ammonium compounds (QAC) levels in betaine fraction of the flag leaves (Fig. 6.5) were found to range from as low as 0.1 μmol g^{-1} fr wt to as much as 33 μmol g^{-1} fr wt (Yang et al., 2003). This variation could be attributed almost exclusively to genetic variability for GB level. Genetic differences in GB

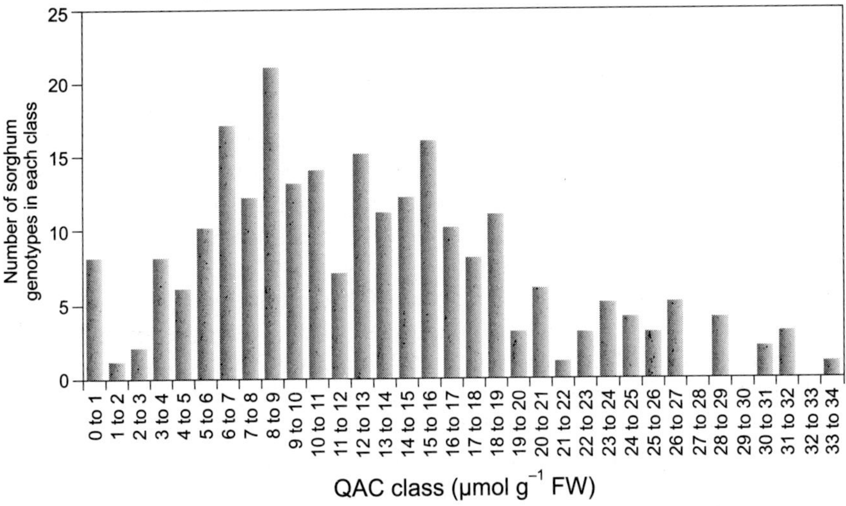

Fig. 6.5 Frequency distribution of total quarternary ammonium compound (QAC) levels (determined by the periodic assay) from the flag leaves of 240 sorghum genotypes. (After Yang et al., 2003).

accumulation exist in sorghum as they do in maize, and that the level of GB in GB-accumulating lines is developmently and environmentally regulated. A recessive allele of a single locus is the cause of this non-accumulation in GB-nonaccumulating sorghum genotypes (Grote et al., 1994).

In controlled environment tests, 339 barley genotypes were grown under well watered conditions and analyzed for shoot GB levels; 145 genotypes from these were also tested under water stress conditions (Ladyman et al., 1983). There were significant differences in GB levels among genotypes, both without and with water stress (approximate ranges of GB levels were 10-40 µmol g^{-1} dry wt in well watered conditions and 35 – 90 µmol g^{-1} dry wt under water stress). Thirteen cultivars which were high and low accumulaters in controlled environments were grown in the field under simulated dryland conditions, and GB was analyzed in the upper leaves. Genotypic variability in GB levels was expressed in both controlled environments and in the field.

Genetics/Inheritance

Low GB accumulation in barley is a dominant trait (Grumet et al., 1985). In maize, a range of 0-5 µmol g^{-1} fr wt of GB has been identified with a single gene controlling GB synthesis (Rhodes and Rich, 1988). The GB deficiency is governed by a single recessive gene. Further, Grote et al. (1994) determined the mode of inheritance of GB-deficient sorghum lines. Crosses were made between three high GB lines (>20 μmol g^{-1} fr wt) and three GB-deficient (<1 μmol g^{-1} fr wt) varieties from which F_1, F_2, and BC progenies were generated. Approximately one-fourth of the F_2 plants in each cross exhibited GB deficiency, supporting the theory of single gene inheritance. Backcrosses to deficient plants segregated 1:1 for deficiency, further supporting this hypothesis. The gene action was primarily additive based on generation mean analysis of high and low parents, F_1 progenies, and backcrosses to the respective parents. A small degree of dominance for GB production was exhibited in the F_1 generation.

Pairs of homozygous near-isogenic GB-containing (*Bet1/Bet1*) and –deficient (*bet1/bet1*) F_8 lines of maize were tested for differences in salt (150 mM NaCl or 127.25 mM NaCl + 22.5 mM $CaCl_2$) tolerance (Saneoka et al., 1995). The *Bet1/Bet1* lines exhibited less shoot growth inhibition (as measured by dry matter accumulation, leaf area expansion rate and/or plant height extension rate) under salinized conditions in comparison to their near-isogenic *bet1/bet1* sister lines. These growth differences were associated with maintenance of a significantly higher leaf relative water content, a higher rate of carbon assimilation, and a greater turgor in *Bet1/Bet1* lines than in *bet1/bet1* lines under salinized conditions. These results strongly suggest that a single gene conferring GB accumulation (and/or a tightly linked locus) plays a key role in osmotic adjustment in maize.

Metabolic engineering

Certain crop plants such as rice, soybean and potatoes lack significant amounts of betaines. This deficiency is the rationale for recent interest in using metabolic engineering technology to install the synthesis of osmoprotectants in such crops in order to improve their tolerance to drought, salinity, and other stresses. The levels of betaine typically rise during exposure to stresses because the biosynthetic enzymes are stress induced. The osmoprotectants are largely confined to the cytoplasm (including organelles) and are almost absent from the vacuole, which generally occupy 90 percent of the cell volume. Isolated chloroplasts of various species have also been shown to contain high concentrations of GB, particularly when isolated from salt-stressed plants.

The idea that introducing GB pathway into plants that lack it will enhance their stress tolerance is based both on comparative physiology and on genetic evidence from a mutation in maize that abolishes GB synthesis and reduces salt and heat tolerance (Saneoka et al., 1995; Yang et al., 1996). Wide crossing work on *Lophopyrum elongatum* and wheat has provided further physiological-genetic evidence that GB accumulation contributes to salt tolerance (Colmer et al., 1995). Nuccio et al. (1998) expressed choline-oxidizing enzymes from spinach-choline monooxygenase into tobacco, a GB nonaccumulater. The transgenic plants produced little

GB and showed small but significant increases in tolerance to various stresses. The GB levels obtained in transgenic plants are only a small percentage of those in spinach, sugarbeet, and other plants that are natural GB accumulaters. The main constraint in GB production in transgenic plants appears to be the endogenous choline supply, because providing choline exogenously leads to a massive increase in GB synthesis (Nuccio et al., 1998). It will therefore most likely be necessary to up-regulate the *de novo* synthesis of choline in order to increase GB synthesis in nonaccumulaters expressing choline oxidizing enzymes.

The first step in GB biosynthesis (choline → betaine aldehyde) is catalyzed by choline monooxygenase (CMO), a ferridoxin-dependent stromal enzyme. The requirement for reduced Fd makes CMO activity light dependent. The second step is mediated by betaine aldehyde dehydrogenase (BADH). Consistent with this, plants such as tobacco that lack GB (nonaccumulaters) show some BADH activity. Such plants do not catabolize GB appreciably. This simplifies engineering as it makes it possible to install a GB pathway in non-accumulaters by expressing choline monooxygenase alone. For GB biosynthesis, free choline is derived from phosphotidylcholine or phosphocholine, and then partitioned between a small active pool (chloroplast) and a large storage (vacuole) pool. Glycinebetaine synthesis is limited by the low choline supply to chloroplast, which can be increased by engineering methods.

Nuccio et al. (1998) constructed transgenic tobacco plants that constitutively express a spinach choline monooxygenase cDNA. Choline monooxygenase from spinach was expressed in tobacco without BADH, there being some endogenous BADH activity in tobacco. The CMO protein was correctly compartmented in chloroplasts and was enzymatically active, showing that its (2Fe-2S) cluster was successfully introduced. On stress testing, Nuccio et al. found that salinization increased the CMO protein levels, apparently via a post-transcriptional mechanism to as high as 10 percent (20-50 nmol fr wt) of that in salinized spinach. However, GB level rose at least 30-fold when 5 mM choline was added to the medium, pointing to a limited internal choline supply. Thus they concluded that in GB-deficient plants engineered with choline-oxidizing genes, size of the free choline pool and the metabolic flux to choline need to be further increased to attain GB levels as high as those in natural accumulaters.

Besides a low capacity for choline synthesis, the import of choline into the chloroplast is a major constraint. Then Nuccio et al. (2000 a) coexpressed a CMO modified to enhance translation and sugarbeet BADH in tobacco, still GB level was not changed significantly. Modeling of ^{14}C-choline labeling data on these double transformants showed that the observed ^{14}C-flux to GB in CMO/BADH plants could be accounted for only by interposing a low-flux choline transport step upstream to CMO.

A single enzyme, phoshoethanolamine N-methyltransferase (PEAMT) mediates all three methylation steps, and this enzyme is only 1-3 percent as active in tobacco as in the GB accumulater spinach (Nuccio et al., 2000 b). Spinach PEAMT was therefore overexpressed already harboring CMO and BADH (Mc Neal et al., 2001). This increased the level of free choline 50-fold, and that of GB 30-fold, without perturbing phosphatidylcholine level or growth. Nevertheless, the flux of GB in the CMO/BADH/PEAMT transcripts is still limited by choline to produce ethanolamine, the precursor of choline (Fig. 6.6). Ethanolamine limitation was identified by an increased choline level when ethanolamine was added to the medium. This limitation prompted cloning of serine decarboxylase (SDC), a soluble, pyridoxal phosphate-dependent enzyme that appears to be the main source of ethanolamine in plants (Rontein et al., 2001). The cloning of SDC opens the way to another round of engineering to boost ethanolamine supply (USDA-Glycinebetaine project No. 2001-25100-10620).

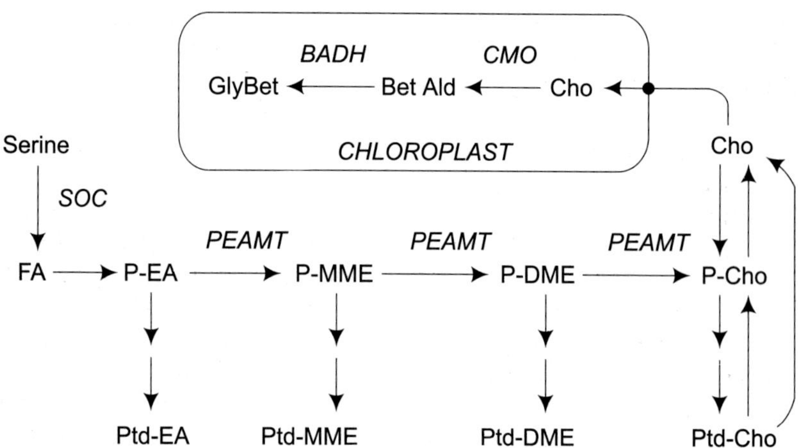

Fig. 6.6 Biosynthesis of glycinebetaine in chloroplasts by a two step oxidation from choline. (USDA-Glycinebetaine project No. 2001-35100-10620). (CMO= Choline monooxidase; BADH= Betaine aldehyde dehydrogenase; PEAMT= Phosphoethanolamine N-methyltransferase; SDC= Serine decarboxylase; P-Cho = Phosphocholine; P-EA = Phosphoethanolamine; Ptd EA = Phosphotidylethanolamine; Ptd Cho = Phosphotidylcholine; GlyBet =Glycinebetaine; BetAld = Betainealdehyde).

Ishitani et al. (1995) reported cloning and characterization of BADH cDNA from barley, and the expression pattern of a BADH transcript. The transgenic tobacco plants harboring the clone expressed high levels of both BADH protein and its enzymatic activity. The leaves and roots of barley plants grown in high-salt conditions, showed that these levels decreased upon release of the stress, whereas they did not decrease under continuous salt stress. BADH transcripts also accumulate in response to drought, indicating a common response of the plant to osmotic changes that affect its water status. Addition of ABA to plants during growth also increased the levels of BADH transcripts dramatically, although the response was delayed when compared to that found for salt-stressed plants.

The enzymes from the monocot barley is localized in the peroxisomes of the transgenic plants (Nakamura et al., 1997). This observation stands in marked contrast to the localization in chloroplasts of BADH from chenopods. However, all BADHs from monocots reportedly contain a signal for targeting to microbodies at the carboxyl-terminus, so this result is easily explained. Nevertheless, it remains unclear whether microbodies/peroxisomes are the site of GB synthesis in Poaceae because the choline-oxidizing enzyme has not been identified (Sakamoto and Murata, 2000).

Huang et al. (2000) installed the metabolic step for oxidation of choline to GB in *Brassica napus* and *Nicotiana tabacum* by constitutive expression of a bacterial choline oxidase gene. The highest levels of GB in independent transgenics were 12.8 and 13 μmol g^{-1} dry wt, respectively i.e. about 15-fold lower values than the values found in natural GB accumulaters. However, the choline-fed transgenic plants synthesized substantially more betaine (Table 6.2). Increasing choline supplementation further enhanced GB synthesis, up to 250 μmol g^{-1} dry wt in *B. napus* and 80 μmol g^{-1} dry wt in tobacco. A moderate stress tolerance was noted in some but not all GB-accumulating transgenic lines based on relative shoot growth. Huang et al. found that the relationship of GB accumulation to precursor (choline)

Table 6.2 ^{13}C composition (δ ^{13}C) of choline-supplimented wild-type (WT) and transgenic plants under osmotic stress (n=3) (Adapted from Huang et al., 2000).

Plant line	δ ^{13}C		
	Control	NaCl[a]	Drought
Brassica napus		← Percentage →	
WT	-33.0	-29.9	-30.8
1916	-33.2	-30.8	-30.9
1928	-33.0	-30.6	-32.2
1929	-32.6	-31.0	-30.7
Tobacco			
WT	-22.2	-17.6	ND[b]
993-1	-23.7	-19.8	ND
993-8	-21.4	-20.0	ND
993-9	-19.4	-20.0	ND

The LSD 0.05 values are 0.82 for *B. napus* and 3.50 for tobacco.
a = Concentrations of NaCl were 300 mM for *B. napus* and 150 mM for tobacco.
b = ND, Not determined.

supply was particularly evident in *B. napus* lines showing progressively higher levels of GB with an increasing supply of choline. Thus, choline supplementation increases the in plant accumulation of GB.

Tomato plants, which normally do not accumulate GB, are susceptible to chilling stress. Exposure to temperatures below 10°C causes various injuries and greatly decreases fruit set in most cultivars. Park et al. (2004) transformed tomato cv Moneymaker with a chloroplast-targeted *codA* gene of *Arthrobacter globiformis,* which encodes choline oxidase to catalyze the conversion of choline to GB. These transgenic plants express *codA* and synthesize choline oxidase, while accumulating GB in their leaves and reproductive organs up to 0.3 and 1.2 μmol g^{-1} fresh weight, respectively. Their chloroplasts contain up to 86 percent of total leaf GB. Over various developmental phases, from seed germination to fruit production, these GB-accumulating plants are more tolerant of chilling stress than their wild-type counterparts. During reproduction, they yield, on average, 10-30 percent more fruit following chilling stress. Exogenous GB contents as low as 0.1 μmol g^{-1} fresh weight are apparently sufficient to confer high levels of tolerance in tomato plants, as achieved via transformation with the *codA* gene. Exogenous application of either GB or H_2O_2 improves both chilling and oxidative tolerance concomitant with enhanced catalase activity. These moderately increased levels of H_2O_2 in *codA* transgenic plants, as a byproduct of choline oxidase-catalyzed GB synthesis, might activate the H_2O_2–inducible protective mechanism, resulting in improved chilling and oxidative tolerances in GB-accumulating *codA* transgenic plants. Thus, introducing the biosynthetic pathway of GB into tomato through metabolic engineering is an effective strategy for improving chilling tolerance.

Kumar et al. (2004) reported high level expression of betaine aldehyde dehydrogenase in cultured cells, roots, and leaves of carrot via plastid genetic engineering. Homoplasmic transgenic plants exhibiting high levels of salt tolerance were regenerated from bombarded cell cultures via somatic embryos was very high, with one transgenic event per approximately seven bombarded plates under optimal conditions. *In vitro* transgenic carrot cells transformed with the *badh* transgene were visually green in color when compared to untransformed carrot cells, and this offered a visual selection for transgenic lines. The betaine aldehyde dehydrogenase enzyme activity was enhanced 8-fold in transgenic carrot cell cultures, grew 7-fold more, and accumulated 50- to 54-fold more betaine (93-101 μmol g^{-1} dry weight of β-alaninebetaine and glycinebetaine) than untransformed cells grown in liquid medium containing 100 mM NaCl. Transgenic carrot plants expressing betaine aldehyde dehydrogenase grew in the presence of high concentrations of NaCl (up to 400 mM), the highest level of salt tolerance reported so far among genetically modified crop plants. The betaine aldehyde dehydrogenase expression was 74.8 percent in non-green edible parts (carrot roots) containing chloroplasts, and 53 percent in proplastids of cultured cells when compared to chloroplasts (100 percent) in leaves.

PRACTICAL APPLICATION AND FUTURE PERSPECTIVE

With sophisticated control of the expression of transferred genes and enhancement of the availability of the substrate in a specific subcellular compartment, it may be possible to generate plants that produce elevated levels of GB and that are more tolerant than the available cultivars to various stresses in the agricultural environment. A second extremely attractive approach is a combination of different strategies that are each, individually, effective in enhancement of stress tolerance. For instance, targets for this approach include genes involved in the biosynthesis of compatible solutes other than GB and regulatory proteins, such as the stress-inducible transcription factor. After engineering the synthesis of GB, subsequent introduction of other genes that have been implicated in stress tolerance might have a significant impact on the efforts aimed at increasing stress tolerance in crop plants (Sakamoto and Murata, 2001).

REFERENCES

Alia, K.Y.; A. Sakamoto; H. Nonaka; et al. 1999. Enhanced tolerance to light stress of transgenic *Arabidopsis* plants that express the *codA* gene for a bacterial choline oxidase. Plant Mol. Biol.40: 279-88.

Allard, F.; M. Houda; M. Kröl; et al. 1998. Betaine improves freezing tolerance in wheat. Plant, Cell Physiol. 39: 1194-1202.

Arakawa, K.; M. Katayama and T. Takabe, 1990. Levels of betaine and betaine aldehyde dehydrogenase activity in the green leaves and etiolated leaves and roots of barley. Plant, Cell Physiol. 31: 797-803.

Bourot, S.; O. Sire; A. Trautwetter; et al. 2000. Glycinebetaine-assisted protein folding in a *lysA* mutant of *Escherichia coli* . J. Biol. Chem. 275: 1050-56.

Colmer,T.D.; E. Epstein and J. Dvorak,1995. Differential solute regulation in leaf blades of various ages in salt-sensitive wheat and a salt-tolerant wheat × *Lophopyrum elongatum* (Host) A. Love amphidiploid. Plant Physiol. 108: 1715-24.

Grote, E.M.; G. Ejeta and D. Rhodes, 1994. Inheritance of glycinebetaine deficiency in sorghum. Crop Sci. 34: 1217-20.

Grumet, R.; T.G. Isleib and A.D. Hanson, 1985. Genetic control of glycinebetaine level in barley. Crop Sci. 25: 618-22.

Hare, P.D.; W.A. Cress and J. van Staden, 1998. Dissecting the roles of osmolyte accumulation during stress. Plant, Cell Environ. 21: 535-53.

Havaux, M. 1992. Stress tolerance by photosystem II *in vitro* : antagonistic effects of water, heat and photoinhibition stresses. Plant Physiol. 100: 424-32.

Holmström, K.O.; S. Somersalo; A. Mandal; et al. 2000. Improved tolerance to salinity and low temperature in transgenic tobacco producing glycinebetaine. J. Exp. Bot. 51(343): 177-85.

Huang, J.; R. Hirji; L. Adam; et al. 2000. Genetic engineering of glycinebetaine production toward enhancing stress tolerance in plants : Metabolic limitations. Plant Physiol. 122: 747-56.

Ishitani, M.; K. Arakawa; K. Mizuno; et al. 1993. Betainealdehyde dehydrogenase in the Graminae : levels in leaves of both betaine-accumulating and nonaccumulating cereal plants. Plant, Cell Physiol. 34: 493-95.

Ishitani, M.; T. Nakamura; S.Y. Han; et al. 1995. Expression of the betainealdehyde dehydrogenase gene in barley in response to osmotic stress and abscisic acid. Plant Mol. Biol. 27: 307-15.

Jolivert,Y.; F. Larher and J. Hamelin, 1982. Osmoregulation in halophytic higher plants : the protective effect of glycinebetaine against the heat destabilization of membranes. Plant Sci. Lett. 25: 193-201.

Jolivert, Y.; J. Hamelin and F. Larher, 1983. Osmoregulation in halophytic higher plants: the protective effects of glycinebetaine and other related solutes against the oxalate destabilization of membranes in beet root cells. Z. fur Pflanzenphysiol. 1098: 171-80.

Kishitani, S.; K. Watanabe; S. Yasuda; et al. 1994. Accumulation of glycinebetaine during cold acclimation and freezing tolerance in leaves of winter and spring barley plants. Plant, Cell Environ. 17: 89-95.

Krall, J.P.; G.E. Edwards and C.S. Andreo, 1989. Protection of pyruvate, Pi dikinase from maize against cold liability by compatible solutes. Plant Physiol. 89: 280-85.

Krishnamurthy, R. and K.A. Bhagwat, 1990. Accumulation of choline and glycinebetaine in salt-stressed wheat seedlings. Curr. Sci. 59: 111-12.

Kumar, S.; A. Dhingra and H. Daniell, 2004. Plastid-expressed betaine aldehyde dehydrogenase gene in carrot cultured cells, roots, and leaves confers enhanced salt tolerance. Plant Physiol. 136: 2843-54.

Ladyman, J.A.R.; K.M. Ditz; R. Grumet; et al. 1983. Genotypic variation of glycinebetaine accumulation by cultivated and wild barley in relation to water stress. Crop Sci. 23: 465-68.

Lilius, G.; N. Holmsberg and L. Bülow, 1996. Enhanced NaCl stress tolerance in transgenic tobacco expressing bacterial choline dehydrogenase. Bio-Technology 14: 177-80.

Mc Neil, S.D.; M.L.. Nuccio and A.D. Hanson, 1999. Betaines and related osmoprotectants : targets for metabolic engineering for stress tolerance. Plant Physiol. 120: 945-49.

Naidu, B.P.; L.G. Paleg; D. Aspinall; et al. 1991. Amino acid and glycinebetaine accumulation in cold stressed wheat seedlings. Phytochemistry 30: 407-09.

Nakamura, T.; S. Yokota; Y. Muramoto; et al. 1997. Expression of a betainealdehyde dehydrogenase gene in rice, a glycinebetaine nonaccumulater, and possible localization of its protein in peroxisomes. Plant J. 11: 1115-20.

Nomura, M.; M. Ishitani; T. Takabe; et al. 1995. *Synechococcus* sp PCC 7942 transformed with *Escherichia coli bet* genes produces glycinebetaine from choline and acquires resistance to salt stress. Plant Physiol. 107: 703-08.

Nuccio, M.L.; B.L. Russell; K.D. Nolte; et al. 1998. The endogenous choline supply limits glycinebetaine synthesis in transgenic tobacco expressing choline monooxygenase. Plant J. 16: 487-96.

Nuccio, M.L; S.D. Mc Neil; M.J. Ziemak; et al. 2000 a. Choline import into chloroplasts limits glycinebetaine synthesis in tobacco : analysis of plants engineered with a chloroplastic or a cytosolic pathway. Metab. Eng. 2: 300-11.

Nuccio, M.L.; M.J. Ziemak; S.A. Henry; et al. 2000 b. cDNA cloning of phosphoethanolamine N-methyltransferase from spinach by complementation in *Schizosaccharomyces pombe* and characterization of the recombinant enzyme. J. Biol. Chem. 275: 14095-101.

Papageorgiou, G.C. and N. Murata, 1995. The unusually strong stabilizing effects of glycinebetaine on the structure and function of the oxygen-evolving photosystem II complex. Photosynthesis Res. 44: 243-52.

Park, E.J.; Z. Jekni; A. Sakamoto; J. De Noma; et al., 2004. Genetic engineering of glycinebetaine synthesis in tomato protects seeds, plants, and flowers from chilling damage. Plant J. 40(4): 474

Parvanova, D.; S. Ivanov; T. Konstantinova; et al. 2004. Transgenic tobacco plants accumulating osmolytes show reduced oxidative damage under freezing stress. Plant Physiol. Biochem. 42: 57-63.

Quan, R.; M. Shang; H. Zhang; et al. 2004. Engineering of enhanced glycinebetaine synthesis improves drought tolerance in maize. Plant Biotechnol. J. 2(6): 477.

Rajashekar, C.B.; H. Zhou; K.B. Marcum; et al. 1999. Glycinebetaine accumulation and induction of cold tolerance in strawberry (*Fragaria × ananassa* Duch.) plants. Plant Sci. 148: 175-83.

Rajendrakumar, C.S.V.; T. Suryanarayana and A.R. Reddy, 1997. DNA helix destabilization by proline and betaine : possible role in the salinity tolerance process. FEBS Lett. 410: 201-205.

Rathinasabapathi, B.; K.F. McCue; D.A. Gage; et al. 1994. Metabolic engineering of glycinebetaine synthesis : plant betainealdehyde dehydrogenases lacking typical transit peptides are targeted to tobacco chloroplasts where they confer betainealdehyde resistance. Planta 193: 155-62.

Rhodes, D. and P.J. Rich, 1988. Preliminary genetic studies of the phenotype of betaine deficiency in *Zea mays* L. Plant Physiol. 88: 102-108.

Rhodes, D.; P.J. Rich; D.G. Brunk; 1989. Development of betaines by fast atom bombardment mass spectroscopy : identification of glycinebetaine-deficient genotypes of *Zea mays*. Plant Physiol. 89: 781-88.

Rontein, D.; I. Nishida; G. Tasshiro; et al. 2001. Plants synthesize ethanolamine by direct decarboxylation of serine using a pyridoxal phoshate enzyme. J. Biol Chem. 276: 35523-29.

Russell, B.L.; B. Rathinasabapathi and A.D. Hanson, 1998. Osmotic stress induces expression of choline monooxygenase in sugarbeet and amaranth. Plant Physiol. 116: 859-65.

Sakamoto, A. and N. Murata, 2000. Genetic engineering of glycinebetaine synthesis in plants : current status and implications for enhancement of stress tolerance. J. Exp. Bot. 51: 81-88.

Sakamoto, A. and N. Murata, 2001. The use of bacterial choline oxidase, a glycinebetaine-synthesizing enzyme, to create stress resistant transgenic plants. Plant Physiol. 125: 180-88.

Sakamoto, A.; M.N.. Alia and A. Murata, 1998. Metabolic engineering of rice leading to biosynthesis of glycinebetaine and tolerance to salt and cold. Plant Mol. Biol. 38: 1011-19.

Sakamoto, A.; r. Valverde; M.N. Alia; et al. 2000. Transformation of *Arabidopsis* with the *codA* gene for choline oxidase enhances freezing tolerance of plants. Plant J. 22: 449-53.

Saneoka, H.; C. Nagasaka; D.T. Hahn; et al. 1995. Salt tolerance of glycinebetaine- deficient and –containing maize lines. Plant Physiol. 107: 631-38.

Takabe, T.; Y. Hayashi; A. Tanaka; et al. 1998. Evaluation of glycinebetaine accumulation for stress tolerance in transgenic rice plants. In: Proc. Int. Workshop on Breeding and Biotechnology for Environmental Stress in Rice. Hokkaido National Agric. Exp. Stn. and Japan Int. Sci. and Technol. Exchange Center, Sapporo, Japan, pp 63-68.

Williams, W.P.; A.P.R. Brain and P.J. Dominy, 1992. Induction of nonbilayer lipid phase separations in chloroplast thylakoid membranes by compatible co-solutes and its relation to the thermal stability of photosystem II. Biochem. Biophys. Acta 1099: 137-44.

Yang, G.; D. Rhodes and R.J. Joly, 1996. Effects of high temperature on membrane stability and chlorophyll fluorescence in glycinebetaine-deficient and glycinebetaine-containing maize lines. Aust. J. Plant Physiol. 23: 437-43.

Yang, W.J.; A. Nadolska-Orczyk; K.V. Wood; et al. 1995. Near-isogenic lines of maize differing for glycinebetaine. Plant Physiol. 107: 621-30.

Yang, W.J.; P.J. Rich; J.D. Axtell; et al. 2003. Genotypic variation for glycinebetaine in sorghum. Crop Sci. 43: 162-69.

Yang, X.H. and C.M. Lu, 2005. Photosynthesis is improved by exogenous glycinebetaine in salt-stressed maize plants. Physiol. Plant. 124: 343-52.

Yang, X.; Z. Liang and C. Lu, 2005. Genetic engineering of the biosynthesis of glycinebetaine enhances photosynthesis against high temperature stress in transgenic tobacco plants. Plant Physiology previev, July 15,2005; 10.1104/pp 105.063164.

Zhao, Y.; D. Aspinall and L.G. Paleg, 1992. Protection of membrane integrity in *Medicago sativa* L. by glycinebetaine against the effects of freezing. J. Plant Physiol. 140: 541-43.

SUGGESTED READING

Chen, T.H. and N. Murata, 2002. Enhancement of tolerance of abiotic stress by metabolic engineering of betaines and other compatible solutes. Plant Biol. 5: 250-57.

Holmström, K.O.; S. Somersalo; A. Mandal; T.E. Palva and W. Welin, 2000. Improved tolerance to salinity and low temperature in transgenic tobacco producing glycinebetaine. J. Exp. Bot. 52(343): 177-85.

Huang, J.; R. Hirji; L. Adam; K.L. Rozwadowski; J.K. Hammerlindl; W.A. Keller and G. Selvaraj, 2000. Genetic engineering of glycinebetaine production toward enhancing stress tolerance in plants : Metabolic limitations. Plant Physiol. 122: 747-56.

Mc Neil, S.D.; M.L. Nuccio and A.D. Hanson, 1999. Betaines and related osmoprotectants. Targets for metabolic engineering of stress tolerance. Plant Physiol. 120: 945-49.

Sakamoto, A. and N. Murata, 2000. Genetic engineering of glycinebetaine synthesis in plants: current status and implications for enhancement of stress tolerance. J. Exp. Bot. 51(342): 81-88.

Sakamoto, A. and N. Murata, 2001. The use of bacterial choline oxidase, a glycinebetaine-synthesizing enzyme, to create stress-resistant transgenic plants. Plant Physiol. 125: 180-88.

7

POLYAMINES

Introduction

Polyamines, triamine spermidine, tetramine spermine and diamine precursor putrescine are positively charged aliphatic amines. Putrescine $[NH_2(CH_2)_4NH_2]$ and spermidine $[NH_2\ (CH_2)_4NH(CH_2)_3NH_2]$ occur in almost all plants, and spermine $[NH_2(CH_2)_3NH\text{–}(CH_2)_4NH\ (CH_2)_3NH_2]$ is less common in procaryotes. Since their primary and secondary amino groups are all protonated at physiological pH, putrescine is divalent, spermidine trivalent and spermine tetravalent organic cation. Polyamines form stronger and more specific interactions than the inorganic cations. Generally, spermidine and spermine are present in millimolar concentrations, whereas putrescine is slightly lower. Most polyamines within cells are bound to nucleic acids and other negatively charged structures. Thus, free and potentially 'reactive' concentration is much lower than the total concentration.

The simplest polyamine, putrescine, is synthesized from the amino acids arginine or ornithine by the enzymes arginine decarboxylase (ADC) and ornithine decarboxylase (ODC), respectively. Additional steps convert putrescine into spermidine and spermine. These steps are catalyzed by spermidine synthetase and spermine synthetase, respectively, which add propylamino groups generated from salenosylmethionine (SAM) by SAM decarboxylase (SAMDC). Polyamines play a role in a wide range of biochemical processes in plants including DNA replication, transcription, protein synthesis, membrane stabilization, and RNA and protein turnover. The most important single characteristic of polyamines is their ability to bind nucleic acids, especially DNA. Polyamines neutralize the charges on phosphate groups of DNA, interact with nucleic acid bases and dock into the major or minor grooves of the double helix. They regulate transcription by inducing bending of DNA after binding to the major grooves. They promote DNA bending by neutralizing the negative charges on DNA phosphate, reducing the energy requirement for bending, and thus facilitating enhanced protein DNA interactions. DNA bending itself is a major pathway for transcriptional regulation of genes expression. Polyamines are also involved in signaling pathway that regulate synthesis of a transcription factors or modulate their bending activity via phosphorylation. Nuclei isolated from the polyamine-depleted cells synthesize 70-80 percent less DNA than the nuclei from control cells. Such cells may accumulate short DNA fragments. The above mentioned effects may be explained due to the ability of polyamines to bind and influence secondary structures of mRNA, tRNA, and rRNA. Only recently with the use of transgenic plants and mutants in which specific enzymes in the polyamine biosynthesis pathway have been specifically overexpressed or inhibited, has it been possible to address the precise roles of polyamines and to determine whether those roles are direct or indirect. The manipulation of polyamine levels in transgenic plants has been made possible through the cloning of genes involved in polyamine biosynthesis.

Putrescine and derived polyamines (spermidine, spermine, long-chain polyamides) have several functions during environmental challenges. In several systems (except during hypoxia, and chilling tolerance of wheat and rice) an induction of polyamines (spermidine, spermine) not putrescine accumulation, may confer stress tolerance (Bouchereau et al., 1999). In several cases stress tolerance is associated with the production of conjugated and bound polyamines and stimulation of polyamine oxidation. In several environmental challenges (osmotic-stress, salinity, hypoxia,, environmental pollutants), results indicate that both arginine decarboxylase and ornithine decarboxylase are required for the synthesis of putrescine and polyamines (spermidine and spermine).

BIOSYNTHESIS

The diamine putrescine is synthesized from the amino acids arginine or ornithine, by the participation of a single cDNA. Capell et al. (2000) transformed transgenic cell lines with a heterologus cDNA derived from the arginine decarboxylase gene of oat, which exhibited significant down-regulation of the activity of the endogenous arginine and ornithine decarboxylases. The changes in enzyme activity were reflected in a marked decrease in the level of putrescine and spermidine but not spermine in the majority of cell lines analyzed. They confirmed that the cell lines with low levels of polyamines exhibited normal morphogenetic responses. In vegetative tissue at the whole plant level no significant variation in polyamine levels was observed. However, significant reduction in putrescine levels in seeds derived from three out of five plants were observed. Thus, simultaneous reduction of the activity of the two alternative enzymes in the early steps of the polyamine pathway results in significant reduction in the end-product accumulation in the seeds of transgenic plants. Further, Lepri et al. (2001) investigated how overexpression of a cDNA for human ornithine decarboxylase (odc) affects the polyamine pools in transgenic rice. They further investigated the tissue-specific expression patterns and product accumulation levels of the transgene driven by either constitutive or seed specific promotors. Their results indicate that : whereas the expression of a heterologus arginine decarboxylase (adc) cDNA in rice resulted in increased putrescine and spermine levels only in seeds, plants engineered to express ornithine decarboxylase (odc) cDNA exhibited significant changes in the levels of all three major polyamines in seeds and also in vegetative tissues (leaves and roots). Lepri and associates suggest that odc rather than adc is responsible for the regulation of putrescine synthesis in plants.

Franceschetti et al. (2004) produced transgenic tobacco plants overexpressing the *Datura stramonium* spermidine synthase cDNA in order to understand the role of this gene in the polyamine metabolism and in particular in affecting spermidine endogenous levels. Determinations of enzyme activities indicated an increased spermidine synthase and S-adenosylmethionine decarboxylase activities. Ornithine decarboxylase was the most active enzyme and its activity was equally distributed between the soluble and the particulate fractions, while the arginine decarboxylase activity in the transgenic plants did not particularly change with respect to the controls. In comparison to the controls, the transformed plants displayed an increased spermidine to putrescine ratio, while the total polyamine content remained almost unchanged. These findings suggest a high capacity of the transformed plants to tightly regulate polyamine endogenous levels and provide evidence that spermidine synthase is not a limiting step in the biosynthesis of polyamines. However, the polyamine level varies with plant species and environmental conditions they are exposed to. For instance, the cereal leaves subjected to osmotic stress, show a rapid increase in arginine decarboxylase activity, a massive accumulation of putrescine, and slow conversion of putrescine to higher polyamines, spermidine and spermine (Tiburcio et al., 1986). The mesophyll protoplasts from leaves which have a high putrescine : polyamine ratio, do not undergo sustained division. By contrast, in *Nicotiana, Capsicum, Vigna* and *Datura,* all dicot genera that readily regenerate plants from mesophyll protoplasts, the response of leaves to osmotic stress is opposite to that in cereals. Putrescine titer as well as arginine and ornithine decarboxylase activities decline in these osmotically stressed dicot leaves, while spermidine and spermine titers increase. Thus, the putrescine : polyamine ratio in *Vigna* protoplasts, which divide readily, is 4-fold lower than in oat protoplasts, which divide poorly. Tiburcio and coworkers suggest that this differing response to polyamine metabolism to osmotic stress may account, in part, for the failure of cereal mesophyll protoplasts to develop readily *in vitro.* Also the environmental stress conditions, such as drought, salinity, and air pollution increase polyamine levels in plants, but the physiological functions of elevated polyamine levels is pooly understood.

Not only the different plant species, but even in different organs of a plant, and at different stages of plant development, accumulated polyamine level varies. Scoccianti et al. (1990) measured the diamine oxidase activity and polyamine content in the shoot apex, leaves, epicotyl, cotyledons, hypocotyl and roots of light-grown bean (*Phaseolus vulgaris* L. cv Lingot) and soybean (*Glycine max* L. cv Sakal) seedlings at three different stages of germination (5, 8 and 14 days) as well as in embryos and cotyledons from soaked seeds. No diamine oxidase activity was detected in embryos and cotyledons of either crops. In bean seedlings diamine oxidase activity was only detectable in the shoot apex, primary leaves and cotyledons, while in soybean the activity was only detectable in the hypocotyl and roots. During seedling growth, in both crop species, a different pattern of diamine oxidase activity was observed. In both the species spermidine and spermine were the most abundant polyamines in embryos and cotyledons. Cadvarine, absent in bean, was only detected in soybean embryos. In the seedlings of both crop species, increasing gradients of putrescine, spermidine and spermine from base to shoot apex were found.

Liu et al. (2000) report that polyamines regulate the voltage dependent inward K^+ channel in the plasma membrane of guard cells and modulate stomatal aperture, a plant "sensor" to environmental changes. All natural polyamines, including spermidine, spermine, cadvarine, and putrescine, strongly inhibit opening and induce closure of stomata. Whole-cell patch-clamp analysis showed that intracellular application of polyamines inhibit the inward K^+ current across the plasma membrane of guard cells. Single-channel recording analysis indicated that polyamine regulation of the K^+ channel requires unknown cytoplasmic factors. In an effort to identify the target channel at the molecular level, Liu and associates found that spermidine inhibited the inward K^+ current carried by KAT1 channel that was functionally expressed in a plant cell model. These findings suggest that polyamines target KAT1-like inward K^+ channels in guard cells and modulate stomatal movements, providing a link between stress conditions, polyamine levels, and stomatal regulation.

POLYAMINES AND ROOT GROWTH

Ben-Hayyim et al. (1994) showed that both α-DL-difluoromethylornithine (an inhibitor of putrescine biosynthesis) and the *rolA* gene (from the root inducing transferred DNA of *Agrobacterium rhizogenes)* stimulate overall root growth and cause a conversion in the pattern of root system formation, producing a dominant or "tap" root. These morphological changes in root were correlated with the depression in accumulation of polyamines and their conjugates. Thus, polyamines play an important role in interactions between environment and development, especially in roots, where development is a crucial part for seedling establishment and/or stress tolerance. *Pringlea antiscorbutica,* Kerguelen cabbage, is the sole endemic cruciferous species from the Kerguelen province in the subantarctic zone characterized by a permanent cold temperature with an annual mean around 4°C, strong and permanent wind and very high precipitation (Hummel et al., 2002). Thus, this species tolerates a variety of adverse conditions such as salt spray and flood, high diurnal temperature amplitude and frequent freeze-thaw cycles. This plant exhibits a high flexibility of polyamine metabolism depending on culture conditions and high contents of polyamines in comparison with other crucifers. The roots and shoots of this plant accumulate a massive amount of agmentine (Hennion and Martin-Tanguy, 2000). Hummel et al. (2002) further found a positive correlation between agmentine, spermidine and spermine content and the primary root growth. Putrescine biosynthesis in *P. antiscorbutica* was found to depend on the arginine decarboxylase pathway. They modified the free polyamine levels by inhibitors of polyamine biosynthesis [D-arginine, difluoromethylornithine (DFMO), cyclohexylammonium (CHA), and methylglyoaxal-bis-guanylhydrazone (MGBG)] and variations of the endogenous pools were compared to changes in root growth. Seedlings treated with DFMO or DFMO + putrescine exhibited significant modifications of polyamine levels (Fig. 7.1), but very little effect on the content of spermidine and spermine in leaves. In addition, DFMO + putrescine treatment led to putrescine accumulation in roots but not in shoots; thus suggesting efficient uptake in roots. Total polyamine content in roots was enhanced by DFMO treatment in comparison with control. The major effect of DFMO treatment was a 10-fold higher spermine level in roots. Roots from seedlings treated with DFMO + putrescine also exhibited an overall increase in polyamines, but the spermine pool only increased 3-fold. Putrescine levels seem negatively linked with root growth rate whereas the levels of spermidine and spermine appear positively linked with this trait.

Methylglyoxal-bis-guanylhydrazone (MGBG) treatment led to an overall decrease in polyamines, mainly in roots (Fig. 7.2). Thus, spermidine content in MGBG-treated roots was 2.3-fold lower than in controls. Moreover, MGBG also led to a depression of agmantine and putrescine pools in roots. In addition, MGBG treatement, while inducing smaller modifications of proline content than cyclohexylammonium (CHA) treatment, led to more drastic phenotypic alterations than CHA. Thus, the authors conclude that the arginine decarboxylase pathway, rather than the ornithine decarboxylase, seems to play a major role in polyamine synthesis in

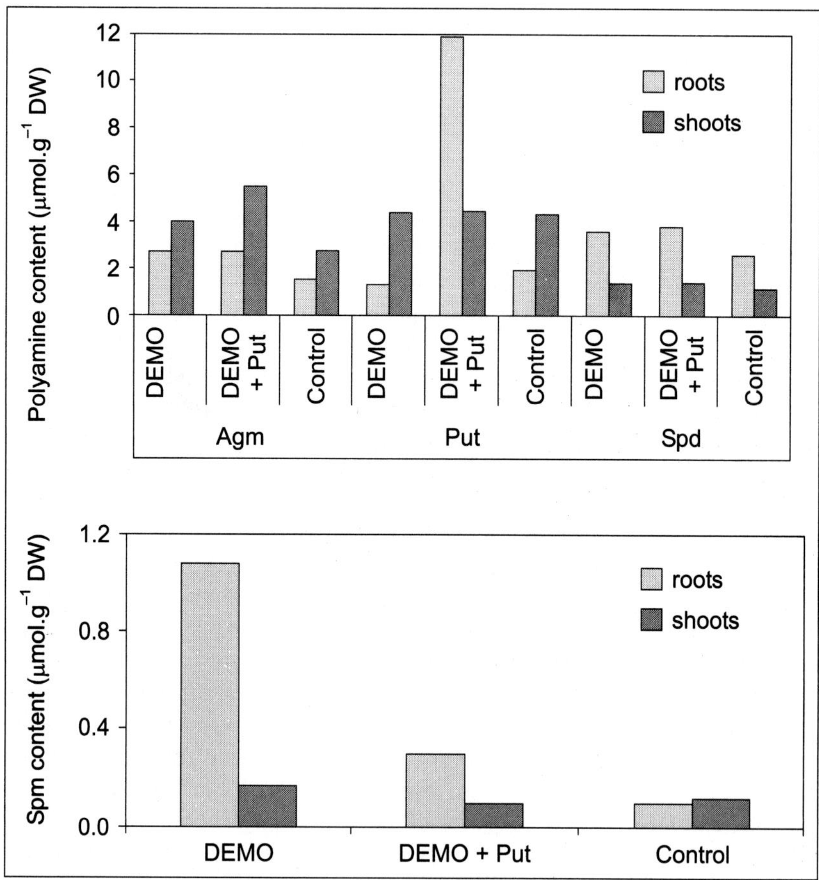

Fig. 7.1 Effects of difluoromethylornithine (DFMO) and DFMO plus putrescine treatments on free polyamine contents in roots and shoots. Levels of agmatine, putrescine and spermine were determined after 15 days of growth in different organs of the *Pringlea antiscorbutica* seedlings. Treatments consisted of 2 mM DFMO or 2 mM DFMO + 0.5 mM putrescine. (After Hummel et al., 2002).

P. antiscarbutica seedlings. Root, but not shoot, phenotypes were greatly affected by these treatments, which modified polyamine endogenous levels. A positive correlation was found between agmantine level and growth rate of the primary root. Spermidine and spermine contents also showed positive correlations with primary root growth whereas the putrescine level showed neutral or negative effects on this trait. Free polyamines were, therefore, found to be differentially involved in the phenotypic plasticity of root architecture. A comparison of developmental effects and physiological concentrations suggest that agmantine and spermine in particular may play a significant role in the control of root development.

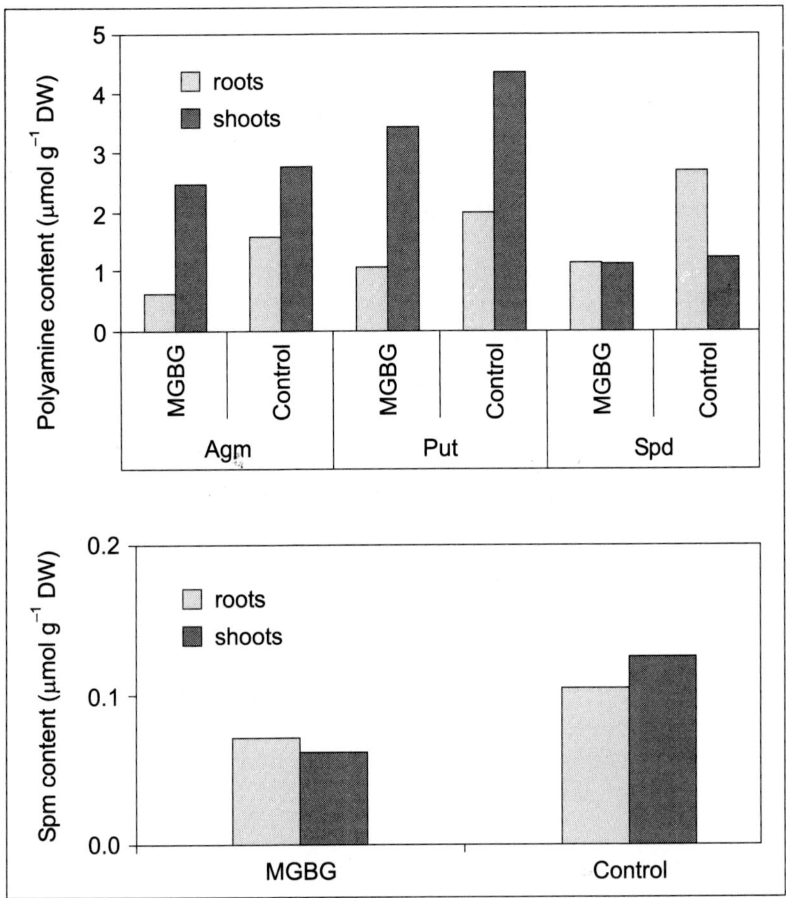

Fig. 7.2 Effects of methylglyoxal-bis-guanylhydrazone (MGBG) treatment on free polyamine contents in roots and shoots. Levels of agmatine, putrescine, spermidine and spermine were determined after 15 days of growth in different organs of the *Pringlea antiscorbutica* seedlings. Treatment consisted of 2 mM MGBG. (After Hummel et al., 2002).

STRESS TOLERANCE

Polyamines, particularly spermidine and spermine, are involved in regulation of gene expression by enhancing DNA-binding activities to particular transcription factors; and are believed to have an osmoprotectant function in plant cells under water deficit (Capell et al., 2004). Polyamines play an important role in plant defence to abiotic stresses. Kasukabe et al. (2004) cloned spermidine synthase cDNA from *Cucurbita ficifolia* and the gene was introduced to *Arabidopsis thaliana* under the control of cauliflower mosaic virus ^{35}S promoter. The transgene was stably integrated and actively transcribed in the transgenic plants. As compared with the wild-type plants, the T_2 and T_3 transgenic plants exhibited a significant increase in spermidine synthase activity and spermidine content in leaves together with enhanced tolerance to various stresses including chilling, freezing, salinity, and drought. During exposure to chilling stress (5°C), the transgenics displayed a remarkable increase in arginine decarboxylase activity and conjugated spermidine contents in leaves compared to the wild-type. A cDNA microarray analysis revealed that several genes were more abundantly transcribed in the transgenics than in the wild-type under chilling stress. These results strongly suggest an important role for spermidine as a signaling regulator in stress signaling pathway, leading to build-up of stress tolerance mechanisms in plants under stress conditions.

Arginine decarboxylase is the key enzyme involved in synthesis of polyamines involved in stress tolerance (Hummel et al., 2002). Mo and Pua (2002) cloned two cDNAs in mustard (*Brassica juncea)* designated *MADC2* and *MADC3,* encoding ADC. *Madc2* and *MADC3* encode polypeptides of 692 and 680 amino acid residues, respectively. Results revealed that ADC transcripts in mustard were generally more abundant in stem and root but were barely detectable in leaf. However, ADC expression in the leaf was up-regulated differentially in response to stress such as chilling, salt and mannitol and to treatments with exogenous polyamines. While chilling-induced expression of all three ADC genes, salt predominantly resulted in increased accumulation of *MADC3* transcript. Leaves exhibited a similar response to exogenous putrescine, spermidine and spermine, all of which stimulated accumulation of *MADC2* and *MADC3* transcripts but not *MADC1.*

The potential role of uncommon polyamines such as thermine and caldvpentamine, in stabilizing cell membranes and nucleic acids against extreme environmental conditions, such as drought and high temperature in crop plants is only now beginning. The genetically susceptible strains of some crop plants do not produce these uncommon polyamines, while abiotic stress induces biosynthesis and accumulation of uncommon polyamines in related, genetically tolerant strains (Prof. G. C. Phillips of the University of Kentucky). Aminopropyltransferase, polyamine oxidase, and schiff-base reductase have been identified as the biosynthetic enzymes responsible for uncommon polyamine synthesis in drought tolerant alfalfa plants.

SALINITY STRESS

Accumulation of spermidine and spermine in salt tolerant rice cultivars and of putrescine in sensitive rice cultivars has been observed in response to salinity stress. Chattopadhyay et al. (1997) compared the effect of salinity stress on the activity of arginine decarboxylase and its transcript level in salt sensitive (M-1-48) and salt tolerant (Pokkali) rice cultivars. Treatment of 72 hour-grown seedlings either with increasing concentrations of NaCl or with 15 mM NaCl for different time periods, showed a gradual increase in activity in Pokkali. In M-1-48 an intermediate increase followed by a sharp decrease was observed on prolonged treatment. While in Pokkali, at least a 20-fold accumulation of *OAD1* (a region of oat genomic DNA containing a coding sequence for 166 amino acids of the C-terminal part of the ADC enzyme) homologus transcript was detected after treatment with 200 mM NaCl; only a 7-fold increase in transcript level was found in M-1-48 after 150 mM NaCl treatment. Results suggest that in the salt tolerant rice cultivar Pokkali, ADC enzyme activity increases and its transcript also accumulates during the prolonged salinity stress; this mechanism is absent in the salt sensitive rice cultivar M-1-48 where a prolonged period of salinity stress down-regulates both ADC activity and its transcript level. The salt tolerant Pokkali rice plants accumulate higher polyamines such as spermidine and spermine in response to salinity stress, while the sensitive cultivar M-1-48 is unable to maintain higher titers of these polyamines under similar conditions (Chattopadhyay et al., 2002). At physiological concentrations spermidine and spermine significantly prevented the leakage of electrolytes and amino acids from roots and shoots induced by salinity stress. To different degrees, they also prevented chlorophyll loss, inhibition of photochemical reactions of photosynthesis, as well as down-regulation of chloroplast-encoded genes like *psbB, psbE* and *rbcL,* indicating a positive correlation between salt tolerance and accumulation of higher polyamines in rice. The inhibitory effect of salinity stress and its reversal by exogenous polyamines were more pronounced in the salt sensitive M-1-48 plants than in the tolerant Pokkali plants.

Later Maiale et al. (2004) observed that ADC and S-sdenosyl-L-methionine decarboxylase activities were reduced in the two rice cultivars differing in salt tolerance, as a consequence of salt treatment. However, the spermidine synthase activity was reduced in the salt tolerant cultivar (Giza) but not in the salt sensitive cultivar (El Paso), while no polyamine oxidase activity was detected. During the salinization period, putrescine and spermidine levels decreased in both cultivars, although less dramatically in Giza. Simultaneously, spermine accumulation occurs in both cultivars, while proline accumulation was major in the sensitive one.

Accumulation of polyamines in rice seeds treated with polyamines and/or NaCl was studied by Upadhyay et al. (2001) to work out a relationship in polyamine levels and salt tolerance in rice. Putrescine levels in untreated seeds were higher in the salt tolerant variety CSR-1, as compared to the salt sensitive variety Manhar. But after salt treatment, putrescine content of Manhar seeds was higher than that in CSR-1 seeds.

Spermidine levels increased in CSR-1 when treated with spermidine alone or in combination with NaCl, while in Manhar, seeds treated with spermidine, putrescine + NaCl or spermine + NaCl recorded an increase in spermidine levels. In Manhar, spermine levels were generally higher than CSR-1 both in treated and untreated seeds. The authors suggest that exogenous application of polyamines on rice plants may help in overcoming NaCl toxicity.

Liu and Liu (2004) measured putrescine, spermidine and spermine and their different forms — free, perchloric acid soluble, and insoluble conjugated together with polyamine oxidase activities in barley cv KP7 seedlings under salt stress. The results show that the free polyamines are the dominant forms among the three polyamines and accounted about two-third of the total polyamines. All three polyamines decreased in roots and leaves after 8 days under NaCl 200 mM/L stress. The free polyamine content decreased dramatically as well as soluble conjugated polyamine content, while the insoluble ones changed little. Different levels of salt stress increased markedly the ratios of (spermidine + spermine)/ putresine and decreased significantly the total polyamine content in roots and leaves.

With commonly cultivated *Glycine max* (cv salt-tolerant Lee 68) and *Glycine soja* (cv salt-sensitive N 23232) as reference, Yu et al. (2004) studied the polyamine contents and polyamine oxidase activities in the highly salt-tolerant cv BB 52 (*G. soja*) seedlings, which showed that under 150 mM NaCl L^{-1} for 2 days, the decrease in putrescine and spermidine contents was more significant, but that of spermine content was less significant in roots of BB 52 than in those of Lee 68 and N 23232. For leaves, the decrease in putrescine and increase in spermidine contents were markedly observed in BB 52. The ascent of (spermine + spermidine)/ putresine ratios and descent of putrescine/polyamines ratios showed a positive relation to their salt tolerance. The polyamine oxidase activity in roots and leaves was all increased, and most obviously in cv N 23232.

Ali (2000) studied the effects of putrescine on germination and seedling growth of *Atropa belladona* (a medicinal plant priced for its alkaloids) plants under the influence of NaCl. Rate of germination, seedling growth, endogenous putrescine and alkaloids content decreased when seeds were subjected to salt stress. But pre-soaking the seeds in 10^{-2} mM putrescine and treatment with NaCl reduced net accumulation of Na^{+} and Cl^{-} ions in different organs of the plant. Thus Ali reported that presoaking seeds in 10^{-2} mM putrescine can alleviate the adverse effect of NaCl during germination and early seedling growth of *A. belladona* and increase alkaloids as well as endogenous putrescine contents.

Santa-Cruz et al. (1997) determined short-term changes in free polyamine induced by salt (200 mmol/L NaCl) and osmotic (400 mmol/L mannitol) stresses in leaf discs of the salt sensitive cultivated tomato species (*Lycopersicon esculentum*) and its wild salt-tolerant relative *L. pennellii*, with the objective of checking whether these metabolites could be used as early traits of tolerance. In *L. pennellii*, putrescine and spermidine accumulation was found in the osmotically-stressed leaf discs from 30 minutes until the end of the experiment (2 hours). Under salt stress, the free polyamine accumulation in *L. pennellii* only took place during the first 15 minutes, mainly due to putrescine, whereas its levels decreased after this time. These results suggest that the initial polyamine accumulation was due to the osmotic effect induced by salinity, but levels rapidly began to decrease as the saline ions were accumulated from the medium. In a similar manner to the wild species, the leaf discs of *L. esculentum* submitted to salt stress also accumulated Na^+ from the beginning of the experiment, although the accumulation was slower. Nevertheless, the patterns of free polyamine were similar under both salt and osmotic stresses in *L. esculentum*, which suggests that the results only show the osmotic shock. In this species, the polyamine levels tended to decrease at the beginning of the experiment, whereas the opposite response was observed at the end of the period (between 1 and 2 hours). The results obtained show that the polyamine levels change in a very different way in the leaf discs of both tomato species. Moreover, NaCl stress and non-ionic osmotic stress caused by mannitol are separable only in the salt-tolerant species in terms of their effects on free polyamine levels.

Polyamines accumulated in wild-type *Arabidopsis thaliana* plants (Col-0 and Ler-0) that were pretreated with 100 mM NaCl before transfer to 125 mM NaCl, but not in plants that were directly transferred to 125 mM NaCl without prior treatment with 100 mM NaCl (Kasinathan and Winger, 2004). The results demonstrate that decreased polyamine formation due to lower arginine decarboxylase activity leads to reduced salt tolerance.

DROUGHT STRESS

Galiba et al. (1993) compared the osmotic and salinity-induced polyamine accumulation in callus cultures of drought and salt tolerant wheat cultivars and in disomic substitution lines. Putrescine, spermidine and spermine occurred in all cultures. Mannitol-induced osmotic stress increased putrescine in all, and cadaverine in two varieties, while salt stress increased spermidine titer, the accumulation rate being higher in sensitive than in tolerant varieties. Erdei et al. (1990) also compared polyamine accumulation under short-term non-ionic osmotic stress in leaf segments of 8 wheat varieties differing in drought and salt tolerance in order to test whether various stress conditions impair different steps of polyamine biosynthesis. The highest putrescine accumulation occurred in the known salt and drought tolerant varieties under non-ionic osmotic stress (400 mOSm L^{-1} sorbitol for 10 h), while in the presence of NaCl of the same osmolarity, only the known salt tolerant variety responded with increased putrescine level. Spermidine titer was slightly influenced by both stress conditions and spermine content increased only under salinity. The putrescine precursor agmentine accumulated under osmotic stress, while the end product 1,3-diaminopropane appeared only under salt stress. These results suggest that with respect to the onset of polyamine biosynthetic and degradative processes, osmotic and salt stresses are two different conditions.

Gonzalez de Mejia et al. (2003) investigated the effect of drought on the metabolism of proline, ornithine and polyamine, and the enzymes involved in their synthesis and degradation, in tepary and common bean seeds. Their results suggest that water stress induced ornithine amino transferase activity and ornithine concentration. Proline also increased under drought conditions. Genotypes L-242-24 and G-400-35, which had the highest yield, also had a major concentration of polyamines under water stress. They also suggest that arginase activity participated in amino acid synthesis such as ornithine, and consequently, in the increased protein concentration under water stress.

Capell et al. (2004) generated transgenic rice plants expressing the *Datura stramonium adc* gene and investigated their response to drought stress. They monitored the steady-state mRNA levels of genes involved in polyamine biosynthesis (*Datura adc,* rice *adc,* and rice *samdc)* and polyamine levels. The wild-type plant responded to the onset of drought stress by increasing endogenous putrescine levels, but this was insufficient to trigger the conversion of putrescine into spermidine and spermine. In contrast, transgenic plants expressing *Datura adc* produced much higher levels of putrescine under stress, promoting spermidine and spermne synthesis and ultimately protecting the plants from drought. Capell and associates demonstrated that the manipulation of polyamine biosynthesis in plants can produce drought-tolerant germplasm, and proposed a model consistent with the role of polyamines in the protection of plants against abiotic stress.

COLD STRESS

Shen et al. (2000) investigated the possible involvement of polyamines in chilling tolerance of cucumber cv Jinchun No. 3 and cv Suyo. Plants with the first expanded leaves were exposed to 3°C or 15°C in the dark for 24 hour chilling, and then transferred to 28°C/22°C under a 12-hour photoperiod for another 24 hour rewarming. The chilling-tolerant cultivar Jinchun No. 3 showed a marked increase of free spermidine in leaves, once during chilling and again during rewarming. Putrescine increased significantly during rewarming, but the increase of spermine was slight. Any of these polyamines did not increase in chilling sensitive cv Suyo during either period. The polyamine-biosynthetic enzyme activities appear to mediate these differences between cultivars. Pretreatment of spermidine to cv Suyo prevented chill-induced increases in the contents of hydrogen peroxide in leaves and activities of NADPH oxidases and NADPH-dependent superoxide generation in microsomes and alleviated chilling injury. Pretreatment of methylglyoxal-bis-guanylhydrazone, a polyamine biosynthesis inhibitor, to chilled cv Jinchun No. 3 prevented spermidine increase and enhanced microsomal NADP oxidase activity and chilling injury. The results suggest that spermidine plays important roles in chilling tolerance of cucumber, probably through prevention of chill-induced activation of NADPH oxidases in microsomes.

Pillai and Akiyama (2004) investigated regulation of the rice gene *OsSAMDC,* which encodes an S-adenosyl-l-methionine decarboxylase (SAMDC) involved in polyamine biosynthesis. Clones representing genes differentially expressed at 5°C and 20°C were isolated from a cDNA library prepared from the chilling-tolerant japonica-type cultivar Yukihikari. Upon exposure to cold stress (5°C) the level of *OsSAMDC* transcripts in the cold-resistant Yukihikari genotype continued to increase for up to 72 hours. In contrast, there was no change in *OsSAMDC* transcription in the susceptible indica cultivar TKM9 under the same conditions. Examination of polyamine levels in the cold-resistant Yukihikari genotype revealed that spermidine levels were elevated during the course of cold treatment. These results suggest that induction of the *OsSAMDC* gene in response to cold may be used as a molecular marker for the ability of rice seedlings to withstand exposure to low temperatures.

In the same year, Hummel et al. (2004) reported that the seedlings of *Pringlea antiscorbutica* (Kerguelen cabbage) show optimal root growth at low temperature conditions of subantarctic zone (5/10°C) and the temperature conditions caused the early arrest of root growth and they found cold tolerance to be associated with increased levels of polyamines or with maintenance of high levels of polyamines. Hummel and coworkers suggested that polyamine levels can serve as a significant marker of chilling tolerance in seedlings.

HEAT STRESS

Basra et al. (1997) subjected the germinating seeds of mungbean (*Vigna radiata*) cv ML 311 with a radicle length of 5 mm to a heat-shock episode of 50°C for 2 hours followed by transfer to the normal temperature (28°C) for 3 days in the dark. Exogenous effects of polyamines (putrescine, spermidine, and spermine) on the recovery growth and membrane integrity of seedling tissues were studied. Application of polyamines, either as a pretreatment at 28°C for 2 hours prior to heat-shock or as a co-treatment (50°C, 2 h) during the heat-shock period itself, enhanced the recovery growth of both roots and hypocotyls but especially the former with the order of effectiveness being putrescine, spermidine, and spermine. Treatment with polyamine biosynthetic inhibitors, i.e. D,L α-diflouromethylarginine and D,L α-diflouromethylornithine resulted in thermosensitization, making seedlings vulnerable to heat-shock. This effect could be ameliorated by putrescine application. An important role of polyamines in heat-shock protection is thus indicated.

MINERAL ION STRESS

The effects of cadmium on putrescine, spermidine and spermine titers were studied in oat and bean leaves (Weinstein et al., 1986). Treatment with cadmium for up to 16 hours in the light or dark resulted in a large increase in putrescine titer, but had little or no effect on spermidine or spermine. The activity of arginine decarboxylase followed the pattern of putrescine accumulation, and experiments with α-difluoromethylarginine established that arginine decarboxylase was the enzyme responsible for putrescine increase. Concentrations of Cd^{2+} as low as 10 micromolar increased putrescine titer in oat segments. In bean leaves, there was a Cd^{2+} -induced accumulation of putrescine in the free and soluble conjugated fractions, but not in the insoluble fraction. This suggests a rapid exchange between putrescine that exists in the free form and putrescine found in acid soluble conjugate forms. The authors conclude that Cd^{2+} can act like certain other stresses to induce substantial increases in putrescine in plant cells. Later, Watson and Malmberg (1996) investigated the mechanism for induction of arginine decarboxylase activity during potassium deficiency stress in *Arabidopsis thaliana* and showed that the plant responds to K-deficiency stress by increasing arginine decarboxylase activity by up to 10-fold over unstressed plants with a corresponding increase in putrescine levels of up to 20-fold. Spermidine and spermine levels did not increase proportionately. The endogenous polyamines, in particular spermine, cause block and modulation of a number of types of ion-channels. Intracellular spermine is responsible for intrinsic gating and rectification of strong inward rectifier K^{2+} channels by directly plugging the ion-channel pore.

UV-B STRESS

The activities of arginine decarboxylase, ornithine decarboxylase and S-adenosylmethionine decarboxylase increased by 165.74, 104.60 and 89.60 percent, respectively, in the leaves of rice cultivar Shan You 63, and by 59.91, 41.30 and 23.68 percent in leaves of the cultivar Nancheum when exposed to 7-14 days of UV-B radiation (Lin et al., 2002). Activities of ADC and ODC increased by 115.95 and 14.45 percent, respectively, but the activity of SAMDC decresed by 33.01 percent in leaves of rice cultivar IR 65600–85. However, by 21-28 days exposure to UV-B radiation, the activities of ADC and ODC increased by 89.72 and 3.71 percent, respectively, in the leaves of cultivar Shan You 63, but by 73.95 and 27.38 percent in the leaves of the cultivar Nancheum. But in the leaves of cultivar IR 65600-85, the activity of ADC increased by 94.41 percent but that of ODC decreased by 13.57 percent. As far as SAMDC is concerned, the activity decreased by 40.06, 19.20 and 38.21 percent in cultivars Shan You 63, Nancheum and IR 65600-85, respectively, by 21-28 days UV-B exposure. The above observations of Lin et al. show that UV-B exposure effectively increases the activities of the three important enzymes concerned in polyamine metabolism. Thus the concentrations of stress-protective polyamines will decrease, making the plants susceptible to stress.

TRAUMATIC STRESS

Cowley and Walters (2005) reported the effect of mechanical wounding on putrescine biosynthesis and catabolism in oilseed rape (*Brassica napus* ssp. *oleifera*). Mechanical wounding of the first leaf led to significant, but transient increases in arginine decarboxylase activity and levels of free putrescine in the wounded first leaf and in the unwounded second leaf. The increased putrescine appeared to be the result of a combination of increased ADC activity, coupled with reduced putrescine catabolism, as activity of the oxidative enzyme diamine oxidase was significantly reduced following wounding, both locally and systemically.

IMPROVEMENT

Noury et al. (2000) introduced the oat *adc* cDNA into rice under control of the constitutive maize ubiquitin 1 promoter. They studied molecularly and biochemically sixteen independent transgenic plant lines. Significant increases in mRNA levels, ADC enzyme activity and polyamines were measured in transgenic callus. Out of 16, only in one lineage the increase was maintained. This particular lineage showed very significant increases in putrescine preferentially in seeds (up to 10-times compared to wild-type and controls transformed with the *hpt* selectable marker alone). In rice, overexpression of the oat *adc* cDNA resulted in increased accumulation of polyamines at different stages of development. They further demonstrated that strong constitutive promoters, such as the maize ubiquitin 1 promoter, are sufficient to facilitate heritable high-level polyamine accumulation in seed. The results demonstrate that by screening adequate numbers of independently derived transgenic plants, it is possible to identify those individuals which express a desired phenotype or genotype. Further, Roy and Wu (2001) observed that the expression of ADC transgene under the control of an ABA-inducible promoter led to stress-induced up-regulation of ADC activity and polyamine accumulation in transgenic rice plants. Second generation transgenic rice plants showed an increase in biomass under salinity stress conditions, as compared to the non-transformed control plants.

Anderson et al. (1998) compared the metabolism of arginine, ornithine, and putrescine in a non-transgenic and a transgenic cell line of carrot (*Daucus carota)* expressing a mouse ornithine decarboxylase cDNA. They fed ^{14}C-Arg, ^{14}C-Orn, and ^{14}C-Put to cells and determined their rates of decarboxylation, uptake, metabolism into polyamines, and incorporation into acid-insoluble material. Transgenic cells showed higher decarboxylation rates for labeled ornithine than the non-transgenic cells. This correlated positively with higher amounts of labeled putrescine production from labeled ornithine. Polyamines have been implicated in a myriad of physiological and developmental processes in many organisms, but their *in vivo* functions remain to be determined (Mehta et al., 2002). They expressed an yeast S-adenosylmetionine decarbosylase gene (*ySAMdc; Spe2*) fused with a ripening-inducible E8 promoter to specifically increase levels of the polyamines spermidine and spermine in tomato fruit during ripening. Independent transgenic plants and their segregating lines were evaluated after cultivation in the greenhouse and in the field for five successive generations. The enhanced expression of the *ySAMdc* gene resulted in increased conversion of putrescine into higher polyamines and thus to ripening-specific accumulation of spermidine and spermine. This led to an increase in lycopene, prolonged vine life, and enhanced fruit juice quality. Lycopene levels in cultivated tomatoes are generally low, and increasing them in the fruit enhances its nutritive value. These results provide direct evidence for a physiological role of polyamines and demonstrate an approach for improving nutritional quality, juice quality, and vine life of tomato fruit.

Capell et al. (1998) recovered transgenic rice cell lines and plants expressing an oat arginine decarboxylase DNA under the control of *CaMV* ^{35}S promoter using particle bombardment. Molecular analyses confirmed stable integration of the transgene and active transcription (mRNA). A 4- to 7-fold increase in arginine decarboxylase activity was observed in transformed plants compared to wild-type controls. Biochemical analysis of cellular polyamines indicated up to 4-fold increase in putrescine levels in transgenic callus and regenerated plants. As the amount of polyamines increase under environmental stress conditions, Wi and Park (2002) used transgenic technology in an attempt to evaluate their potential for mitigating the adverse effects of several abiotic stresses in plants. Due to a metabolic competition for S-adenosylmethionine as a precursor between polyamine and ethylene biosyntheses, it is expected that the antisense-expression of ethylene biosynthetic genes could result in an increase in polyamine biosynthesis. Antisense constructs of cDNAs for senescence-related l-aminocyclopropane l-carboxylic acid (ACC) synthase and ACC oxidase were isolated from carnation flowers that were introduced into tobacco by *Agrobacterium*-mediated transformation. Several transgenic lines showed higher polyamine contents than wild-type plants. The number and weight of seeds also increased. Stress-induced senescence was attenuated in these transgenic plants in terms of total chlorophyll loss and phenotypic changes after oxidative stress with hydrogen peroxide, high salinity, acid stress (pH 3.0) and ABA treatment. These results suggest that the transgenic plants with antisense ACC synthase and ACC oxidase cDNAs are more tolerant to abiotic stresses than wild-type plants. This shows a positive correlation between polyamine content and stress tolerance in plants.

Galiba et al. (1993) studied the genetics of osmotic stress induced spermidine accumulation in wheat. The specific chromosome (5A and 7A) involvement revealed that mannitol was the most effective stress agent and only spermidine titer of the wheat variety Chinese Spring was significantly changed as a consequence of chromosome substitution. The A genome of Cappelle Desprez (donor) substituted into Chinese Spring (recipient) appears to carry genes involved in the control of osmotic stress induced spermidine accumulation, and the genes controlling cadaverine biosynthesis may be localized in chromosome 5 B. Urano et al. (2003) later characterized the genes for enzymes involved in the biosynthesis of polyamines in *Arabidopsis thaliana* . The genes *AtADC2* and *AtSPMS* mRNAs encoding arginine decarboxylase and spermine synthase were found to increase in response to NaCl and dehydration and ABA treatments. Stress-inducible accumulation of *AtADC2* mRNA correlated with putrescine accumulation under NaCl and dehydration treatments. In a cold condition, *AtSAMDC2* mRNA was expressed in sexual organs such as flowers, buds and immature siliques. Polyamines also accumulated in sexual organs.

REFERENCES

Ali, R.M. 2000. Role of putrescine in salt tolerance of *Atropa belladona* plant. Plant Sci. 152: 173-79.

Anderson, S.E.; D.R. Bastola and SC. Minocha, 1998. Metabolism of polyamines in transgenic cells of carrot expressing a mouse ornithine decarboxylase cDNA. Plant Physiol. 116: 299-307.

Basra, R.K.; A.S. Basra; C.P. Malik and I.S. Grover, 1997. Are polyamines involved in the heat-shock-protection of mungbean? Bot. Bull. Acad. Sin. 38: 165-69.

Ben-Hayyim, G.; J.P. Damon; J. Martin-Tanguy and D. Tepfer, 1994. Changing root system architecture through inhibition of putrescine and feruloyl putrescine accumulation. FEBS Lett. 342: 145-48.

Bouchereau, A.; A. Aziz; F. Laher and J. Martin-Tanguy, 1999. Polyamines and environmental challenges : recent development. Plant Sci. 40: 103-25.

Capell, T.; C. Escobar; H. Liu; et al. 1998. Overexpression of the arginine decarboxylase cDNA in transgenic rice (*Oryza sativa* L.) affects normal development patterns *in vitro* and results in putrescine accumulation in transgenic plants. Theo. Appl. Genet. 97: 246-54.

Capell, T.; L. Bassie; L. Topsom; et al. 2000. Simultaneous reduction of the activity of two related enzymes involved in early steps of the polyamine biosynthetic pathway by a single antisense cDNA in transgenic rice. Mol. Gen. Genet. 264: 470-76.

Capell, T.; L. Bassie and P. Christou, 2004. Modulation of the polyamine biosynthetic pathway in transgenic rice confers tolerance to drought stress. Proc. Nat. Acd. Sci., U.S.A. 101 (26): 9909-14.

Chattopadhyay, M.K; S. Gupta; D. N. Sengupta and B. Ghosh, 1997. Expression of arginine decarboxylase in seedlings of indica rice (*Oryza sativa* L.) cultivars as affected by salinity stress. Plant Mol. Biol. 34: 477-83.

Chattopadhyay, M.K.; B.S. Tiwari; G. Chattopadhyay; et al. 2002. Protective role of exogenous polyamines on salinity-stressed rice (*Oryza sativa)* plants. Physiol. Plant. 116: 192-99.

Cowley, T. and D.R. Walters, 2005. Local and systemic changes in arginine decarboxylase activity, putrescine levels and putrescine catabolism in wounded oilseed rape. New Phytol. 165 : 807-11.

Erdei, L.; S. Trivedi; K. Takeda and H. Matsumoto, 1990. Effects of osmotic and salt stresses on the accumulation of polyamines in leaf segments from wheat varieties differing in salt and drought tolerance. J. Plant Physiol. 137: 165-68.

Franceschetti, M. ; S. Fornale; A. Tassonia; et al. 2004. Effects of spermidine synthase overexpression on polyamine biosynthetic pathway in tobacco plants. J. Plant Physiol. 161 : 989-1001.

Galiba, G.; G. Kocsy; R. Kaursawhney; et al. 1993. Chromosomal localization of osmotic and salt stress-induced differential alterations in polyamine content in wheat. Plant Sci. 92: 203-11.

Gonzalez de Mejia; V. Martinez Resendiz; E. Castano and G. Loarca Pina, 2003. Enzymes related to polyamine metabolism in beans. J. Sci. Food Agric. 83 (10): 1022-30.

Hennion, F. and J. Martin-Tanguy, 2000. Amines of the subantarctic crucifer *Pringlea antiscorbutica* are responsive to temperature conditions. Physiol. Plant. 109: 232-43.

Hummel, I.; I. Couee; A. El-Amrani; et al. 2002. Involvement of polyamines in root development at low temperature in the subantarctic cruciferous species *Pringlea antiscorbutica.* J. Exp. Bo. 53 (373): 1463-73.

Hummel, I.; A. El-Amrani and G. Gouesbet, 2004. Involvement of polyamines in the interacting effects of low temperature and mineral supply on *Pringlea anticarbutica* (Kerguelen cabbage) seedlings. J. Exp. Bot. 55 (399): 1125-34.

Kasinathan, V. and A. Wingler, 2004. Effect of reduced arginine decarboxylase activity on salt tolerance and on polyamine formation during salt stress in *Arabidopsis thaliana.* Physiol. Plant. 121 : 101-107.

Kasukabe, Y.; L. He; K. Nada; et al. 2004. Overexpression of spermidine synthase enhances tolerance to multiple environmental stresses and upregulates the expression of various stress-regulated genes in transgenic *Arabidopsis thaliana.* Plant Cell Physiol. 45: 712-22.

Krishnamurthy, R. and K.A. Bhagwat, 1989. Plant Physiol. 91: 500-04.

Lepri, O.; L. Bassie; G. Safwat; et al. 2001. Overexpression of a cDNA for human ornithine decarboxylase in transgenic rice plants alters the polyamine pool in a tissue-specific manner. Mol. Genet. Genomics 266: 303-12.

Lin, W.; X. Wu; K. Liang; et al. 2002. [Effect of enhanced UV-B radiation on polyamine metabolism and endogenous hormone contents in rice (*Oryza sativa* L.)] (In Chinese). Ying Yong Sheng Tai Xue Bao 13 (7): 807-13.

Liu, J. and Y.L. Liu, 2004. [The relations between polyamine types and forms and polyamine oxidase activities in barley seedlings under salt stress] (In Chinese). Zhi Wu Sheng Li Yu Fen Zi Sheng Wu Xue Xue Bao 30 (2): 141-46.

Liu, K.; H. Fu; Q. Bei and S. Luan, 2000. Inward potassium channel in guard cells as a target for polyamine regulation of stomatal movements. Plant Physiol. 124: 1315-26.

Maiale, S.; D.H. Sanchez; A. Guirado; et al. 2004. Spermine accumulation under salt stress. J. Plant Physiol. 161: 35-42.

Mehta, R.A.; T. Cassol; N. Li; et al. 2002. Engineered polyamine accumulation in tomato enhances phytonutrient content, juice quality, and vine life. Nat. Biotechnol. 20: 613-18.

Mo, H. and E.C. Pua, 2002. Upregulation of arginine decarboxylase gene expression and accumulation of polyamines in mustard (*Brassica juncea)* in response to stress. Plant Physiol. 114: 439-49.

Noury, M.; L. Bassie; O. Lepri; et al. 2000. A transgenic rice cell lineage expressing the oat arginine decarboxylase (*adc*) cDNA constitutively accumulates putrescine in callus and seeds but not in vegetative tissues. Plant Mol. Biol. 43: 537-44.

Pillai, M.A. and T. Akiyama, 2004. Differential expression of an S-adenosyl-L-methione decarboxylase gene involved in polyamine biosynthesis under low temperature stress in japonica and indica rice genotypes. Mol. Genet. Genomics 27: 141-49.

Roy, M. and R. Wu, 2001. Arginine decarboxylase transgene expression and analysis of environmental stress tolerance in transgenic rice. Plant Sci. 160: 869-75.

Santa-Cruz, A.; M.T. Estan; A. Rus; et al. 1997. Effects of NaCl and mannitol iso-osmotic stresses on the free polyamine levels in leaf discs of tomato species differing in salt tolerance. J. Plant Physiol. 151: 754-58.

Scoccianti, V.; P. Torrigiani and N. Bagni, 1990. Distribution of diamine oxidase activity and polyamine pattern in bean and soybean seedlings at different stages of germination. Physiol. Plant. 80: 515-19.

Shen, W.; K. Nada and S. Tachibana, 2000. Involvement of polyamines in the chilling tolerance of cucumber cultivars. Plant Physiol. 124: 431-40.

Tiburcio, A.F.; M.A. Masdeu; F.M. Dumortier and A.W. Galston, 1986. Polyamine metabolism and osmotic stress. I. Relation to protoplast viability. Plant Physiol. 82: 369-74.

Upadhyay, T.; S. Chavan and D.P. Misra, 2001. Polyamine accumulation under salt stress in salt resistant and salt sensitive varieties of rice. Physiol. Mol. Biol. Plants 7: 55-59.

Urano, K.; Y. Yoshiba; T. Nanjo; et al. 2003. Characterization of *Arabidopsis* genes involved in biosynthesis of polyamines in abiotic stress responses and developmental stages. Plant Cell Environ. 26 (11): 1917.

Watson, M.B. and R.L. Malmberg, 1996. Regulation of *Arabidopsis thaliana* (L.) Heynh arginine decarboxylase by potassium deficiency stress. Plant Physiol. 111:1077-83.

Weinstein, L.H.; R. Kaur-Sawhney; M.V. Rajan; et al. 1986. Cadmium-induced accumulation of putrescine in oat and bean leaves. Plant Physiol. 82: 641-45.

Wi, S.J. and K.Y. Park, 2002. Antisense expression of carnation cDNA encoding ACC synthase or ACC oxidase enhances polyamine content and abiotic stress tolerance in transgenic tobacco plants. Mol. Cells 13: 209-20.

Yu, B.; X. Ji; J. Liu, 2004. [Changes in polyamines level in *Glycine soja* and *Glycine max* seedlings under NaCl stress] (In Chinese). Ying Yong Sheng Tai Xue Bao 15(7): 1223-26.

SUGGESTED READINGS

Capell, T.; O. Lepri; P. Hang; L. Bassie; N. Nogueira; g. El-Hussieng; J. Newman and P. Christou, 1999. Modulation of polyamine biosynthesis in rice by constitutive expression of oat *adc* cDNA. Rice Genetics Newsl. 16: 140-43.

Hummel, I.; I. Couee; A. El Amrani; J. Martin-Tanguy and F. Hennion, 2002. Involvement of polyamines in root development at low temperature in the subantarctic cruciferous species *Pringlea antiscorbutica* . J. Exp. Bot. 53 (373): 1463-73.

8

TREHALOSE

Introduction

Trehalose is a non-reducing disaccharide in which the two glucose units are linked in an α,α-1,1-glycosidic linkage. This sugar is present in a wide variety of plants where it serves as a source of energy and carbon. It also serves as a signaling molecule to direct or control certain metabolic pathways or even to affect growth. In addition, trehalose can protect proteins and cellular membranes from inactivation or denaturation caused by a variety of stress conditions, including desiccation, dehydration, heat, cold, and oxidation. There are at least three different pathways described for the synthesis of trehalose. The best known and most widely distributed pathway involves the transfer of glucose from UDP-glucose (or GDP-glucose in some cases) to glucose-6-phosphate to form trehalose-6-phosphate and UDP. This reaction is calalyzed by the trehalose-P synthase (TPS). Organisms that use this pathway usually also have a trehalose-P phosphatase (TPP).

BIOSYNTHESIS AND SUGAR METABOLISM

The trehalose producing transgenic plants have improved photosynthesis under drought conditions and reduced photooxidative damage (Garg et al., 2002). The transgenic plants also exhibit approximately 20 percent higher carbohydrate levels than those of the corresponding non-transformed control plants, suggesting that trehalose may be involved in sugar sensing and modulating carbon metabolism. Paul et al. (2001) have also demonstrated the ability of trehalose to modulate photosynthetic capacity in transgenic tobacco plants expressing *Escherichia coli* trehalose biosynthetic genes. At light saturation, rates of photosynthesis in the transgenic plants was 5-15 percent higher than in the nontransformed controls. The elevated levels of light-saturated photosynthesis in the transgenic plants supports the hypothesis that trehalose acts as a regulator of sugar sensing, and thus, the expression of genes associated with carbon metabolism. The higher capacity for photosynthesis before stress provides a larger sink for the products of photosynthesis during stress, thus limiting the extent of excess-

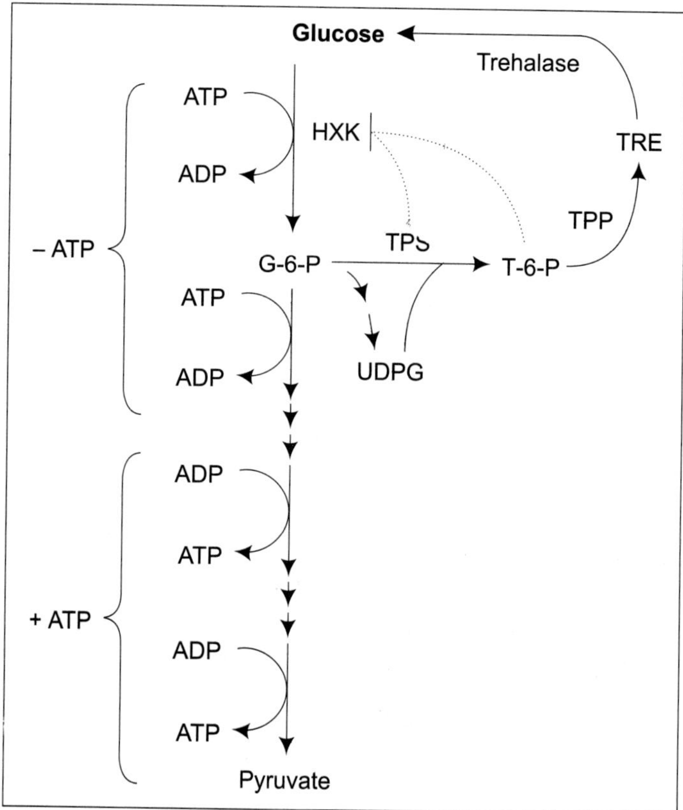

Fig. 8.1 A schematic diagram of the pathway of trehalose metabolism and its relationship to glycolysis in *Saccharomyces cerevisiae.* (After Eastmond et al., 2002). [G-6-P is glucose-6-phosphate, UDPG is uridine-5-diphosphoglucose, T-6-P is trehalose-6-phosphate, TRE is trehalose-6-phosphate synthase, TPP is trehalose-6-phosphate phosphatase, and HXK is hexokinase].

light-induced photooxidative damage and accounting, in part, for the more vigorous growth of the transgenic lines during stress.

Müller et al. (1998) also studied the role of trehalose in sucrose metabolism. Their experiments with soybean plantlets grown under sterile conditions demostrated that trehalose, added to the growth medium, strongly induced sucrose synthase and, to a lesser extent, alkaline invertase. A comparative study showed that nodules with high levels of trehalose had significantly higher levels of sucrose synthase and alkaline invertase, and significantly lower levels of sucrose than nodules with low levels of trehalose. Thus the trehalose synthesized by symbiotic microorganisms may have a regulatory role in sucrose metabolism in the host plants. Müller et al. (2001) further reported that trehalose activity is highest in the floral organs of the host plants, particularly in the anthers (approximately 700 nkat g^{-1} protein) and maturing siliques (approximately 250 nkat g^{-1} protein) and much lower in leaves, stems and roots (>5 nkat g^{-1} protein). Inhibition of trehalose *in vivo* by validamycin A leads to accumulation of an endogenous substance that has all the properties of trehalose, and a strong reduction in sucrose and starch contents in flowers, leaves and stems. Thus trehalose and trehalase appear to play a role in regulating carbohydrate allocation in plants.

INCREASED PRODUCTION BY UNSTRESSED PLANTS

Increased photosynthesis and sugar metabolism lead to increased production in unstressed plants. The improved photosynthetic performance is achieved by modifying sugar signalling mechanisms that control the expression of genes for whole pathways and processes that determine photosynthetic capacity and source-sink balance, rather than by directly targeting individual 'key' enzymes. Paul et al. (2001) support for the hypothesis that genetic modification of trehalose metabolism through its interaction with sugar signalling pathways can enhance photosynthetic capacity. Genes encoding functional enzymes for the synthesis and degradation of trehalose have been detected (Wingler et al., 2000). Wingler and associates analyzed how trehalose affects metabolism and development of seedlings. They report that the exogenously applied trehalose (25 mM) strongly reduced elongation of roots, and concomitantly induced a strong accumulation of starch in shoots, whereas the contents of sugars were not increased. When the seedlings were grown on trehalose plus sucrose, root elongation was restored, but starch still accumulated to a much larger extent than during growth on sucrose alone. Accumulation of starch in the shoots of trehalose-treated seedlings was accompanied by an increased activity of ADP-glucose pyrophosphorylase and an induction of the expression of ADP-glucose pyrophosphorylase gene, *ApL3*. Even in presence of 50 mM sucrose, which itself also slightly induced *ApL3*, trehalose (5 mM) led to a further increase in *ApL3* expression. The results suggest that trehalose interferes with carbon allocation to the sink tissues by inducing starch synthesis in the source tissues. Furthermore, trehalose induced the expression of the β-amylase gene, *AT.β-Amy*, in combination with sucrose but not when trehalose was supplied alone, indicating that trehalose can modulate sugar-mediated gene expression.

Increased trehalose also promotes growth of dicot crops by producing increased nodulation and nitrogen fixation. To confirm this, Müller et al. (1992) infected soybean cv Maple Arrow with *Bradyrhizobium japonicum*, strain 61-A-101, and planted under semisterile conditions for purification and characterization of trehalose. They extracted trehalase from the nodules under acidic conditions (pH 3.7) and purified as an apoplastic glycoprotein. Its native molecular weight was close to 54 kDa. The enzyme had a broad pH optimum (pH 3.5 to 7.0) and a temperature optimum of 59°C. Validamycin, a fungitoxic antibioticum isolated from *Streptomyces,* competitively inhibited trehalase biosynthesis. Later, Aeschbacher et al. (1999) purified trehalase to homogeneity and cloned the full length cDNA encoding the enzyme, named *GMTRE 1* (*G. max trehalase 1*). The amino acid sequence derived from the open frame of *GMTRE 1* showed strong homology to known trehalases. *GMTRE 1* is a single-copy gene and is expressed at a low but constant level in many tissues.

Müller et al. (1994 a) analyzed the N_2 –fixing (effective) nodules from various legume-*Rhizobium* combinations for trehalose and other soluble carbohydrates. Whereas the bacterial disaccharide trehalose was present only in the minority of the nodules, trehalase activity was found in all of them. Extracts from determinate nodules had a higher trehalase activity than extracts from indeterminate nodules. Detailed studies were conducted on soybean nodules formed in interactions with 2 effective and 5 ineffective *Bradyrhizobium japonicum* strains. Only in effective soybean nodules colonized by the strain 61-A-101 was trehalose a major soluble carbohydrate. Irrespective of the wild-type strains used, effective soybean nodules contained about 10 nkat trehalase g^{-1} fresh weight, whereas the ineffective nodules colonized by mutant strains derived from these wild-type strains contained 2-30-times less trehalase.

As nitrate ion is known to inhibit nodulation and N_2 fixation, Müller et al. (1994 b) exposed soybean cv Maple Arrow plants to various amounts of nitrate either from the beginning or after completion of nodulation to study its effects on accumulation of trehalose and other carbohydrates and on trehalase activity in soybean root nodules. Presence of 5 mM or more nitrate during nodulation caused considerable reduction in the number and biomass of nodules per plant, and the nitrogenase activity per nodule fresh weight. The level of the disaccharide trehalose, produced by the microsymbiont, was 50 percent lower in nodules formed in presence of 20 mM nitrate than in control nodules formed in its absence. Nodules formed in presence of high amounts of nitrate contained about 75 percent less starch but 3- to 4-fold higher levels of sucrose and pinitol than control nodules. When plants with fully established nodules grown in absence of nitrate were shifted to 20 mM nitrate, levels of trehalose and starch decreased over a period of 3 weeks while the level of sucrose increased, until the carbohydrate level attained similar values as found in the nodules established in presence of nitrate. The activity of trehalase, an enzyme induced in nodules, was about 75 percent lower in nodules formed in presence of nitrate than in the control nodules. However, the trehalase activity did not change in established nodules during a 3-week exposure to 2 mM nitrate.

As early as in 1985, Streeter studied the accumulation of trehalose by *Rhizobium* bacteria and bacteriods in 4 strains of *R. japonicum* (61A76 USDA 110, 123 and 138) grown in 8 different defined media. Regardless of the carbon or nitrogen source supplied, trehalose was the major carbohydrate accumulated by all the 4 strains. After 7-9 days of growth, trehalose generally accounted for 90-100 percent of the mono- and disaccharides detected. Its concentration varied widely; the highest concentration recorded was 41 $\mu g\ mg^{-1}$ of dry weight. In all but six strains, trehalose accounted for greater than 80 percent of the mono- and disaccharides in cells. *Rhizobium japonicum* bacteriods also synthesized trehalose; the quantity in nodules varied in approximate correspondance to accumulation of trehalose by cultured bacteria. In young soybean nodules (29 days after planting), 45-80 percent of the trehalose was recovered in the cytosol. There were differences among *R. japonicum* strains in the retention of trehalose, and the proportion of trehalose retained by bacteriods enhanced with increasing plant age for all the strains.

Possibly trehalose, a stress-protectant synthesized by many microorganisms including rhizobia is important in their survival. To obtain more information about the survival of rhizobia and carbohydrate metabolism during nodule senescence, Müller et al. (2001) treated nodulated soybean plants with the photosynthesis inhibitor DCMU, a specific inhibitor of PSII, was taken up rapidly and with good systemic mobility. Nodules were then harvested over a period of 14 days and the levels of the major plant non-structural carbohydrates, sucrose and starch, as well as the major bacterial carbohydrate, trehalose, were monitored. At the beginning of DCMU treatment, pool sizes of trehalose did not differ significantly from the control values. Only 10 days after adding DCMU, trehalose values were significantly lower than in controls, but during the experiment, they did not decline to the background. Even after 14 days, trehalose values were still about 60 percent of the control values (Fig. 8.2). In contrast to plant-borne carbohydrates, trehalose was not depleted in senescing nodules, and with more than 4 $mg\ g^{-1}$ dry weight became the most abundant non-structural carbohydrate at the end of the experiment; other non-structural carbohydrates accounted for less than 1 $mg\ g^{-1}$ dry weight. The trehalose accumulated in bacteria could protect them against membrane injuries and/or serve as an intermediate energy reserve. Trehalose seems to be at least partially maintained during senescence. It is expected to be degraded and used as an energy source upon re-initiation of growth by the bacteria. The number of reisolated viable bacteria was not significantly decreased in senescent nodules as compared to control nodules. These results indicate that during terminal senescence of nodules an appreciable part of the bacteria conserve their trehalose pools and survive. Taken together, it can be postulated that nodule senescence does not necessarily lead to a decay of the microsymbiont population. A quite substantial amount of the population may still be able to grow and thus be able to transmit the selected traits.

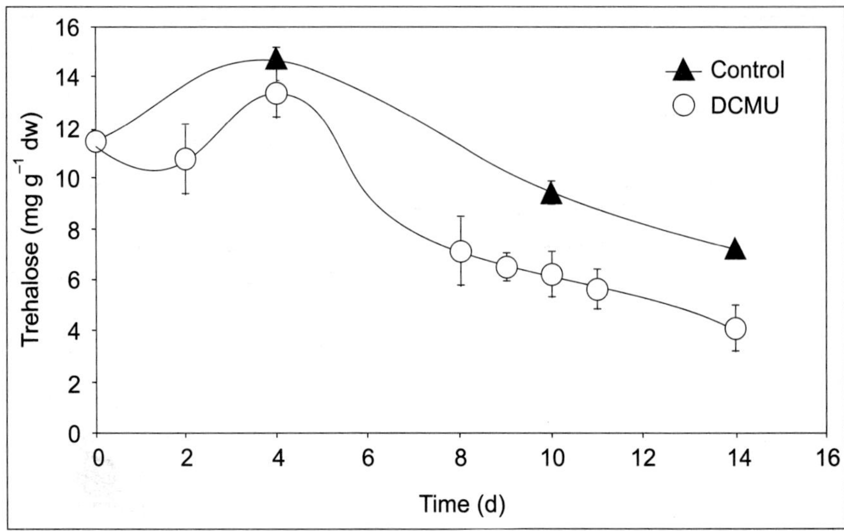

Fig. 8.2 Trehalose in soybean nodules. [Soybean cv Maple Arrow plants were nodulated by *Bradyrhizobium japonicum* 61-A-101. At an age of 4 weeks, some plants were treated with 0.5 ppm DCMU (O) or kept as controls (-▲-). Trehalose was assayed using GC. Mean values ± SE are given for three independent samples for each time point]. (After Müller et al., 2001).

Avonce et al.(2005) have shown that overexpression of the *Arabidopsis* trehalose-6-phosphate synthase gene (*AtTPS1*) in *Arabidopsis* promotes trehalose and trehalose-6-phosphate (T6P) accumulation. The gene-expression analysis in transgenic plants showed up-regulation of several genes involved in sugar signalling and metabolism. These data suggest that *AtTPS1* and accordingly T6P and trehalose play an important role in the regulation of glucose sensing and signalling genes during plant development.

TREHALOSE VS STRESS TOLERANCE

Trehalose serves not only as a carbohydrate reserve or osmoprotectant, but also as a protective agent against a variety of physical and chemical stresses. Trehalose is known to have high water retention capacity, which maintains the fluidity of membranes under dry conditions (Leslie et al., 1995). Thus this sugar allows desert (resurrection) plants to tolerate naturally occurring stresses during cycles of dehydration and rehydration (Müller et al., 1995). Exogenously applied trehalose causes a decrease in NaCl accumulation and growth inhibition (Garcia et al., 1997). Thus Goddijn and van Smeekens (1998) and Goddijn and van Dun (1999) overproduced trehalose-6-phosphate synthase genes in plants but observed that this resulted in significant morphological growth defects and altered metabolism due to changes in carbon allocation between source and sink tissues which may be implicated in sugar signalling. Tobacco plants expressing *E. coli* trehalose-6-phosphate synthase accumulated more trehalose-6-phosphate and displayed increased rates of photosynthesis per unit leaf area under saturating light, whereas those expressing trehalose-6-phosphate phosphatase displayed reduced photosynthetic rates. These data suggest that trehalose-6-phosphate either directly or indirectly controls carbon assimilation. Exogenous trehalose has also been shown to affect plant metabolism and gene expression (Müller et al., 1998; Wingler et al., 2000). Although trehalose-6-phosphate synthase is clearly important, it remains to be ascertained about the physiological role of trehalose synthesis.

Encouraged with the observation that trehalose in resurrection plants, *Myrothamnus flabillifolius* and *Silaginella lepidophylla,* confers the ability to desiccate almost to dryness, and then pop up whenever water becomes available; attempts were made to incorporate trehalose producing genes in crop plants. Early attempts with the dicot crops, potato and tomato did not yield encouraging results. Then Seo et al. (2000) made an important contribution after testing the effect of physical proximity of the two enzymes, trehalose-6-phosphate synthase (T-6-PS) and trehalose-6-phosphate phosphatase (T-6-PP) catalyzing sequential reactions; a bifunctional fusion enzyme; TPSP was constructed by fusing the *E. coli* genes for trehalose-6-phosphate (T-6-P) synthase (TPS) and trehalose-phosphate phosphatase (TPP) . TSPS catalyzes the sequential reaction in which T6P is formed and then dephosphorylated, leading to the synthesis of trehalose. The increased catalytic efficiency of TPSP results from the proximity of TPS and TPP in the TPSP fusion enzyme. Then two groups of scientists, Garg et al. (2002) in USA and Jang et al. (2003) in Korea made successful attempts with the staple cereal crop, rice. Trehalose stablized dehydrated enzymes, proteins, and lipid membranes efficiently, as well as protected the biological structures from damage during desiccation. Engineering trehalose overproduction in the indica rice was achieved by stress-inducible or tissue specific expression of the bifunctional trehalose-6 phosphate synthase/phosphatase fusion enzyme without any detrimental effect on plant growth or grain yield (Garg et al., 2002). During the abiotic stress, the transgenic plants (coding regions of the *E. coli otsA* and *otsB* genes-encoding trehalose-6-phosphate synthase and trehalose-6-phosphate phosphatase, respectively) accounted increased amounts of trehalose and showed high levels of tolerance to salt, drought, and low-temperature stresses, as compared with the nontransformed plants.

Jang et al. (2003) generated trehalose producing, transgenic rice plants by introducing a gene encoding a bifunctional fusion trehalose-6-phosphate synthase/phosphatase (TPSP) of the *E. coli* under control of the maize ubiquitin promoter (*Ubi 1*). Trehalose levels in the leaf and seed extracts from *Ubi1:: TPSP* plants were increased up to 1.076 mg g^{-1} fresh weight. This level is 200-fold higher than that of transgenic tobacco plants transformed independently with either *TPS* or *TPP* expression cassettes. The *Ubi1 :: TPSP* plants showed no growth inhibition or visible phenotypic alterations despite the high level production of trehalose. Moreover, trehalose accumulation in *Ubi1 :: TPSP* plants resulted in increased tolerance to drought, salt, and cold. Results suggest that trehalose acts as a global protectant against abiotic stress, and that rice is more tolerant to trehalose synthesis than dicots.

The catalytic efficiency of *TPSP* was 3.5-4.0-fold higher than that of a mixture of the individual enzymes, which demonstrates the kinetic advantage of the fusion enzyme (Seo et al., 2000). The resultant transgenic plants produced trehalose levels that were up to 0.1 percent of the fresh weight, and the plants showed no visible growth inhibition. The *Ubi1 :: TPSP* plants showed normal vegetaive phenotype and fertility as compared with untransformed control plants (Jang et al., 2003). A slight delay in germination of *Ubi1 :: TPSP* seeds was observed at 3 days after the start of germination; but the growth rates converged at later stages without notable differences in shoot and root growth. Also the chemical composition of the rice grains of transgenic plants remained unchanged (Garg et al., 2002).

Attempts of incorporating trehalose overproduction in legumes have not proved beneficial under drought conditions. One of the reasons is failure of *Rhizobium* growth under drought stress and the rhizobial inoculation is also ineffective due to the same reason. Streeter (2003) thought that trehalose might improve the survival of *Rhizobium japonicum* and thus Streeter added trehalose to cultures just prior to desiccation, or supplied to bacteria during the 6-day growth period. Trehalose added to cultures at the time of desiccation improved survival slightly, but trehalose loading during growth was much more effective in protection against desiccation. Growth of bacteria with 3 mmol L^{-1} trehalose increased trehalose concentration in cells by about 3-fold and increased survival of cells placed on soybean seeds by 2- to 4-fold after 2 or 24 hours, respectively. These results indicate that growth of bacteria with trehalose in the medium resulted in a 294 percent increase in survival after 24 hours of desiccation. The concentration of trehalose in cells was highly correlated with survival of bacteria. The authors expect that it may be possible to engineer greater trehalose accumulation in rhizobia. Further, the trehalose concentration in cells should be a useful predicator of survival during desiccation .

Salinity stress

After prolonged exposure to salt stress, almost all the transgenic rice plants survived and displayed vigorous root and shoot growth (Garg et al., 2002). In contrast, all of the non-transformed-stressed plants were either dead or nearly dead because of severe salt damage to the leaves and concomitant loss of chlorophyll. The transgenic plants developed longer and thicker roots than the non-transformed-stressed plants after salt stress. Salt stress severely inhibited growth of shoot and roots of non-transformed stressed plants. After continuous salt stress (100 mM NaCl) for 4 weeks, non-transformed stressed plants showed a very large increase in Na^+ content in both shoots and roots campared with non-transformed control, whereas the increase in shoots of all the transgenic plants was much smaller. Sodium ion content of transgenic plant shoots was only 30-35 percent of the non-transformed-stressed plants after salt stress. Garg et al. argued that trehalose might have played a direct or indirect role in maintaining ion selectivity and thus, fascilitating cellular Na^+ exclusion. Garcia et al. (1997) have earlier noted that exogenous trehalose application significantly reduced salt-induced accumulation of Na^+ in the leaves which supports Garg's contention. On the other hand, the transgenic lines R80 and A05 maintained shoot to root K^+ homeostasis both under non-stress and salt-stress conditions. After salt stress, the levels of shoot and root K^+ content in transgenic plants was similar to the non-stressed controls, while a 4-fold decrease in root K^+ in the non-transformed-stressed plants was observed. Thus the transgenic trehalose producing rice plants were able to maintain a higher level of selectivity for K^+ over Na^+ uptake in roots and Na^+ exclusion from the shoots compared with the non-transformed-stressed plants. Maintenance of the Na^+/K^+ ratio in both shoot and roots of transgenic plants correlated with nearly normal plant growth and may be the basis for minimizing Na^+ toxicity under salt stress. Also, the salt stress led to a significant increase in root and shoot Ca^{2+} content in the non-transformed-stressed plants, whereas in the transgenic plants, this Na-mediated increase in Ca^{2+} content was only observed in the shoots and not in the roots. Significantly higher levels of shoot Fe content was also observed in the transgenic plants compared with the non-transformed control plants.

Trehalose-producing, transgenic rice plants were generated by the introduction of a gene encoding a bifunctional fusion (*TPSP*) of the trehalose-6-phosphate synthase and trehalose-6-phosphate phosphatase (*TPP*) of *Escherichia coli* , under the control of maize ubiquitin promoter (*Ubi1*) (Jang et al., 2003). To investigate the increased tolerance of *Ubi1 :: TPSP* rice plants against salinity, Jang et al. measured growth during germination of 5 homozygous T_2 seedlings in hydroponic solutions that contained 100 mM NaCl. In the absence of NaCl, *Ubi1 :: TPSP* seedlings grew similarly to non-transgenic seedlings during 13 days after germination. In presence of NaCl, in contrast, both shoot and seminal root growth of the *Ubi1 :: TPSP* seedlings was much faster than that which occurred in those of the non-transgenic seedlings (Fig. 8.3).

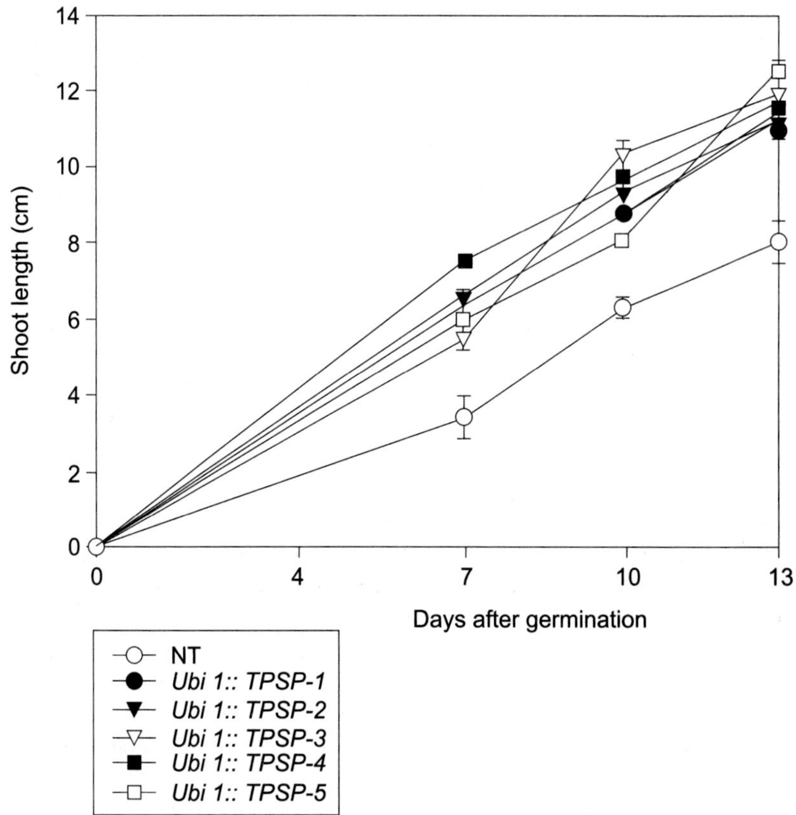

Fig. 8.3 Salt tolerance of non-transgenic and *Ubi 1:: TPSP* seedlings grown in the presence of 100 mM NaCl. Ten seeds from each of the five (1-5) *Ubi 1:: TPSP* lines and the non-transgenic (NT) plants were germinated and grown in hydroponic solutions that contained 100 mM NaCl under continuous 150 mmol m^2 s^{-1} light conditions. The shoot length was scored at various intervals. Each data point represents the mean ± SE of triplicate experiments (n=10). (After Jang et al., 2003).

Drought stress

A low but significant amount of trehalose is present in the shoots (17 µg g^{-1} fr wt) of non-transformed control rice plants; but these levels increase significantly under drought stress (Garg et al., 2002). The transgenic plants grown under control conditions exhibit trehalose levels comparable with non-transformed-stressed plants. After drought stress, trehalose levels in the transgenic lines increased 3- to 9-fold (Fig. 8.4). The transgenic lines showed considerable protection at much lower trehalose concentrations during drought stress. There was no obvious relationship between trehalose accumulation and stress tolerance among the transgenic lines evaluated. However, the difference in trehalose levels between the transgenic and non-transgenic lines clearly correlated with increased tolerance to drought stress.

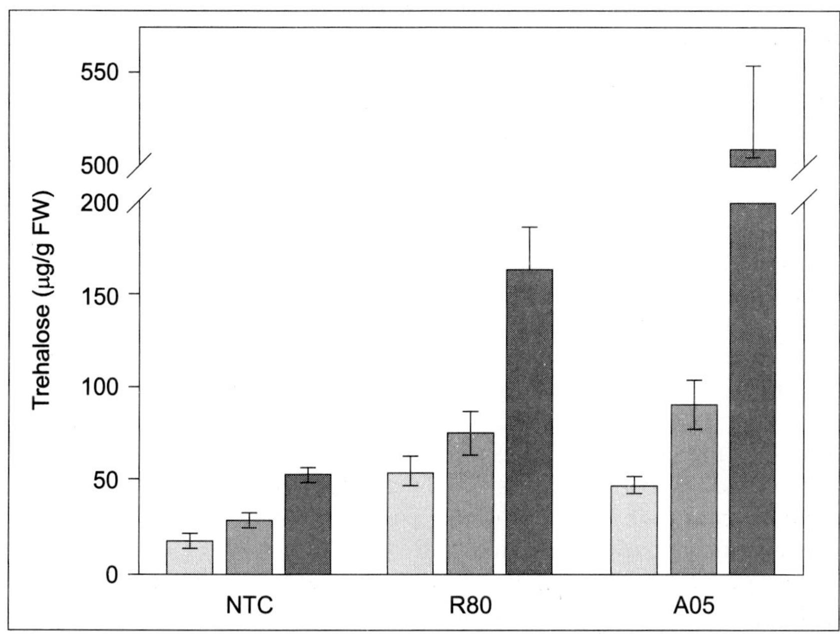

Fig. 8.4 Trehalose content in shoots of transgenic (R80 and A05) and nontransgenic plants with or without stress. Trehalose accumulation under nonstressed (white bars), salt-stressed (100 mM NaCl for 4 weeks, hatched bars), or drought-stressed (100 h, black bars) conditions. (After Garg et al., 2002).

Five-week old non-transformed and trehalose producing transgenic indica rice seedlings were subjected to 2 cycles of 100 hours of drought stress. The non-transgenic plants exhibited rolling of leaves within 48 hours of stress as compared with the considerably fewer visual symptoms in transgenic plants during the same time period. After 2 cycles of 100-hour drought stress and subsequent watering for 3 weeks, growth of both the transgenic lines, R80 and A05 were almost identical to non-stressed control plants. In contrast, growth of the drought stressed non-transformed-stressed plants was severely inhibited.

After the first cycle of 100 hours of drought stress, the quantum yield of PSII photochemistry in non-transformed-stressed plants decreased by approximately 68 percent, whereas the activity of the two best performing transgenic lines (R80 and A05) decreased only by 29-37 percent compared with the non-stressed controls. Similarly, drought-induced decreases in the fluorescence parameters Fv/Fm, which is a measure of accumulated photooxidative damage to PSII, were considerably smaller in the transgenic lines than in the non-transformed-stressed plants.

After prolonged exposure to drought stress, the *Ubi1 :: TPSP* rice plants survived and displayed vigorous root and shoot growth; over the same treatment period, the untransformed plants were nearly dead because of severe damage to leaves and concomitant loss of chlorophyll (Jang et al., 2003). A decrease in the Fv/Fm ratio was observed after the plants were subjected to dehydration, salt or low temperature stresses. The Fv/Fm ratios were 15 percent to 19 percent higher in *Ubi1 :: TPSP* plants than in the untransformed control plants. Avonce et al. (2004) further showed that the *Arabidopsis* trehalose-6-phosphate synthase (*AtTPS1*) gene also regulates glucose, abscisic acid and stress signalling. *AtTPS 1* gene results in the development of ABA-insensitive phenotypes. Thus, the constitutive expression of *TPSP* in transgenic plants leads to increased levels of trehalose accumulation, which correlates with enhanced tolerance against drought, salinity, and low temperature, suggesting that trehalose acts as a global protectant against abiotic stress in rice. The same strategy should also work in a range of other crops, including maize, wheat, millet, soybeans, and sugarcane. For the critics of GM crops, it is important to note that chemical composition of the edible parts of plants, such as rice grains, remains unchanged (Garg et al., 2002).

The gene *TPS* was isolated and characterized from cultivated cotton cv ZETA 2, using a chromosome-walking technique (Kosmas et al., 2005). *TPS* has three exons comprising the coding region. Southern blot analysis indicated that the *Gossypium* genomes (A and D) contain a single copy of *TPS* per genome. They studied the expression of this gene in different plant tissues. Plants of the Australian cotton variety Siokra L23, known for its drought tolerance, were subjected to drought stress (using PEG 6000 solution, for four hours during the dark period of the day and for four consecutive days); leaves, stems and roots were collected after the end of the stress period. Total extracted RNA was examined for the presence of transcripts, in the above-mentioned tissues of stressed and well-watered plants, by reverse transcription polymerase chain reaction (RT-PCR). The expression levels, determined semi-quantitatively, indicated that the gene was expressed in all plant tissues under both water availability conditions. However, increased expression levels of *TPS* were observed mainly in stressed leaves and root compared to those of the well-watered control. This finding is in agreement with the fact that *TPS* participates in trehalose biosynthesis, known for its participation in stress signal transduction in higher plants.

The *TPS* gene of *Saccharomyces cerevisiae* was engineered under the control of the *CaMV* ^{35}*S* promoter for constitutive expression in transgenic potato plants by Ti-plasmid of *Agrobacterium*-mediated transformation (Yeo et al., 2000). The resulting *TPS1* transgenic potato plants exhibited various morphological phenotypes in culture tubes, ranging from normal to severely retarded growth, including dwarfish growth, yellowish lencet-shaped leaves, and aberrant root development. However, the plants recovered from these negative growth effects when grown in a soil mixture. The *TPS1* transgenic potato plants showed significantly increased drought resistance. These results suggest that the production of trehalose not only affects plant development but also improves drought tolerance.

IMPROVEMENT

Chloroplast vs nuclear engineering

Lee et al. (2003) introduced yeast trehalose phosphate synthase (*TPS1*) gene in tobacco chloroplast or nuclear genomes to study resultant phenotypes, especially with regard to their drought tolerance. The chloroplast transformant expressed 169-fold more *TPS1* transcript than the best surviving nuclear transgenic plant. Although both the chloroplast and nuclear transgenic plants showed significant *TPS1* enzyme activity, no significant trehalose was observed in T_0/T_1 nuclear transgenic plants whereas the chloroplast transgenic plants showed 15-25-fold higher accumulation of trehalose than the best surviving nuclear transgenic plants. The nuclear transgenic plants (T_0) that showed even small amounts of trehalose accumulation showed stunted phenotype, sterility and other pleiotropic effects whereas the chloroplast transgenic plants (T_1, T_2, T_3) showed normal growth and no pleiotropic effects. Transgenic chloroplast thylakoid membranes showed high integrity under osmotic stress as evidenced by retention of chlorophyll even when grown in 6 percent PEG whereas chloroplasts in untransformed plants were bleached. After 7-hour drying, chloroplast transgenic seedlings (T_1, T_2) successfully rehydrated while the control plants died. There was no difference between control and transgenic plants in water loss during dehydration but dehydrated leaves from transgenic plants (not watered for 24 days) recovered upon rehydration turning green while the control leaves dried out. These observations suggest that trehalose functions by protecting biological membranes rather than regulating water potential. In order to prevent escape of drought tolerance trait to weeds and associated pleiotropic traits to related crops, it may be desirable to engineer crop plants for drought tolerance via the chloroplast genome instead of the nuclear genome.

FUTURE PERSPECTIVE

Plant adaptation to environmental stresses is dependent upon the activation of cascades of molecular networks involved in stress perception, signal transduction, and the expression of specific stress-related genes and metabolites. Consequently, engineering genes that protect and maintain the function and structure of cellular components can enhance tolerance to stress (Vinocur and Altman , 2005). Unraveling additional stress-associated gene resources, from both crop plants and highly salt- and drought-tolerant model plants, will enable future molecular dissection of salt-tolerance mechanisms in important crop plants, possible. Upon induction of trehalose-6-phosphate synthase, the elevated levels of intracellular trehalose correlates not only with increased tolerance to heat shock but also with resistance to freezing and thawing, dehydration, osmostress, and toxic levels of ethanol, indicating that trehalose may be the stress metabolite underlying the overlap in induced tolerance to these stresses. Among the isogenic strains transformed with this construct, one in which the gene coding for the trehalose-hydrolyzing enzyme, neutral trehalase was disrupted, accumulated trehalose to a greater extent and was more resistant to the above stresses. Increased trehalose concentration is thus a major determinant of the general stress protection response. Trehalose is an active stabilizer of enzymes, proteins and biomass. Trehalose has been accepted as a safe food ingredient by the European regulation system following approval by the US Food and Drug Administration. The wide range of applications of this sugar has increased the interest of many research groups into the development of novel and economically feasible production systems.

REFERENCES

Aeschbacher, R.A.; J. Müller; T. Boller; et al. 1999. Purification of the trehalose *GMTRE1* from soybean nodules and cloning of its cDNA , *GMTRE1* is expressed at a low level in multiple tissues. Plant Physiol. 119: 489-96.

Avonce, N.; B. Leyman; J.O. Mascorro-Gallardo; et al. 2004. The *Arabidopsis* trehalose-6-P synthase *AtTPS1* gene is a regulator of glucose, abscisic acid, and stress signaling. Plant Physiol. 136: 3649-59.

Avonce, N.; B. Leyman; J. Thevelein; et al. 2005. Trehalose metabolism and glucose sensing in plants. Biochem. Soc. Trans. 33 (Pt 1): 276-79.

Bradford, M. 1976. A rapid and sensitive method for the quantitation of microgram quantities of protein utilizing the principle of protein-dye binding. Anal. Biochem. 72: 248-54.

Eastmond, P.J.; A.J. van Dijken; M Spielman; et al. 2002. Trehalose-6-phosphate synthase1, which catalyzes the first step in trehalose synthesis, is essential for *Arabidopsis* embryo maturation. Plant J. 29: 225-35

Garcia, A.B.; J de E. Engler; S. Iyer et al. 1997. Effects of osmoprotectants upon NaCl stress in rice. Plant Physiol. 115: 159-69.

Garg, A.K; J.K. Kim; T.G. Owens; et al. 2002. Trehalose accumulation in rice plants confers high tolerance levels to different abiotic stresses. Proc. Nat. Acad. Sci. 99: 15898-903.

Goddijn, O.J. and K. van Smeekens, 1998. Sensing trehalose biosynthesis in plants. Plant J. 14: 143-46.

Goddijn, O.J. and K. van Dun, 1999. Trehalose metabolism in plants. Trends in Plant Sci. 4: 315-19.

Jang, I.C.; S.J. Oh; J.S. Seo; et al. 2003. Expression of a bifunctional fusion of the *Escherichia coli* genes for trehalose-6-phosphate synthase and trehalose-6-phosphate phosphatase in transgenic rice plants increases trehalose accumulation and abiotic stress tolerance without stunting growth. Plant Physiol. 131: 416-24.

Kosmas, S.A.; A. Argyrokastritis; M.G. Loukas; et al. 2005. Isolation and characterization of drought-related trehalose-6-phosphate synthase gene from cultivated cotton (*Gossypium hirsutum* L.). Planta Aug. 6; 1-11 (E. pub. ahead of print).

Lee, S.B.; H.B. Kwon; S.J. Kwon; et al. 2003. Accumulation of trehalose within transgenic chloroplasts confers drought tolerance. Molecular Breeding 11: 1-13.

Leslie, S.B.; E. Israel; B. Lighthart; et al. 1995. Trehalose and sucrose protect both membranes and proteins in intact bacteria during drying. Appl. Environ. Microbiol. 61: 3592-97.

Müller, J.; C. Staehelin; R.B. Mellor; et al. 1992. Partial-purification and characterization of trehalase from soybean nodules. J. Plant Physiol. 140: 8-13.

Müller, J.; Z.P. Xie; C. Staehelin; et al. 1994 a. Trehalose and trehalase in root nodules from various legumes. Physiol. Plant. 90: 86-92.

Müller, J.; Z.P. Xie, C. Staehelin; et al. 1994 b. Effects of nitrate on accumulation of trehalase and other carbohydrates and on trehalase activity in soybean root nodules. J. Plant Physiol. 143: 153-60.

Müller, J.; T. Boller and A. Wiemken, 1995. Trehalose and trehalase in higher plants : new developments. Plant Sci. 112: 1-8.

Müller, J.; T. Boller and A. Wiemken, 1998. Trehalose affects sucrose synthase and invertase activities in soybean (*Glycine max* (L.) Merr.) roots. J. Plant Physiol. 153: 255-57.

Müller, J.; R.A. Aeschbacher; A. Wingler; et al. 2001. Trehalose and trehalase in *Arabidopsis.* Plant Physiol. 125: 1086-93.

Paul, M.; T. Pellny and O.J. Goddijn, 2001. Enhancing photosynthesis with sugar signals. Trends in Plant Sci. 6: 197-200.

Seo, H.S.; Y.J. Koo; J.Y. Lim; et al. 2000. Characterization of a bifunctional fusion enzyme between trehalose-6- phosphate synthase and trehalose-6- phosphate phosphatase of *Escherichia coli.* Appl. Environ. Microbiol. 66: 2484-90.

Streeter, J.G. 1985. Accumulation of α-α-trehalose by *Rhizobium* bacteria and bacteriods. J. Bacteriol. 164: 78-84.

Streeter, J.G. 2003. Effects of trehalose on survival of *Bradyrhizobium japonicum* during desiccation. J. Appl. Microbiol. 95: 484-91.

Vinocur, B. and A. Altman, 2005. Recent advances in engineering plant tolerance to abiotic stress : achievements and limitations. Curr. Opin. Biotechnol. 16(2): 123-32.

Wingler, A.; T. Fritzius; A. Wiemken; et al. 2000. Trehalose induced the ADP-glucose pyrophosphorylase gene, *ApL3* , and starch synthesis in *Arabidopsis.* Plant Physiol. 124: 105-14.

Yeo, E.T.; H.B. Kwon; S.E. Han; et al. 2000. Genetic engineering of drought resistant potato plants by introduction of the trehalose-6-phosphate synthase (*TPS1*) gene from *Saccharomyces cerevisiae.* Mol. Cells 10(3): 263-68.

Zhou, J.; B. Yang and X. Dai, 2001. [Detection of trehalose in transgenic tobacco by HPLC with ELSD]. Wei Sheng Wu Xue Bao 41 (3): 378-80 (In Chinese).

SUGGESTED READING

Garg, A.K.; J.K. Kim; T.G. Owens; A.P. Ranwala; Y.D. Choi; L.V. Kochian and R.J. Wu, 2002. Trehalose accumulation in rice plants confers high tolerance levels to different abiotic stresses. Published online before print November 27, 2002, 10.1073/pnas.252637799. PANS 99 (25): 15898-903.

Jang, I.C.; S.J. Oh; J.S. Seo; W.B. Choi; S.I. Song; et al. 2003. Expression of a bifunctional fusion of the *Escherichia coli* genes for trehalose-6-phosphate synthase and trehalose-6-phosphate phosphatase in transgenic rice plants increases trehalose accumulation and abiotic stress tolerance without stunting growth. Plant Physiol. 131: 516-24.

9

FRUCTAN

Introduction

Fructans are polyfructose molecules that function as non-structural storage carbohydrates in about 15 percent of all flowering plant species (over 40,000 species) which are linear and branched polymers of fructose. Some important crops (wheat, barley, oats), vegetables (onion, lettuce, chichory, artichoke, asparagus), forage grasses (*Festuca, Lolium*) and ornamental plants (dahlia, tulip) store fructans. Small fructans have a sweet taste, whereas longer fructan chains form emulsions with a fat-like texture and a neutral taste. In addition to serving as a reserve carbohydrate, fructans have been implicated in protecting plants against water-deficit caused by drought, salinity and low temperature (Pilon-Smits et al., 1995; 1999; Puebla et al., 1997; Demel et al., 1998; Albrecht et al., 1993; Amiard et al., 2003). Fructan metabolism is initiated by intermittent drought, as shown by Hendry and Wallace (1993). Fructans protect membranes or other cellular components from the adverse effects of drought, or perhaps influence the growth processes directly (Pilon-Smits, 1995).

The substrate for fructan synthesis is sucrose, and like sucrose, fructans are also stored in vacuoles. Sucrose is synthesized in cytoplasm, but the fructans are produced in the vacuole by the action of specific enzymes (fructosyl-transferases) that transfer fructose from sucrose to the growing fructan chain. Fructan synthesis is modulated by light. In the vacuole, sucrose can be converted into fructan by fructosyltransferases or hydrolysed into glucose and fructose by invertase. The reserve metabolism of natural fructan accumulators is distinct from natural starch accumulators in a number of important respects. Fructan is synthesized directly from sucrose as the sole precursor, apparently without the involvement of phosphorylated sugars or nucleotide co-factors. Its synthesis is extrachloroplastic, it is water soluble, and it accumulates in the vacuole of both photosynthetic and storage cells. In addition to its reserve role, it also confers tolerance to abiotic stresses.

The fructosyl chain length in plants, as opposed to bacteria, varies greatly and is much shorter than that of bacterial fructan. In general, the degree of polymerization of 30-50 fructosyl residues are found, but occasionally, the degrees of polymerization may exceed 200. Furthermore, plant fructans have a greater variety in the linkage of the fructosyl residues. In higher plants five major classes of structurally different fructans can be distingushed (Fig. 9.1); inulin, levan (bacterial fructans), mixed levan, inulin neoseries, and levan neoseries.

Fructans have a much stronger effect on different lipid systems than other (poly)saccharides, which appears to be related to their hydrophobic properties (Vereyken et al., 2001). Fructans are able to stabilize the liquid-crystalline lamellar phase, which is consistent with a drought-protecting role in plants. The inulin-type fructan has a more profound interaction with the membrane than the levan-type fructan (Vereyken et al., 2003 a). Fructans are thus involved in membrane preservation during dehydration by interacting with the membrane lipids (Vereyken et al., 2003 b). Fructan increases the mobility of acyl chains, but immobilizes the lipid headgroup region. Most likely, fructans insert between the headgroups of lipids, thereby spacing the acyl chains. This results in a much lower phase transition temperature. The headgroup is immobilized by the interaction with fructan.

BIOSYNTHESIS

Fructan is synthesized from sucrose by the action of two or more different fructosyltransferases. According to the clasical model of Edelman and Jefford (1968), the enzymes are involved in the synthesis of the most simple form of fructan, inulin. The first enzyme 1-sucrose: sucrose fructosyltransferase (-SST) initiates *de novo* fructan synthesis by catalyzing the transfer of a fructosyl residue from sucrose to another sucrose molecule, resulting in the formation of the trisaccharide, 1-ketose (Fig 9.2). The second enzyme 1-fructan: fructan fructosyltransferase (1-FFT), transfers fructosyl residues from a fructan molecule with a degree of polymerization of $\supseteq 3$ to another fructan molecule or to sucrose. The action of 1-SST and 1-FFT results in the formation of a mixture of fructan molecules with different chain lengths.

Fructan synthesis controls sucrose concentration in the vacuole, fructan thus reduces sucrose concentration in the cell and prevents sugar-induced feedback inhibition of photosynthesis. Fructan accumulation in the vacuole can reach levels as high as 70 percent of dry weight inhibiting photosynthesis, though unlike starch, fructans are water soluble. Fructans are synthesized abundantly in temperate plants with seasonal drought and frost, but are almost absent in tropical plants. At temperatures below 10°C, starch synthesis decreases in temperate plants but photosynthesis and fructan production are less sensitive to low temperature. Encouraged by the reports of Pilon-Smits et al. (1995) and others that fructans have a role in increasing drought

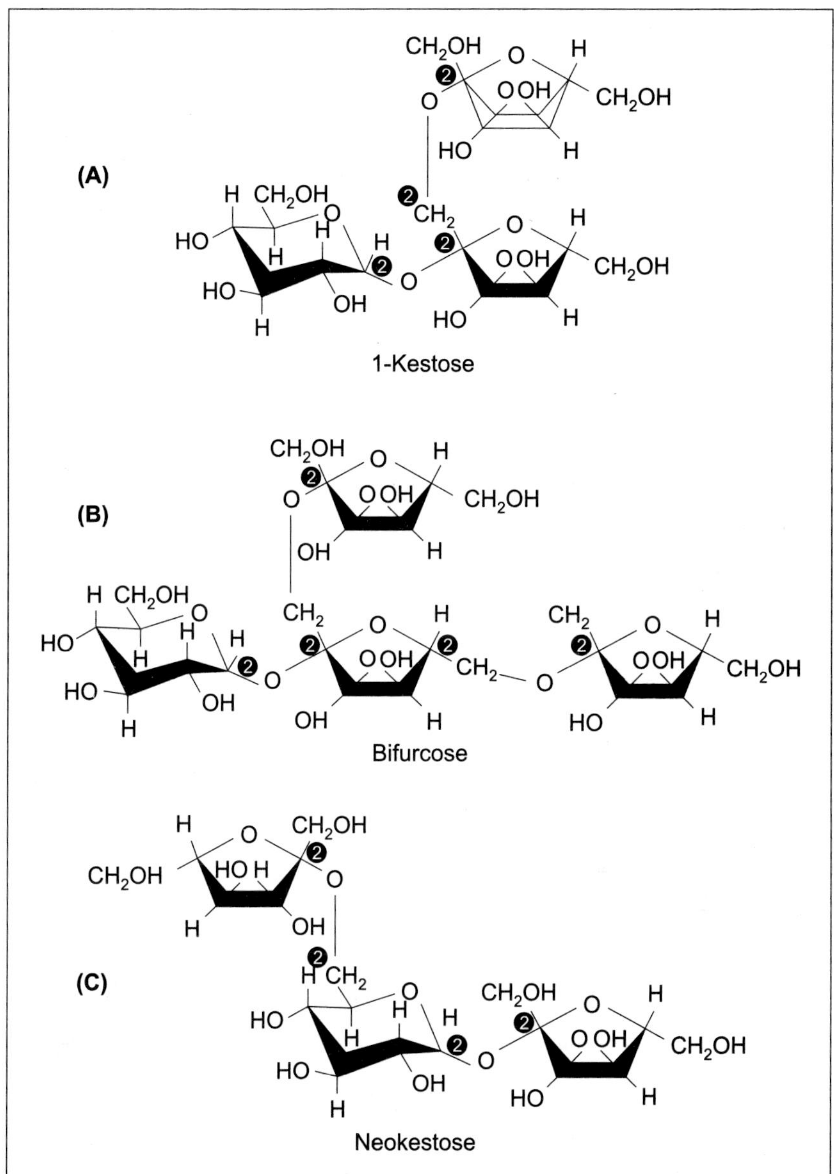

Fig. 9.1 Some examples of structurally different functan molecules found in plants: (A) The trisaccharide 1-ketose consists of a (2-1)-linked b-D-fructosyl unit to sucrose and is the shortest inulin molecule. (B) The tetrasaccharide bifurcose is an example of a mixed-type levan and consists of a (2-1)- and a (2-6)-linked b-D-fructosyl unit to sucrose. (C) Neoketose is the smallest inulin neoseries molecule, and in this molecule a b-D-fructosyl unit is linked to the C6 of the glucose moiety of sucrose. The numbers encircled in black represent the numbers of the carbon atoms in the sugar molecule. (After Vijn and Smeekens, 1999).

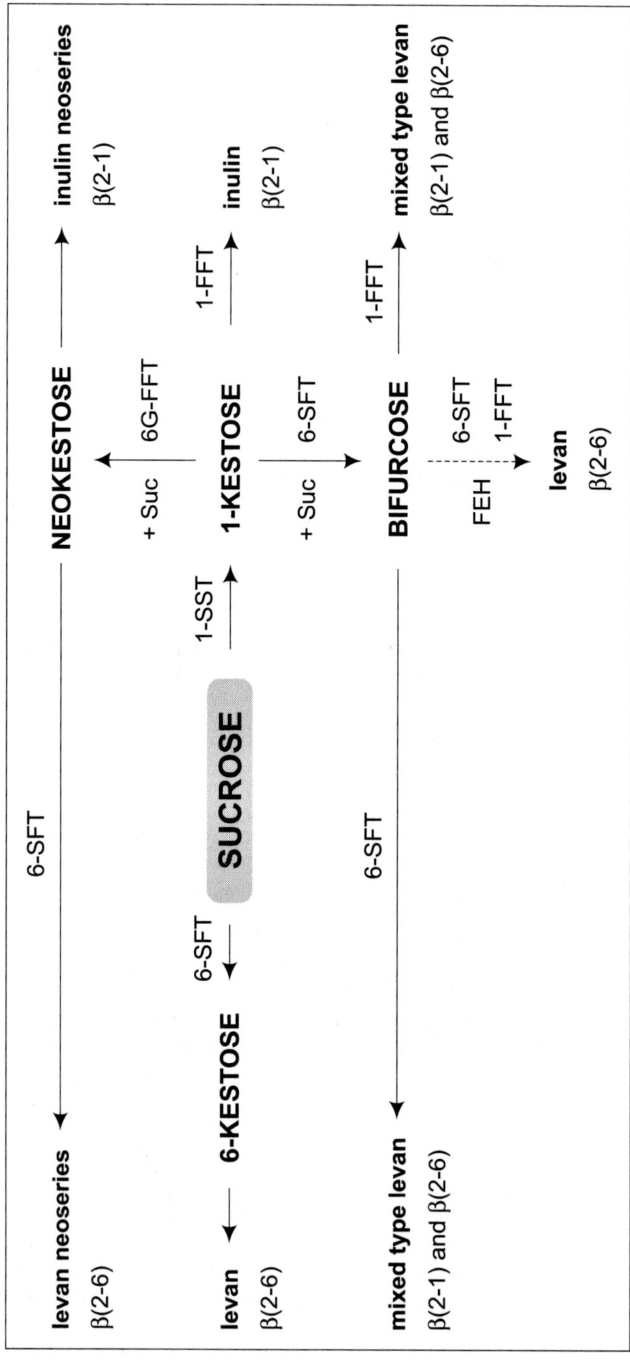

Fig. 9.2 Model of fructan biosynthesis in plants. Starting from sucrose, structurally different fructan molecules can be produced by the concerted action of different fructosyltransferases. The dotted arrow shows an elternative route for the production of levan. FEH = Fructan exohydrolase; FFT = 1-fructose : fructose fructosyltransferase , SFT = 1-sucrose : fructose 6-fructosyltransferase (After Vijn and Smeekens, 1999).

and cold tolerance in crop plants, efforts have been made to engineer fructan producing gene in non-fructan producing crops.

In plants at least two fructosyltransferases are needed to synthesize fructans. One enzyme synthesizes the fructan triglyceride 1-ketose, the next enzyme uses 1-ketose for elongation and/or modification, producing longer fructans. The specificity of fructosyltransferases determines the type of glycosidic bond formed and the donor and acceptor substrates used. This enables the synthesis of many structurally diverse fructans (Ritsema and Smeekens, 2003).

Cairns (1993) argued that the enzymological evidence is insufficient to sustain the SST/FFT model for fructan synthesis. The model fails to explain the synthesis of linkage types and branched fructans. There is no demonstration of the enzymatic *de novo* synthesis of the full complement of the tissue fructan. Further, Cairns (1995) observed that at low concentrations of the partially purified fructosyltransferases, only trisaccharide products were detected, but at the higher enzyme concentrations, larger oligofructans were detected. Depending upon the enzyme concentration, results indicated participation of either FFT activity exclusively, or a more complex fructan polymerizing system.

Sucrose:fructose 6-fructosyltransferase (6-SFT), an enzyme catalyzing the formation and extension of β-2,6-linked fructans was purified from barley (Sprenger et al., 1995). It occurred in two closely similar isoforms with indistinguishable catalytic properties with apparent masses of 49 and 23 kDa. Later, sucrose:sucrose 1-fructosytransferase enzyme was purified to homogeneity from the tubers of *Helianthus tuberosus* having a molecular mass of about 67 kDa (Koops and Jonker, 1996). 1-SST preferentially catalyzed the conversion of sucrose into trisaccharide 1-ketose.

Based on the activities observed for fructosyltransferases, Vijn and Smeekens (1999) proposed a model for the biosynthesis of structurally different fructan molecules in plants. Starting from sucrose, 1-SST produces 1-ketose, which can be elongated by 1-FFT, resulting in the formation of inulin. From sucrose and 1-ketose, 6G-FFT produces neoketose, which can be elongated by 1-FFT or 6-SFT (sucrose, fructose 6-fructosyltransferase), resulting in branched, mixed-type levans. When only sucrose is available as a substrate, 6-SFT produces 6-ketose, which can also be elongated by 6-SFT to produce levans. Another possibility for the production of levans, which was proposed by Wiemken et al. (1995), involves removal of the β (2-1)-linked fructosyl residue from bifurcose by either 1-FFT or exohydrolase.

Graminan-type fructans are temporarily stored in wheat stems. Two phases can be distinguished : a phase of fructan biosynthesis (green stems) followed by a breakdown phase (stems turning yellow). Van der Ende et al. (2003) performed cloning, purification and characterization of two fructan 1-exohydrolase cDNAs (*1-FEH* w_1 and w_2) from winter wheat stems. Characterization of the purified enzymes revealed that inulin-type fructans are much better substrates than levan-type fructans. Although both enzymes are highly identical (98 percent identity), they showed different substrate specificity toward branched wheat-stem fructans.

Sevenier et al. (1998) transformed sugarbeet (a non-fructan producing crop, Van der Ende, 2004) into a crop that produces fructans. The gene encoding *1-SST,* which was isolated from *Helianthus tuberosus,* was introduced into sugarbeet. In *H. tuberosus, 1-SST* mediates the first steps in fructan synthesis through the conversion of sucrose into low molecular weight fructans. In the tap root of sugarbeet transformed with the *1-sst* gene, the stored sucrose is almost totally converted into low molecular weight fructans. Caimi et al. (1996) described the transformation of a *Bacillus amyloliquefaciens SacB* gene into maize callus. The tissue-specific expression and targeting of the *SacB* protein into endosperm vacuoles resulted in stable accumulation of high-molecular-weight fructan in mature seeds. Accumulation of fructan in the vacuole had no detectable effect on kernel development or germination. Fructan levels were found to be approximately 9-fold higher in sh_2 mutants compared to wild-type maize kernels.

ABIOTIC STRESS TOLERANCE

Drought

Fructans play a drought-protecting role in plants. Fructans interact strongly with model membranes (Demel et al., 1998). The surface active effect of fructans contrasts strongly with the maximal effects observed for trehalose, sucrose and glucose under comparable conditions. The results demonstrate a profound and specific membrane interaction of the fructans which is probably very different from the effect of the smaller carbohydrates. They suggested that the water-stress-protective effect of fructans may be induced by membrane-fructan interaction which prevents lipid condensation and phase transition to take place. The inulin-type fructan is able to preserve the membrane barrier during dehydration. Using a vesicle leakage assay, Vereyken et al. (2003 a) fould that both levan- and inulin-type fructans protect membrane barrier and suggested that fructans in general, would be able to protect the membrane barrier fusion. Furthermore, both fructan-types inhibit vesicle fusion to a large extent as measured using a liquid-mixing assay. Using X-ray defraction, it was found that in the presence of both inulin- and leavan-type fructans the lamellar repeat distance increased considerably. From this Vereyken and coworkers concluded that fructans are present between the lipid bilayers during drying. The above data support the hypothesis that fructans can have a membrane-protecting role during dehydration, and give insight into the mechanism of protection. Vereyken et al. (2003 b) further studied the consequences of fructan-membrane-lipid interaction for the molecular organization and dynamics in the dry state. Fructan increased the mobility of acyl chains, but immobilized the lipid headgroup region. Most likely, fructans insert between the headgroups of lipids, thereby spacing the acyl chains. This results in a much lower phase transition temperature. The headgroup is immobilized by the interaction with fructan. The location of the interaction with the lipid headgroup is different for the inulin-type fructan compared to the levan-type fructan, since inulin shows interaction with the lipid phosphate group, whereas levan does not.

In order to study the effect of fructans on drought tolerance, Pilon-Smits et al. (1995) transformed tobacco, a non-fructan accumulating plant, to fructan accumulating. The fructan-producing tobacco plants performed significantly better under polyethylene-glycol-mediated drought stress than the wild-type tobacco plants. Growth rate of the transgenic plants was significantly higher (+55%), as were fresh weight (+33%) and dry weight (+59%) yields. The difference in weight was observed in all organs and was particularly pronounced in roots. They concluded that introduction of fructans in this non-fructan-producing species mediates enhanced resistance to drought stress. Further, Pilon-Smits et al. (1999) introduced *SacB* gene from *Bacillus subtilis,* to produce bacterial fructans in sugarbeet. The transgenic sugarbeets accumulated fructans to low levels (max. 0.5 percent of dry weight) in both roots and shoots. Two independent transgenic lines of fructan-producing sugarbeets showed significantly better growth under drought stress than the untransformed beets. The drought-stressed fructan-producing plants attained higher total dry weights (+ 25-35%) than the wild-type sugarbeet due to higher biomass production of leaves (+ 30-33%), storage roots (+ 16-33%) and fibrous roots (+ 37- 60%). Under well-watered conditions, no significant differences were observed between the transgenic and wild-type beets. Thus they concluded that introduction of fructan biosynthesis in transgenic plants is a promising approach to improving crop productivity under drought stress. de Roover et al. (2000) further substantiated the above findings by way of reporting 10-times higher fructan concentration in the roots and leaves of one month-old chichory (*Chichorium intybus)* seedlings subjected to drought for two weeks. The onset of fructan synthesis coincided with the increase in sucrose : sucrose 1-fructosyltransferase, a key enzyme for fructan biosynthesis in roots. Expression of *1-SST* gene could be observed in roots and leaves of stressed plants.

Further to this, Amiard et al. (2003) reported that in perennial ryegrass, fructans and not sucrosyl-galactosides, raffinose or loliose are affected by drought stress. Drought stress increased neither the concentrations of loliose and raffinose nor the activities of loliose synthase and raffinose synthase. Moreover, concentrations of raffinose precursors, myoinositol and galactinol, as well as the gene expressions of myoinositol 1-phosphate synthase and galactinol synthase were either decreased or unaffected by drought stress. On the contrary, drought stress caused fructans to accumulate in leaf tissues, mainly in leaf sheaths and elongating leaf bases. This increase is mainly due to the accumulation of long-chain fructans (degree of polymerization >8) and is not accompanied by a sucrose increase.

Wheat var C306 and WH542 (less and more water requiring, respectively) were subjected to varying degrees of water stress (-12, -16.5 and –22 bars) by withholding irrigation (Narang et al., 2000). The plants were rewatered to attain the original water potential of –7 bars. All carbohydrates except starch and fructan were found to increase with the enhancement of water stress while starch and fructan decreased. On rewatering, the reverse happened. With the progress of water stress, SST and invertase activities decreased, FET remained unaffected, while FEH activity increased sharply. The decrease in SST and increase in FEH activity during water deficit rule out *de novo* synthesis of fructans. Rather, pre-existing fructans are degraded/ depolymerized to sucrose and free hexoses, which in turn contribute to osmotic adjustment during water-deficit, indicating indirect involvement of fructan metabolism in tiding over short-term water stress conditions.

Cold stress

Santoiani et al. (1993) observed that the activity of sucrose : sucrose fructosyltransferase increased more than 25-fold when wheat cv Trigal 800 plants were cooled to 4°C. Sucrose synthase and sucrose-phosphate synthase activities also increased, but low temperatures had no significant effect on invertase or on fructan hydrolase activities. In roots, chilling stimulated the synthesis of fructans of high degree polymerization. Further, Livingston III and Henson (1998) determined the changes in carbohydrate concentrations and activities of carbohydrate-degrading enzymes in the crown tissues of oat cv Wintok during cold hardening. During the second-phase hardening (-3°C for 3 days) levels of fructan, sucrose, glucose, and fructose in the apoplast increased significantly above that in non-hardened and first-phase-hardened plants. The extent of increase in apoplastic fructan during second-phase-hardening varied with the degree of fructan polymerization (e.g. from 3 and 4 to 7 and > 7). The activities of invertase and fructan exohydrolase in the crown apoplast increased approximately 4-fold over non-hardened and first-phase-hardened plants.

In this crop at the grain-filling stage and lower day/night temperature (18/10°C) regime, fructan synthesis increases, but decreases when the plants are shifted to the higher (20/20°C day/night) temperature regime due to higher carbon requirement for grain-filling (Bancal and Triboi, 1993). The SST content of these plants was higher at the low temperature regime, but increased during the higher temperature regime. Then Puebla et al. (1997) instead of shifting the same cultivar to two temperature regimes, selected two *Bromus* species adapted to different climatic conditions to determine their responses to chilling and water stress. Their results indicate that *Bromus pictus* (adapted to cold desert area) has a constitutive fructan synthesis, while the *B. anleticus* (adapted to a warmer climate) only accumulates fructan under cold stress. Presence of fructans in *B. pictus* and their absence in *B. anleticus* supports the hypothesis that fructans play a role in plant tolerance to environmental stress.

Freezing tolerance and carbohydrate changes during cold acclimation of green-type annual bluegrass (*Poa annua* L.) ecotypes was studied by Dionne et al. (2001). The high molecular weight fructans were the most abundant carbohydrates found in plants, cold-acclimated under low non-freezing temperature conditions with low levels up to 170 mg g^{-1} dry weight as compared with 60-70 mg g^{-1} dry weight in non-acclimated plants. Two weeks of low, non-freezing temperature promoted a significant accumulation of both low molecular weight and high molecular weight fructans in the crowns of annual bluegrass ecotypes. In all ecotypes, high molecular weight fructans were the major carbohydrate stored under these conditions with levels more than twice the initial values. In all ecotypes, high molecular weight fructans and sucrose were the major carbohydrates found in the cold-hardened crowns (Fig. 9.3). Exposition to cold temperatures induced a marked accumulation of sucrose from ≅ 10 mg g^{-1} dry weight up to a maximum of 78 mg g^{-1} dry weight. High molecular weight and low molecular weight fructan showed an opposite response with a decrease during fall and winter. A strong and specific interaction between high molecular weight fractions with model membranes suggests that fructans might help prevent lipid condensation and phase transitions to take place in droughted or freeze-desiccated cells (Demel et al., 1998). The hardy winter cereal cultivars generally accumulate higher levels of high molecular weight fructans when compared to the less hardy cultivars (Suzuki and Nass, 1988).

Kawakami and Yoshida (2002) isolated two cDNAs of winter wheat designated *wft1* and *wft2,* which encoded sucrose : fructan 6-fructosyltransferase and sucrose : sucrose 1-fructosyltransferase, respectively, which are involved in the synthesis of fructan in wheat. *wft 1* and *wft 2* were cloned by screening of a cDNA library with probed-cDNA fragments corresponding to plant fructosyltransferase and invertase. The identity of the clones was verified by functional characterization of recombinant proteins expressed in methylotrophic yeast, *Pichiapastoris.*

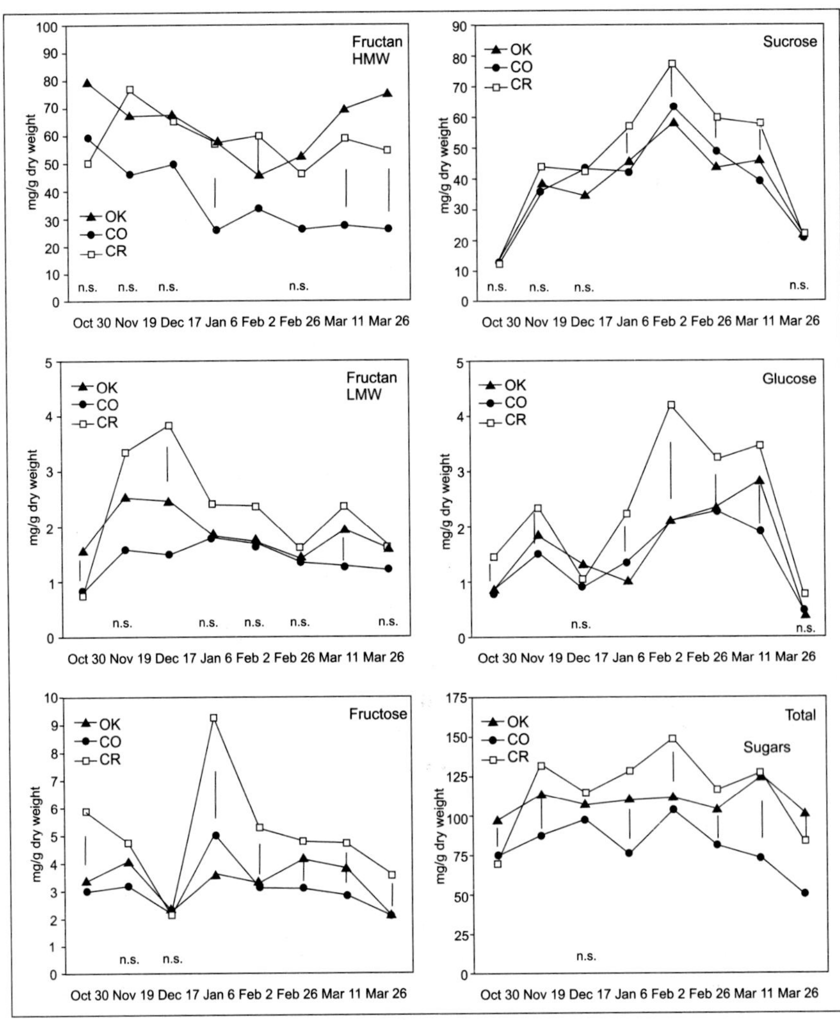

Fig. 9.3 Changes in carbohydrate composition in crowns of three annual bluegrass ecotypes (OK, CO and CR) cold-acclimated in an unheated glasshouse during winter. HMW=High molecular weight, LMW=Low molecular weight, n.s.=Not significant. (After Dionne et al., 2001).

Northern blotting showed that the level of *wft 2* transcripts increased from autumn to early winter in the crown tissues of all field-grown wheat cultivars examined. Higher levels of *wft 1* and *wft 2* transcripts were found in leaf tissues of snow mold-resistant cultivars, which accumulated more fructan than other cultivars. The results show that *Wft1* and *Wft2* are important in fructan accumulation during cold hardening of winter wheat.

Hypoxia

Albrecht et al. (1993) studied fructan accumulation in wheat cv Alcedo seedlings caused by oxygen deficiency around the root. Plants were cultivated in either nitrogen-flushed or aerated nutrient solution. In response to nitrogen treatment there was an accumulation of soluble carbohydrates in the shoots as well as in the roots. The main contribution was due to fructans. The content of fructo-oligosaccharides increased 5-fold in roots and 7-fold in shoots. This is incompatible with the assumption that higher substrate consumption follows enhanced fermentation under oxygen deficiency. Reaeration of the nitrogen-flushed nutrient solution resulted in enhanced consumption of stored carbohydrates, mainly fructans, accompanied by high growth rates of the roots. The possibility of utilizing fructans quickly, if oxygen is available, is a possible advantage for plants adapted to hypoxia. It is suggested that the low energy requirements for SST, allowing fructan synthesis even under oxygen shortage, may improve the fitness of plants to survive temporary hypoxia in the rhizosphere. Further, Albrecht et al. (2000) found that following oxygen shortage in the rooting environment a 2- to 4-fold increase in carbohydrate in all the plant species studied, both in the shoots as well as in the roots. The flooding-tolerant species of the genera *Crepis* and *Senacio* accumulated the highest amounts of carbohydrates, in particular fructans with a degree of polymerization up to 35 compared to 10 under control conditions. Nearly 50 percent of the soluble carbohydrates were fructans compared with 20 percent under aerated conditions. The ability to accumulate fructans seems to be important for plants to cope with oxygen shortage due to the very low energy demand for fructan synthesis in comparison to starch synthesis.

Further, Bertrand et al. (2003) subjected four perennial forage species, lucerne, red clover, timothy and cocksfoot to a progressively developing oxygen deficiency stress. Timothy is reportedly more resistant to oxygen deficiency than the three other species. Non-structural carbohydrates increased and remained at a higher concentration in timothy than in the other three species under low oxygen concentration. Concentrations of sucrose, fructose, glucose and fructans increased in response to oxygen deficiency in timothy, whereas the concentration of soluble sugars decreased under the same conditions in lucerne, red clover, and cocksfoot. The gene expression of glyceraldehyde-3-phosphate dehydrogenase increased in response to low oxygen concentration in oxygen deficiency-sensitive lucerne, while it remained unchanged in the oxygen deficiency-resistant timothy. It is concluded that timothy maintains higher carbohydrate reserves under oxygen deficiency, a specific feature that could favor its winter survival under spring growth.

Mineral nutrient stress

Omission of nitrogen from the medium in which barley was grown under controlled conditions induced a fast decline in relative growth rate (RGR), whereas omission of phosphorus did not affect the RGR during the 156 hours of the experiment (Wang and Tillberg, 2000). The content of fructans, however, increased in both sink and source leaves under both types of starvation. At the end of the experiment, increase in fructan level under nitrogen starvation was about 70 percent higher in the sink leaves and 45 percent higher in the source leaves compared to leaves from phosphorus starved plants. Starch levels in both sink and source leaves increased in parallel with the increase of fructan at nitrogen-starvation but not at phosphorus starvation. The results show that fructan is the major type of reserve carbohydrate accumulated in barley leaves subjected to nitrogen or phosphorus deficiency, but the accumulation rates are different in the two types of mineral deficiency. The rate and level of starch accumulation is different from that of fructan accumulation, and not as sensitive to phosphorus deficiency as the level of fructan.

Morcuende et al. (2005) investigated the inhibition of fructan accumulation by phosphate nutrition in warm-grown and cold-treated barley plants. In warm-grown leaves, phosphate decreased fructan accumulation and total carbon in carbohydrates and did not affect protein content. In cold-treated leaves, however, which before incubation had higher sucrose phosphate synthase activities than the warm-grown leaves, phosphate had no inhibitory effect on fructan accumulation, carbohydrate content or total carbon in carbohydrates. The activities of sucrose phosphate synthase and cytosolic fructose-1,6-bisphosphatase were unaffected by phosphate. The results indicate that phosphate decreases fructan accumulation through an inhibition of sucrose phosphate synthase whenever this activity is not high before a rise in phosphate content.

Salinity stress

Besides sucrose and starch, fructans are the dominant carbohydrates in wheat. Kerepesi et al. (2000) attempted to compare the changes of fructan levels during short-term drought and salt stresses in wheat seedlings. Sugar (glucose, fructose, sucrose) and fructan content and changes in degree of polymerization were determined in seedlings of 4-wheat varieties differing in drought and salt tolerance. Hydroponically-grown seedlings were exposed to consecutive water (PEG) and salinity (NaCl) stresses. Distribution of degree of polymerization (3-9 fructans) due to salt and drought effects were genotype-dependent, but changes in degree of polymerization (3, 4, and 6 fructan concentration) were more characteristic than the others. Fructan content increased in each part of plants including stem, leaf and root. The rate of accumulation in stems of the NaCl-treated plants correlated with salt tolerance. Thus the authors propose that the fructan concentration might be a good marker to predict the degree of salt tolerance of wheat seedlings.

IMPROVEMENT

Over 40,000 species of plants accumulate fructan. The tissue specific expression and targeting of the *SacB* protein to endosperm vacuoles results in stable accumulation of high molecular weight fructan in mature maize kernels. Fructan level was approximately 9-fold higher in sh_2 mutants compared to the wild-type maize kernels (Caimi et al.,1996). Sevenier et al. (1998) transformed sugarbeet into a crop that produces fructans. The gene encoding 1-sucrose : sucrose fructosyltransferase was introduced from sunflower. In the tap root of sugarbeet transformed with the *1-sst* gene, the stored sucrose gets almost totally converted into low molecular weight fructans. The transformants of tobacco, a plant naturally unable to form fructans, synthesized the trisaccharide ketose and a series of unbranched fructans of the phlein type (Sprenger et al., 1997). The normally non-fructan storing potato plant was modified by introducing the microbial fructosyltransferase genes so that it could accumulate fructans (Van der Meer et al., 1994). The regenerated potato plants accumulated high molecular mass fructan molecules in which the degree of polymerization of fructose units exceeded 25,000. The fructan content in the transgenic potato plants tested varied between 1 and 30 percent of dry weight in leaves and 1 and 7 percent of dry weight in microtubers. The transformants of chichory, a plant naturally producing only unbranched fructans of the inulin type, synthesized in addition branched fructans of the graminan type, particularly the tetrasaccharide bifurcose which is also the main fructan in barley leaves.

Van der Meer et al. (1998) isolated two full size cDNA clones encoding the two enzymes responsible for fructan biosynthesis in Jerusalem artichoke (*Helianthus tuberosus*), *1-SST* and *1-FFT*. Both *1-SST* and *1-FFT* were encoded by single copy genes. In tubers, the spatial and developmental expression correlated with the accumulation of fructans. The progeny of the crosses between a ^{35}*S-1 sst-* containing plant and ^{35}*S-1-fft-* containing plant, showed accumulation of high molecular weight fructans in old, senescent leaves. Based on the comparison of the predicted amino acid sequences of *1-sst* and *1-fft* with those of other plant fructosyltransferase genes, Van der Meer and associates postulated that both plant fructan genes have evolved from the plant invertase genes.

Vijn et al. (1997) also transformed chichory plants harboring onion FF6G fructosyltransferase (6G-FFT). The deduced amino acid sequence showed a high homology with plant invertases and 6SFT. Incubation of protein extracts from transgenic tobacco plants with the triglyceride 1-ketose and sucrose resulted in the formation of neoketose and fructans of the inulin neoseries with a degree of polymerization up to six. Introduction of the onion 6G-FFT into chichory resulted in the synthesis of fructan of the inulin neoseries, in addition to the synthesis of linear inulin. Vijn et al. (1998) further cloned 1-SST from onion by screening a cDNA library using acid invertase from tulip (*Tulipa gesneriana)* as probe. Expression assays in tobacco protoplasts showed the formation of 1-ketose from sucrose. In addition, an onion acid invertase clone was isolated from the same cDNA library. The protein extracts of tobacco protoplasts transformed with this clone showed extensive sucrose-hydrolyzing activity. Structurally different fructan molecules could be produced from sucrose by a combined incubation of protein extract of protoplasts transformed with 1-SST and protein extract of protoplasts transformed with either the onion fructan : fructan 6G-fructosyltransferase or the barley sucrose : fructan 6-fructosyltransferase.

Cairns (2003) discussed the biosynthesis of fructan in transgenic plants. Cairns assessed the plants transformed to accumulate fructan in the context of natural concentrations of reserve carbohydrates and natural fluxes of carbon in primary metabolism: transgenic fructan accumulation is universally reported as an instantaneous endpoint concentration. In experimental cases, concentrations of 60-160 mg g^{-1} fresh mass have been reported and compare favorably with naturally occurring maximal starch and fructan content in leaves and storage organs. Generally, values are less than 20 mg g^{-1} for plants transformed with bacterial genes and < 9 mg g^{-1} for plant-plant transformants. Superficially, the results indicate a marked modification of carbon partitioning. However, transgenic fructan accumulation is generally constitutive and involves accumulation over time-scales of weeks or months. When calculated as a function of accumulation period, fluxes into the transgenic product are low, in the range 0.00002- 0.03 nkat g^{-1}. By comparison with an estimated minimum daily carbohydrate flux in leaves for a natural fructan-accumulating plant in field conditions (37 nkat g^{-1}), transgenic fructan accumulation is only 0.00005- 0.08 percent of primary carbohydrate flux and does not indicate radical modification of carbon partitioning, but rather, a quantitatively minor leakage into transgenic fructan. Possible mechanisms for this low fructan accumulation in the transformants include : (i) rare codon usage in bacterial genes compared with eukaryotes, (ii) low transgene mRNA concentrations caused by low expression and/or high turnover, (iii) resultant low expression of enzyme protein, (iv) resultant low total enzyme activity, (v) inappropriate kinetic properties of the gene products with respect to substrate concentrations in the host, (vi) *in situ* product hydrolysis, and

(vii) levan toxicity. Transformants expressing bacterial fructan synthesis exhibit a number of aberrant phenotypes such as stunting, leaf bleaching, necrosis, reduced tuber number and mass, tuber cortex, discoloration, reduction in starch accumulation, and chloroplast agglutination. In severe cases of developmental aberration, potato tubers are replaced by florets. In most instances, the attempted subcellular targeting of the transgene product is not demonstrated. When localization is attempted, the transgene product is generally mis-localized, for example, to the cell perimeter or to the endomembrane system, instead of the intended target, the vacuole. Fructosyltransferases exhibit different product specificities *in planta* than *in vitro,* expression *in planta* generally favoring the formation of larger fructan oligomers and polymers. This implies a direct influence of the intracellular environment on the capacity for polymerization of fructosyltransferases and may have implications for the mechanism of natural fructan polymerization *in vivo.*

The critics of GM crops should bear in mind that the *de novo-* introduction of fructans in crop plants is not harmful, rather it is beneficial to human beings who consume these modified crops or crop products. The human digestive tract does not contain enzymes able to degrade fructans; therefore, there is a great interest from the food industry to use them as low-caloric food ingredients. Moreover, fructans have been implicated in the prevention of osteoporosis by enhanced calcium uptake, a lower blood cholestrol and triglyceride status, and prevention of colon cancer. In plants, fructans have functions other than carbon storage; they have been implicated in protecting plants against water deficit caused by drought or low temperatures. Fructans are produced in the vacuole by the action of specific enzymes. Several genes encoding fructan biosynthetic enzymes have been cloned and the activity of the proteins have been characterized. It has now been detected that very complex branched neo-series fructans are present in highly drought tolerant agave species. Probably the phase transition of lipids is responsible for most of the *in vivo* damage that is experienced by the cells of stressed plants. *In vitro* experiments have provided evidence of the ability of fructans to protect the integrity of lipid bilayer.

REFERENCES

Albrecht, G.; S. Kammerrer; W. Praznik; et al. 1993. Fructan content of wheat seedlings (*Triticum aestivum* L.) under hypoxia and following reaeration. New Phytol. 123: 471-76.

Albrecht, G.; S. Baumgartner; S. Bjemelt; et al. 2000. Accumulation of fructans in plant species with different flooding tolerance following oxygen deficiency stress. 24th Int. Fructan Conf., Basel, Switzerland, Aug., 2000, hosted by Dr. Andres Wiemken.

Amiard, V.; A. Morvan-Bertrand; J.P. Billard; et al. 2003. Fructans, but not the sucrosyl-galactosides, raffinose and loliose are affected by drought stress in perennial ryegrass. Plant Physiol. 132: 2218-29.

Bancal, P. and E. Triboi, 1993. Temperature effect on fructan oligomer contents and fructan related enzyme-activities in stems of wheat (*Triticum aestivum* L.) during grain filling. New Phytol. 123: 247-53.

Bertrand, A.; Y. Castonguay; P. Nadeau; et al. 2003. Oxygen deficiency affects carbohydrate reserves in over-wintering forage crops. J. Exp. Bot. 54 (388): 1721-30.

Caimi, P.G.; L.M. McCole; T.M. Klein; et al. 1996. Fructan accumulation and sucrose metabolism in transgenic maize endosperm expressing a *Bacillus amyloliquefaciens SacB* gene. Plant Physiol. 110: 2355-63.

Cairns, A.J. 1993. Evidence for the *de novo* synthesis of fructan by enzymes from higher plants – A reappraisal of the SST : FFT model. New Phytol. 123 : 15-24.

Cairns, A.J. 1995. Effect of enzyme concentration on oligo-fructan synthesis from sucrose. Phytochemistry 40: 705-08.

Cairns, A.J. 2003. Fructan biosynthesis in transgenic plants. J. Exp. Bot. 54 (382): 549-67.

Demel, R.A.; E. Dorrepaal; M.J. Ebskamp; et al. 1998. Fructans interact strongly with model membranes. Biochem. Biophys. Acta 1375 (1-2): 36-42.

de Roover, J.; Vandenbranden; A. van Laere; 2000. Drought induces fructan synthesis and 1-SST (sucrose : sucrose fructosyltransferase) in roots and leaves of chichory seedlings (*Chichorium intybus* L.) Planta 210: 808-14.

Dionne, J.; Y. Castonguay; P. Nadeau; et al. 2001. Freezing tolerance and carbohydrate changes during cold acclimation of green-type annual bluegrass (*Poa annua* L.) ecotypes. Crop Sci. 41: 443-51.

Edelman, J. and T.G. Jefford, 1968. The mechanism of fructan metabolism in plants as examplified in *Helianthus tuberosus.* New Phytol. 67: 517-31.

Hendry, G.A.F and R.K. Wallace, 1993. The origin, distribution, and evolutionary significance of fructans. pp 119-39. In : Science and Technology of Fructans (eds) M. Suzuki and N.J. Chatterton, CRC Press, London.

Kawakami, A. and M. Yoshida, 2002. Molecular characterization of sucrose : sucrose 1-fructosyltransferase and sucrose : fructan 6-fructosyltransferase associated with fructan accumulation in winter wheat during cold hardening. Biosci. Biotechnol. Biochem. 66 (11): 2297-305.

Kerepesi, J.; G. Galiba and L. Boross, 2000. Fructans in wheat seedlings under drought and salt stresses. 24th Int. Fructan Conf. – Basel, Switzerland, Aug., 2000, hosted by Dr. Andres Wiemken.

Koops, A.J. and H.H. Jonker, 1996. Purification and characterization of the enzymes of fructan biosynthesis in tubers of *Helianthus tuberosus* Colombia (II. Purification of sucrose : sucrose 1-fructosyltransferase and reconstitution of fructan synthesis *in vitro* with purified sucrose : sucrose 1-fructosyltransferase). Plant Physiol. 110 : 1167-75.

Livingston III, D.P. and C.A. Henson, 1998. Apoplastic sugars, fructans, fructan exohydrolases, and invertase in winter oat : Responses to second-phase cold hardening. Plant Physiol. 116: 403-08.

Morcuende, R.; S. Kostadinova; P. Perez; 2005. Fructan synthesis is inhibited by phosphate in warm-grown, but not in cold-treated, excised barley leaves. New Phytol. , Aug. 2005; doi : 10.1111/j.1469-8137.2005.01534.x.

Narang, R.; S. Malhotra and R. Singh, 2000. Fructan metabolism in wheat (*Triticum aestivum* L.) leaves as affected by water deficit. 24th Int. Fructan Conf. – Basel, Switzerland, Aug., 2000, hosted by Dr. Andres Wiemken.

Pilon-Smits, E.A.M.; M.J.M. Ebskamp; M.J. Paul; et al. 1995. Improved performance of transgenic fructan-accumulating tobacco under drought stress. Plant Physiol. 107: 125-30.

Pilon-Smits, E.A.M.; N. Terry; T. Sears; 1999. Enhanced drought resistance in fructan-producing sugarbeet. Plant Physiol. Biochem. 37: 313-17.

Pubela, A.E.; G.L. Salerno and H.G. Pontis, 1997. Fructan metabolism in two species of *Bromus* subjected to chilling and water stress. New Phytol. 136: 123-29.

Ritsema, T. and S.C. Smeekens, 2003. Engineering fructan metabolism in plants. J. Plant Physiol. 160 (7): 811-20.

Santoiani, C.S.; J.A. Tognetti; H.G. Pontis; et al. 1993. Sucrose and fructan metabolism in wheat roots at chilling temperatures. Physiol. Plant. 87: 84-88.

Sevenier, R.; R.D. Hall; I.M. van der Meer; et al. 1998. High level of fructan accumulation in a transgenic sugarbeet. Nat. Biotechnol. 16: 843-46.

Sprenger, N.; K. Bortlik; A. Brandt; et al. 1995. Purification, cloning, and functional expression of sucrose : fructose 6-fructosyltransferase, a key enzyme of fructan synthesis in barley. Proc. Nat. Acad. Sci. 92: 11652-56.

Sprenger, N.; L. Schellenbaum; K. van Dun; et al. 1997. Fructan synthesis in transgenic tobacco and chicory plants expressing barley sucrose ; fructose 6-fructosyltransferase. FEBS Lett. 400 (3): 355-58.

Suzuki, M. and H.G. Nass, 1988. Fructan in winter wheat, triticale, and fall rye cultivars of varying cold hardiness. Canad. J. Bot. 66: 1723-28.

Van der Ende, W.; S. Clerens; R. Vergauwen; et al. 2003. Fructan 1-exohydrolase. β-(2,1)-trimmers during germinan biosynthesis in stems of wheat ? Purification, characterization, mass mapping, and cloning of two fructan 1-exohydrolase isoforms. Plant Physiol. 131: 621-31.

Van der Ende, W.; B. de Coninck and A. van Laere, 2004. Plant fructan exohydrolases : a role in signalling and defence ? Trends in Plant Sci. 9 (11): 523-28.

Van der Meer, I.M.; M. Ebskamp; R. Visser; et al. 1994. Fructan as a new carbohydrate sink in transgenic potato. Plant Cell 6: 561-70.

Van der Meer, I.M.; A.J. Koops; J.C. Hakkert; et al. 1998. Cloning of the fructan biosynthesis pathway of Jerusalem artichoke. Plant J. 15: 489-500.

Vereyken, I.J.; V.. Chupin, R.A. Demel; 2001. Fructans insert between the headgroups of phospholipids. Biochim. Biophys. Acta 1510: 307-20.

Vereyken, I.J.; J. Albert van Kuik; T.H. Evers; et al. 2003 a. Structural requirements of the fructan-lipid interaction. Biophysical J. 84: 3147-54.

Vereyken, I.J.; V. Chupin; F.A. Hoekstra; et al. 2003 b. The effect of fructan on membrane-lipid organization and dynamics in the dry state. Biophysical J. 84: 3759-66.

Vijn, I.; A. van Dijken; M. Lüscher; et al. 1998. Cloning of sucrose : sucrose 1-fructosyltransferase from onion and synthesis of structurally defined fructan molecules from sucrose. Plant Physiol. 117: 1507-13.

Vijn, I.; A. van Dijken; N. Sprenger; et al. 1997. Fructan of the inulin neoseries is synthesized in transgenic chichory plants (*Chichorium intybus* L.) harbouring onion (*Allium sepa* L.) fructan : fructan 6G-fructosyltransferase. Plant J. 11: 387-98.

Vijn, I. and S.C. Smeekens, 1999. Fructan: more than a reserve carbohydrate? Plant Physiol. 120: 351-59.

Wang, C. and J.E. Tillberg, 2000. Fructans in leaves of barley plants starved for phosphorus and nitrogen. 24th Int. Fructan Conf. –Basel, Switzerland, Aug., 2000, hosted by Dr. Andres Wiemken.

Wiemken, A.; N. Sprenger and T. Boller, 1995. Fructan : an extension of sucrose by sucrose. pp 178-89 In: Sucrose Metabolism, Biochemistry, Physiology and Molecular Biology (eds) H.G. Pontis; G.L. Salerno and E.J. Echeverrria; Am. Soc. Plant Physiologists, Rockville, MD.

SUGGESTED READING

Cairns, A.J. 2003. Fructan biosynthesis in transgenic plants. J. Exp. Bot. 54 (382): 549-67.

Pollock, C.J. and A.J. Cairns, 1991. Fructan metabolism in grasses and cereals. Annu. Rev. Plant Physiol. Plant Mol. Biol. 42: 77-101.

Vijn, I. and S. Smeekens, 1999. Fructan : more than a reserve carbohydrate. Plant Physiol. 120: 351-60.

10

ECTOINE

Introduction

The stress-stimulated osmolytes/compatible solutes (inorganic nutrients, sugars, oligosaccharides and proline, etc) result in cell osmotic adjustment at relatively much higher concentrations than the other stress-produced chemical mediators/protectants like trehalose, fructans, citrulline and ectoine, etc. at much lower concentrations to be osmotically effective. These protectants offer protection to cell membranes and the macromolecules. Thus the multi-gene transformed transgenics with osmoregulants and other protectants with the membrane/nucleotide/protein/enzyme-protection ability, are likely to have greater stress tolerance ability. Ectoine (1,4,5,6-tetrahydro-2-methyl-4-pyrimidinecarboxylic acid or 3,4,5,6-tetrahydro-2-methyl-4-pyrimidinecarboxylic acid) is a cyclic amino acid, identified as a compatible solute in moderately halophilic bacteria *Halomonas elongata.** Extrogenously provided ectoine stimulates growth of *Escherichia coli* in media of inhibitory osmotic strength (Jebbar et al., 1992). The stimulation is independent of any specific solute, electrolyte, or nonelectrolyte. Ectoine accumulates in *E. coli* cells proportionately to the osmotic strength of the medium, and it is not metabolized. Its protective ability is as potent as that of glycinebetaine (Jebbar et al., 1992). The intracellular ectoine pool is regulated by both influx and efflux systems. Talibart et al. (1994) further investigated the involvement of ectoine in the osmotic adaptation of *Rhizobium meliloti.* Ectoine appears almost as effective as glycinebetaine in improving growth of *R. meliloti* under salinity stress (0.5 M NaCl). Ectoine also stimulates the growth of rhizobial strains non responsive to glycinebetaine. Ectoine transport proved inducible, periplasmic protein dependent, and distinct from the transport of glycinebetaine. Ectoine, unlike glycinebetaine, is not intracellularly accumulated and, as a consequence, does not repress the synthesis of

---

* This highly protective chemical mediator present in bacteria, if engineered in important crop plants, will have tremendous stress-protective/economic implications. In fact some results have now started coming in and therefore have been included in this volume.

endogenous compatible solutes (glutamate, N-acetylglutaminylglutamine amide, and trehalose).

Moderate halophilic *Halomonas elongata* DSM3043 accumulates ectoine, hydroxyectoine*, glutamate and glutamine in response to salt stress (3 M NaCl). The *H. elogata* mutant CHR 62 could not grow above 0.75 M NaCl, and CHR 63 did not grow above 1.5 M NaCl as these mutants did not synthesize ectoine, but accumulated ectoine precursors (Canovas et al., 1997). The mutant CHR 62 accumulated low levels of diaminobutyric acid, and mutant CHR 63 accumulated high concentrations of N-δ-acetyldiaminobutyric acid. These results suggest that the strain CHR 62 could be defective in the gene for diaminobutyric acid acetyltransferase (*ectB*), and the strain CHR 63 could be defective in the gene for ectoine synthase (*ectC*). Salt sensitivity of the mutants at 1.5 to 2.5 M NaCl could be partially corrected by cytoplasmic extracts of the wild-type strain containing ectoine, and the salt sensitivity of the strain CHR 62 could be partially repaired by the addition of extracts of the strain CHR 63 which contains N-δ-acetyldiaminobutyric acid. Finally, a cosmid from *H. elongata* genomic library was isolated which complemented the *Ect*-phenotype of both mutants, indicating that it carried at least the genes *ectB* and *ectC* of the biosynthetic pathway of ectoine (Canovas et al., 1997). This cosmid was sequenced and three open reading frames were found (Canovas et al., 1998) corresponding to the genes *ectA* (encoding diaminobutyric acid acetyltransferase), *ectB* (encoding diaminobutyric acid aminotransferase) and *ectC* (encoding ectoine synthesis). These three genes were able to restore the salt tolerance of two *H. elongata* mutants defective in the synthesis of ectoine (strains CHR 62 and CHR 53). However, the *H. elongata* ectoine synthesis genes did not confer to *E. coli* the ability to synthesize ectoine.

Louis and Galinski (1997) cloned genes of the biosynthetic pathway of ectoine from the δ-positive moderate halophilic *Marinococcus halophylus* by functional expression in *E. coli.* These genes were not only expressed, but also osmoregulated in *E. coli,* as demonstrated by increasing cytoplasmic ectoine concentration in response to medium salinity. Sequencing of a 4.4 kb fragment revealed four genes designated *ectA, ectB, ectC* and *orfA.* The significance of three of these genes for ectoine synthesis was proved by sequence comparison with known proteins and by physiological experiments. Several deletion derivatives of the sequenced fragments were introduced into *E. coli* and the resulting clones were investigated for their ability to synthesize ectoine or one of the intermediates in its biosynthetic pathway. A DNA region upstream of *ectA* was shown to be necessary for

* Hydroxyectoine has considerable potential for the improvement of desiccation tolerance in sensitive microorganisms, particularly for those recalcitrant to trehalose (Manzanera et al., 2002).

the regulated expression of ectoine synthesis in response to the osmolarity of the medium.

BIOSYNTHESIS

Ectoine occurs in *Halomonas elongata* and other halophilic bacteria, which synthesize it from aspartate semialdehyde in three steps (Fig. 10.1, Canovas et al., 1998). The three *H. elongata* genes (*ectA*, *ectB*, and *ectC*) encoding the enzymes of ectoine synthesis were each placed under the control of the constitutive CaMV ^{35}S promoter in a single construct, which was introduced into cultured tobacco cells (Nakayama et al., 2000}. Transformed cell lines synthesized low levels of ectoine (≤ 79 nmol g^{-1} fresh weight) and showed a small increase in resistance to osmotic stress imposed with mannitol (Nakayama et al., 2000). This work is exceptional in having expressed three genes at once using a single construct, which is difficult in plants, particularly when the same promoter is used for all the genes because this predisposes the gene silencing.

The nucleotide sequence of a DNA coding for ectoine synthase has been determined by isolating a gene coding for an enzyme involved in the biosynthesis of ectoine in the form of a DNA comprising about 4.2 kilobase pairs which is obtained from *Halomonas* species KS-3 by cleaving with restriction endonucleases *EcoRI* and *SalI*. By introducing the obtained gene DNA into *E. coli*, it is possible to provide the capability of biosynthesizing ectoine and the characteristic of high osmotic tolerance, thus permitting the development of an efficient fermentation technique utilizing the resultant transformant and the creation of plants having high resistance to drought.

In *Halomonas elongata* 2,4-diaminobutyrate (DABA) aminotransferase catalyzes reversably the first step of the pathway, conversion of aspartate β-semialdehyde (ASA) to DABA by transamination with L-glutamate (Ono et al., 1999). This enzyme requires pyridoxal 5′-phosphate and potassium ions for its activity and stability and exhibits an optimum pH of 8.6 and an optimum temperature of 25°C, and has Km of 9.1 mM for L-glutamate and 4.5 mM for DL-ASA. DABA acetyltransferase catalyzes acetylation of DABA to δ-N-acetyl-α, δ-diaminobutyric acid (ADABA) with acetyl coenzyme A and exhibits an optimum pH of 8.2 and an optimum temperature of 20°C in presence of 0.4 M NaCl. The molecular mass is 45 kDa by gel filtration. Ectoine synthase catalyzes circulation of ADABA to ectoine and exhibits an optimum pH of 8.5 to 9.0 and an optimum temperature of 15°C in presence of 0.5 M NaCl. This enzyme has an apparent molecular mass of 19 kDa and a Km of 8.4 mM in presence of 0.77 M NaCl. DABA acetyltransferase and ectoine synthase are stabilized in presence of NaCl (>2 M) and DABA (100 mM) at temperatures below 30°C. Further, Kuhlmann and Bremer (2002) investigated the types of compatible

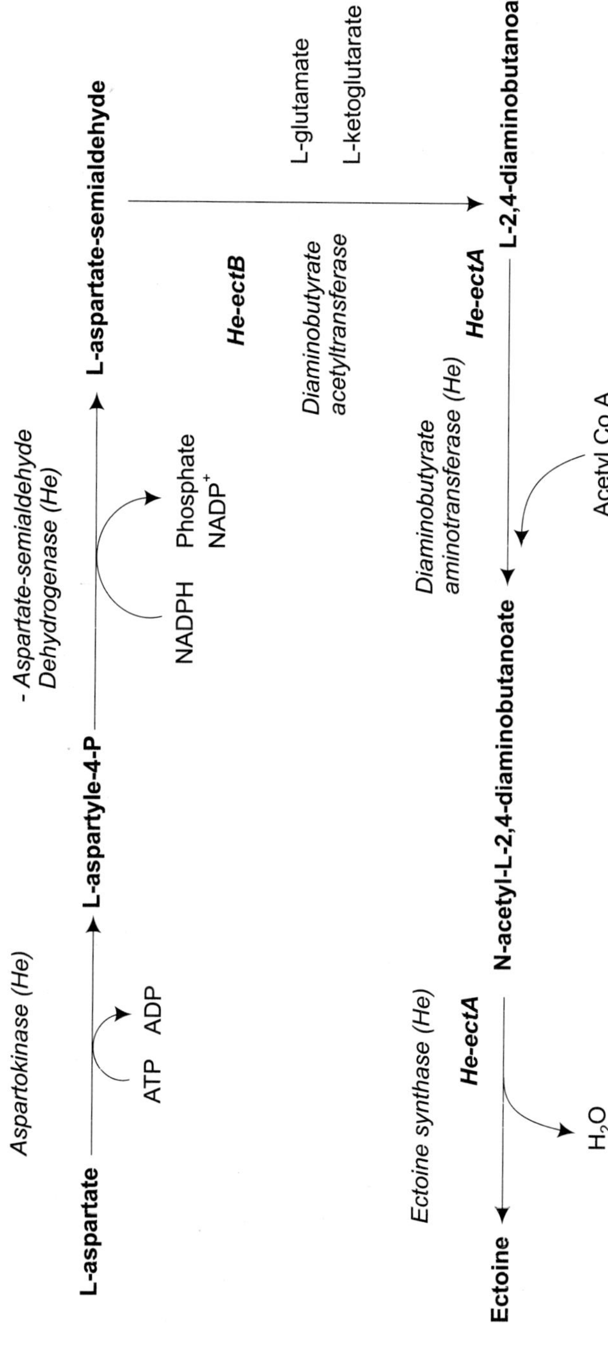

Fig. 10.1 Biosynthesis of ectoine (1,4,5,6-tetrahydro-2-methyl-4-pyrimidine carboxylic acid) (After Canovas et al., 1998).

solutes that are synthesized *de novo* in a variety of *Bacillus* species under high osmolality-growth conditions. Out of the 13 *Bacillus* species studied, *B. alcalophilus, B. psychrophilus,* and *B. pasteurii* synthesized ectoine. One other species synthesized ectoine plus hydroxyectoine, and one another species synthesized both ectoine and proline. Kuhlmann and Bremer cloned and sequenced the ectoine biosynthetic gene (*ectABC*) in *B. pasteurii.* The *ectABC* genes encode the diaminobutyric acid acetyltransferase (*EctA),* and the diaminobutyric acid aminotransferase (*EctC*). Together these proteins constitute the ectoine biosynthetic pathway, and their heterologus expression in *B. subtilis* led to the production of ectoine. The *ectABC* genes are genetically organized as an operon whose expression is strongly enhanced when the osmolality of the growth medium is raised. They pinpointed the osmoregulated promoter of the *B. pasteurii ectABC* gene cluster. The osmotically challenged *B. pasteurii* cells revealed that ectoine production within this bacterium is finely tuned and closely correlated with the osmolality of the growth medium. These observations together with the osmotic control of *ectABC* transcription suggest that the *de novo* synthesis of ectoine is an important facet in the cellular adaptation of *B. pasteurii* to high osmolarity surroundings.

TRANSPORT

Tetsch and Kunte (2002) reported that *TeaABC* is a novel osmoregulated secondary transport system for compatible solutes found in the halophilic proteobacterium *Halomonas elongata* DSM 2581 (T). Sequence comparisons confirmed that *TeaABC* consists of two putetive transmembrane proteins (*TEAb, TeaC*) and a putetive periplasmic substrate-binding protein (*TeaA*). *TeaABC* is the only osmoregulated transporter for ectoine found in *H. elongata* . In *E. coli,* Tetsch and Kunte demonstrated that *TeaA* is processed and transported to the periplasm. Furthermore, it was proven that *TeaA* is indeed a periplasmic ectoine-binding protein. The functionality of the purified protein as an ectoine-binding protein was tested employing a modified binding assay.

Ectoine-induced osmoprotection in *Sinorhizobium meliloti* grown in medium supplemented with ectoine revealed the induction of 10 proteins (Jebbar et al., 2005). Out of these, four genes encode the characteristic components of an ATP-binding cassette transporter named *ehu,* for ectoine/hydroxyectoine uptake. This transporter is coded by four genes (*ehuA, ehuB, ehuC,* and *ehuD*) that form an operon with another gene cluster that contains five genes, named *eutABCDE* for ectoine utilization. *Sinorhizobium meliloti* possesses at least one additional ectoine catabolic pathway as well as a lower affinity transport system for ectoine and hydroxyectoine.

Addition of ectoine or hydroxyectoine in the growth medium exhibited a remarkable growth stimulation of *Bravibacterium* strain JCM 6894, regardless of the range of osmotic shifts (Nagata and Wang, 2001). Growth rates of this strain were dependent on the amount of ectoine taken up. The cells subjected to 1 to >2 M NaCl and 2 to >2 M NaCl transfers took up less ectoine and this resulted in lower growth rates than those of cells with the largest osmotic shift (0 to >2 M NaCl). The *Brevibacterium* epidermis DSM 20659 grows at relatively high salinity, tolerating up to 2 M NaCl (Onraedt et al., 2004). The bacterium synthesizes ectoine and the intracellular content increases with the medium salinity, with a maximum of 0.14 g^{-1} ectoine g^{-1} cell dry weight at 1 M NaCl. Ectoine synthesis is induced by salinity as the stress caused by sugar does not induce the synthesis of ectoine.

ENGINEERING FOR ECTOINE PRODUCTION AND IMPROVEMENT

To investigate the functions of ectoine as a compatible solute in plant cells, the three genes (*ectA, ectB, ectC*) were individually placed under the control of cauliflower mosaic virus ^{35}S promoter and induced together into cultured tobacco cv Bright Yellow 2 (BY2) cells (Nakayama et al., 2000). The transgenic BY2 cells accumulated a small quantity of ectoine (14 – 79 nmol g^{-1} fresh weight) and showed increased tolerance to hyper osmotic shock (900 mOsm). Furthermore, the transgenic BY2 cells exhibited a normal growth pattern even under hyperosmotic conditions (up to 350 mOsm), in which the growth of the untransformed BY2 (wild-type) cells was obviously delayed. Nakayama and associates cloned a 4.1 kb DNA fragment involving the *ectC* gene encoding L-ectoine synthase (ES), which catalyzes the final reaction step of ectoine biosynthesis in *H. elongata* OUT 30018. The 4.1 kb DNA fragment was introduced into *E. coli* and the resulting clones exhibited accumulation of ectoine and increased salt tolerance (Min-Yu et al., 1993). The *ectA* and *ectB* genes, encoding L-2,4-diaminobutyric acid acetyltransferase (DAA) and L-2,4-diaminobutyric acid transaminase (DAT), respectively, were also found in the 4.1 kb DNA fragment. Although the *ect* genes are considered to be powerful tools for the molecular breeding of salt-tolerant plants, there is still no evidence either for ectoine biosynthesis or for the functional role of ectoine in plants. Nakayama et al. report that ectoine confers increased hyperosmotic tolerance in transgenic BY2 cells, and that the extent of hyperosmotic tolerance is correlated with the level of ectoine accumulation, despite a low level of accumulation of the solute. Like trehalose and fructan, ectoine also confers protection to cell membranes in transgenic plants at a very low level, rather than conferring an osmolyte effect. Ectoine synthesis allows maintenance of the normal permeability of cell membranes even upon exposure to hyperosmotic shock. Further, Nakayama and associates suggest the multigene introduction system, in which the accumulation of osmolytes such as proline and mannitol for intracellular osmotic adjustment and ectoine production to protect the cell membrane from initial damage are simultaneously achieved, will be a powerful strategy of metabolic engineering to grow useful salt- and drought-tolerant crops.

Later, Moghaieb et al. (2004) introduced ectoine biosynthetic genes (*ectABC*) from *H. elongata* to tobacco plants using the *Agrobacterium* mediated gene delivery system. Accumulation of ectoine in the transgenic plants under the salt stress condition was higher in roots than in leaves. Moghaieb and coworkers suggest that the accumulation of ectoine and Na^+ ion in transgenic plants improves tolerance to salt stress by maintenance of the cell osmotic potential at the level that allows water uptake under salt stress. Accumulation of ectoine increased with increasing salt concentrations in the media up to approximately 50 µmol g^{-1} fresh weight (Table 10.1). The ectoine concentrations were considerably higher in roots than in leaves; a reverse trend, however, was true for the mRNA amount. These results suggest that although the ectoine gene is expressed at higher levels in leaves than in roots, the ectoine synthesized in leaves may be translocated to roots; consequently, ectoine concentrations increase more considerably in the roots. In this study, the difference in ψ_s and osmotic adjustment appears to be related to the accumulation of ectoine and sodium (Table 10.2). These data indicate that the ψ_p values of the ectoine transgenic lines increased with increasing salt concentrations. The increase in ψ_p values may be responsible for the promotion of transgenic plant growth under salt stress conditions. Ectoine is synthesized in plant leaves and is translocated into roots, thus leading to maintenance of the osmotic potential of the cell at a level that allows water uptake.

Halomonas elongata strain KS 3 grows in the presence of 0.3 to 21 percent NaCl and at temperatures of 5 to 45°C. The optimum concentration of NaCl and temperature for growth are 3 to 5 percent and 37°C, respectively. Under hyperosmotic stress in a glucose-mineral medium, accumulation of several low molecular weight nitrogenous organic compounds is observed (Ono et al., 1998). These compounds are ectoine, hydroxyectoine and L-alanine. Ectoine production was induced immediately by the addition of NaCl and the amount accumulated increased with an increasing external concentration of NaCl and reached approximately 120 mµg mg^{-1} of dry cells at a concentration of 2.56 M NaCl. Hydroxyectoine was detected with a lag in time after the addition of NaCl at a concentration greater than 1.71 M, and the amount accumulated increased and reached 45 mµg mg^{-1} of dry cells at a concentration of 2.56 M NaCl. Thus, hyperosmotic stress induced by salt appears to cause the accumulation of ectoine and hydroxyectoine as compatible solutes in *Halomonas elongata* KS 3.

Jebbar et al. (1997) report that the cyclic amino acid ectoine functions as an osmoprotectant for the soil bacterium *Bacillus subtilis*. Ectoine specifically

Table 10.1 Effect of salinity on Na^+, K^+ (mM g^{-1} fr wt) and ectoine (µmol g^{-1} fr wt) concentrations in ectoine transgenic tobacco and wild-type control plants (After Moghaieb et el. 2004).

Genotpe	NaCl (mM)	Na^+		K^+		Ectoine	
		Leaf	Root	Leaf	Root	Leaf	Root
Control	0	0.00	0.00	0.01	0.02	0.0	0.0
	100	0.14	0.15	0.02	0.00	0.0	0.0
	200	0.22	0.21	0.02	0.01	0.0	0.0
Line 1	0	0.00	0.00	0.02	0.02	8.5	9.4
	100	0.19	0.19	0.02	0.02	31.2	45.9
	200	0.25	0.22	0.02	0.01	33.4	50.1
Line 2	0	0.00	0.00	0.03	0.02	8.1	9.2
	100	0.13	0.13	0.03	0.02	32.8	41.3
	200	0.28	0.30	0.03	0.02	49.9	55.5
Line 3	0	0.00	0.00	0.02	0.02	8.8	9.5
	100	0.21	0.21	0.02	0.02	24.4	38.4
	200	0.26	0.29	0.03	0.03	26.7	52.3

Table 10.2 Effect of salinity on water relations of leaves from ectoine transgenic tobacco plants and the wild-type control plants. (After Moghaieb et al., 2004).

Genotypes	NaCl (mM)	ψ_w (MPa)	ψ_s (MPa)	ψ_p (MPa)	Osmotic adjustment
Control	0	-0.27±0.03	-1.1± 0.02	0.83± 0.05	
	100	-0.25± 0.05	-1.29± 0.07	1.04± 0.04	0.19
	200	-0.5 0±.02	-1.35± 0.05	0.85± 0.07	0.25
Line 1	0	-0.25± 0.03	-0.93± 0.06	0.68± 0.01	
	100	-0.32± 0.01	-1.15± 0.07	0.83± 0.05	0.22
	200	-0.35± 0.04	-1.39± 0.03	1.04± 0.02	0.46
Line 2	0	-0.28± 0.01	-1.00± 0.04	0.72± 0.07	
	100	-0.40± 0.04	-1.18± 0.02	0.78± 0.06	0.18
	200	-0.42± 0.02	-1.61± 0.08	1.19± 0.04	0.61
Line 3	0	-0.24± 0.01	-0.99± 0.08	0.75± 0.02	
	100	-0.40± 0.03	-1.30± 0.03	0.90± 0.05	0.31
	200	-0.40± 0.06	-1.43± 0.06	1.03± 0.09	0.44

accumulates via the ABC-transport system *OpuC*. Competition experiments employing unlabeled ectoine and radiolabeled glycinebetaine showed that the *OpuC* transport system has a low affinity for ectoine with a K_i value of approximately 1.5 mM. Ectoine was identified by ^1H NMR spectroscopy in the solute pool of cells grown in the presence of ectoine. Ectoine could not be used by *B. subtilis* as sole carbon or nitrogen source. The results thus characterize ectoine as a metabolically inert stress compound for *B. subtilis* and established a crucial role for the ABC-transport system *OpuC* for the acquisition of the osmoprotectant ectoine from the environment.

The HPLC, MS, and NMR studies indicate variation in amounts of ectoine in the transgenic tobacco lines ranging from 0.25 to 25 nmol g^{-1} fresh weight (Mamta Rai, 2001). Ninety-eight to one hundred percent seeds of T1 generation of both transgenic and SR1 progenators germinated at NaCl concentrations ranging from 100 to 200 mM in the growth medium. Growth of transgenic seedlings was enhanced at all additional NaCl concentrations as compared to SR1 progenator lines. Forty to forty-five percent germination at 250, and 25-30 percent germination at 300 mM NaCl in the growth medium, was seen in transgenic lines as compared to 2 percent germination at 250 mM, and absence of germination at 300 mM NaCl of SR1 controls. Leaves of the seedlings growing at these NaCl concentrations were bleached and deformed. Transfer of these seedlings to media having 100 and 200 mM NaCl, 20 weeks after germination restored growth in 90 percent of the transgenic seedlings. Cation and amino acid analyses indicated higher amounts of Na^+ , K^+ , Ca^{2+} , proline, glutamine, aspergine, serine and threonine in the transgenic seedlings in concert with increasing NaCl concentrations as compared to controls.

REFERENCES

Canovas, D.; C. Vergas; F. Iglesias-Guerra; et al. 1997. Isolation and characterization of salt sensitive mutants of the moderate halophilic *Halomonas elongata* and cloning of the ectoine synthesis genes. J. Biol. Chem. 272 (41): 25794-801.

Canovas, D.; C. Vargas; M.J. Calderon; et al. 1998. Characterization of the genes for the biosynthesis of the compatible solute ectoine in the moderately halophilic bacterium *Halomonas elongata* DSM3043. Syst. Appl. Microbiol. 21: 487-97.

Jebbar, M; R. Talibart; K. Gloux; et al. 1992. Osmoprotection of *Escherichia coli* by ectoine : uptake and accumulation characteristics. J. Bacteriol. 174 (15): 5027-35.

Jebbar, M.; C. von Blohn and E. Bremer, 1997. Ectoine functions as an osmoprotectant in *Bacillus subtilis* and is accumulated via the ABC-transport system *OpuC*. FEMS Microbiol. Lett. 154 (2): 325-30

Jebbar, M.; L. Sohn-Bösser; E. Bremer; et al. 2005. Ectoine-induced proteins in *Sinorhizobium meliloti* include an ectoine ABC-type transporter involved in osmoprotection and ectoine catabolism. J. Bacteriol. 187: 1293-1304.

Kuhlmann, A.U. and E. Bremer, 2002. Osmotically regulated synthesis of the compatible solute ectoine in *Bacillus pasteurii* and related *Bacillus* spp. Appl. Environ. Microbiol. 68: 772-83.

Louis, P. and E.A. Galinski, 1997. Characterization of genes for the biosynthesis of the compatible solute ectoine from *Marinococcus halophilus* and osmoregulated expression in *Escherichia coli.* Microbiology 143 (Pt 4): 1141-49.

Mamta Rai, 2001. Biosynthesis of 1,4,5,6-tetrahydro-2-methyl-4-pyrimidinecarboxylic acid (ectoine) from *Marinococcus halophilus* into tobacco.DOI: 10.1300/J153v02n01-02.

Min-Yu, L.; H. Ono; M. Takano, 1993. Gene cloning of ectoine synthase from *Halomonas* species. Annu. Rep. Int. Cent. Coop. Res. Biotechnol., Japan 16: 193-200.

Moghaieb, R.E.A.; N. Tanaka; H. Saneoka; et al. 2004. Expression of ectoine biosynthetic genes in tobacco plants (*Nicotiana tabacum*) leads to the maintenance of osmotic potential under salt stress. 4th Int. Crop Sci. Congr., 2004.

Manzanera, M.; A. Garcia de Castro; A. Tondervik; et al. 2002. Hydroxyectoine is superior to trehalose for anhydrobiotic engineering of *Pseudomonas putida* KT2440. Appl. Environ. Microbiol. 68 (9): 4328-33.

Nagata, S. and Y.B. Wang, 2001. Interrelation between synthesis and uptake of ectoine for the growth of the halotolerant *Brevibacterium* species JCM 6894 at high osmolarity. Microbios. 104 (407): 7-15.

Nakayama, H.; K. Yoshida; H. Ono; et al. 2000. Ectoine, the compatible solute of *Halomonas elongata,* confers hyperosmotic tolerance in cultured tobacco cells. Plant Physiol. 122: 1239-47.

Ono, H.; M. Okuda; S. Tongpim; et al. 1998. Accumulation of compatible solutes, ectoine and hydroxyectoine, in a moderate halophyte, *Halomonas elongata* KS3 isolated from dry salty land in Thailand. J. Ferment Bioeng. 85: 362-68.

Ono, H.; K. Sawada; N. Khunajakr; et al. 1999. Characterization of biosynthetic enzymes for ectoine as a compatible solute in a moderately halophilic eubacterium, *Halomonas elongata.* J. Bacteriol. 181: 91-99.

Onraedt, A.; C. de Muynck; B. Walcarius; et al. 2004. Ectoine accumulation in *Brevibacterium* epidermis. Biotechnol. Lett. 26 (19): 1481-85.

Talibart, R.; M. Jebbar; G. Gouesbent; et al. 1994. Osmoadaptation in rhizobia : ectoine induced salt tolerance. J. Bacteriol. 176 (17): 5210-17.

Tetsch, L. and H.J. Kunte, 2002. The substrate-binding protein *TeaT* of the osmoregulated ectoine transporter *TeaABC* from *Halomonas elongata* : purification and characterization of recombinant *TeaT.* FEMS Microbiol. Lett. 211: 213-18.

SUGGESTED READING

Moghaieb, R.E.A.; N. Tanaka; H. Saneoka and K. Fujita, 2004. Expression of ectoine biosynthetic genes in tobacco plants (*Nicotiana tabacum)* leads to the maintenance of osmotic potential under salt stress. 4th Int. Crop Sci. Congr., 2004.

Nakayama, H.; K. Yoshida; H. Ono; Y. Murooka and A. Shinmyo, 2000. Ectoine, the compatible solute of *Halomonas elongata,* confers hyperosmotic tolerance in cultured tobacco cells. Plant Physiol. 122: 1239-47.

11

CITRULLINE

Introduction

In Botswana, Kawasaki et al. (2000) observed that the wild watermelon plants have the ability to survive under severe drought conditions by maintaining their water status (water content and water potential). Under moderate drought stomata normally close and thus the CO_2 inflow in the leaves decreases. In this way more electrons from the solar radiation are directed for the formation of reactive oxygen species (ROS). As the rate of transpiration decreases, so does the amount of heat that can be dissipated safely. Plants suffer from multiconstraints, including injury of cell components by active oxygen and increasing temperature, under these conditions. The osmolytes synthesized *in planta* protect cells and cell components from damage; high concentrations of osmolytes may also function to lower the water potential, to retain cell water and to take up more water from the soil. In the quest of finding the precise mechanism, Akashi et al. (2001) conducted two-dimensional electrophoresis of leaf proteins, and observed that seven spots were newly induced after plants were droughted. One chemical with the molecular mass of 40 kDa (citrulloine $–C_6H_{13}N_3O_3$) accumulated abundantly. As the drought stress progressed, citrulline became one of the major components in the total free amino acids. Eight days after withholding watering, although the lower leaves wilted significantly, the upper leaves still maintained their water status and the content of citrulline reached about 50 percent of the total free amino acids accumulated.

The Botswana-wild watermelon primarily accumulated citrulline, and then glutamate and arginine. The amount of citrulline accumulated reached 24 μmol g^{-1} fresh weight, or a concentration of 0.6 M, which constitutes 50 percent of the total volume of developed mesophyll cells. The concentrations of glutamate and arginine were each one-quarter of that of citrulline.

BIOSYNTHESIS

Carbamoyphosphate, a metabolic precursor of both the pyrimidine and arginine pathway, is a thermolabile intermediate. Carbamoyphosphate is used by ornithine carbamoytransferase, to carbamoylate ornithine, forming citrulline which is converted into arginine in two steps (Fig. 11.1 – Yokota et al., 2002). The side chains of arginine residues in a polypeptide are particularly sensitive to oxidation by hydroxy radicals, implying the citrulline, which is structurally analogous to arginine, may also exhibit high reactivity toward hydroxy radicals. The excellent performance of the wild watermelon plants against drought/oxidative stress, together with a proposed hypothesis that compatible solutes act as radical scavengers, prompted Akashi et al. (2001) to evaluate the role of citrulline as a hydroxy radical scavenger. Earlier, Kawasahi et al. (2000) had observed that drought stress induces an *ArgE*-related polypeptide and causes massive accumulation of the free amino acid citrulline in the leaves of highly drought tolerant wild watermelon plants (Fig. 11.2), also supported his contention. Thus, Akashi et al. (2001) got equally prompted to determine the physiological function of citrulline which accumulates in large amounts under drought conditions.

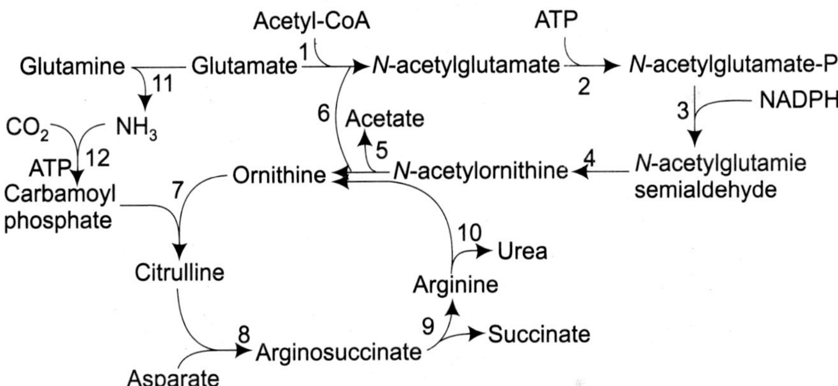

Fig. 11.1 Pathway for citrulline and arginine synthesis from glutamate. The names of enzymes functioning in the metabolic steps from 1 to 12 are as follows : 1, N-acetylglutamate synthase; 2, N-acetylglutamate kinase; 3, N-acetylglutamate phosphate reductase; 4, N-acetylornithine aminotransferase; 5, N- acetylornithine deacetylase (DRIP-1 may function for this step); 6, glutamate N-acetyltransferase; 7, ornithine transcarbamylase; 8, arginosuccinate synthase; 9, arginine synthase; 10, arginase; 11, glutaminase and glutamate dehydrogenase; and 12, carbamoyl phosphate synthetase. (After Yokota et al., 2002).

Out of the four osmolytes, proline, mannitol, glycinebetaine and citrulline examined, citrulline was the most effective hydroxyradical scavenger (Fig. 11.3). Mannitol and proline were less effective and glycinebetaine was virtually ineffective. The rate constant* for citrulline was 3.9×10^{-9} M^{-1} s^{-1},

* The hydroxyl radical scavenging activity of a compound is analyzed by the competitive trapping assay. The second order rate constant for the reaction between the compound and hydroxyl radicals is calculated according to the kinetic competition model for ROS scavengers. The constant for salicylate, 1.2×10^{-10} M^{-1} s^{-1} is used for calculation of the constant for the competitor.

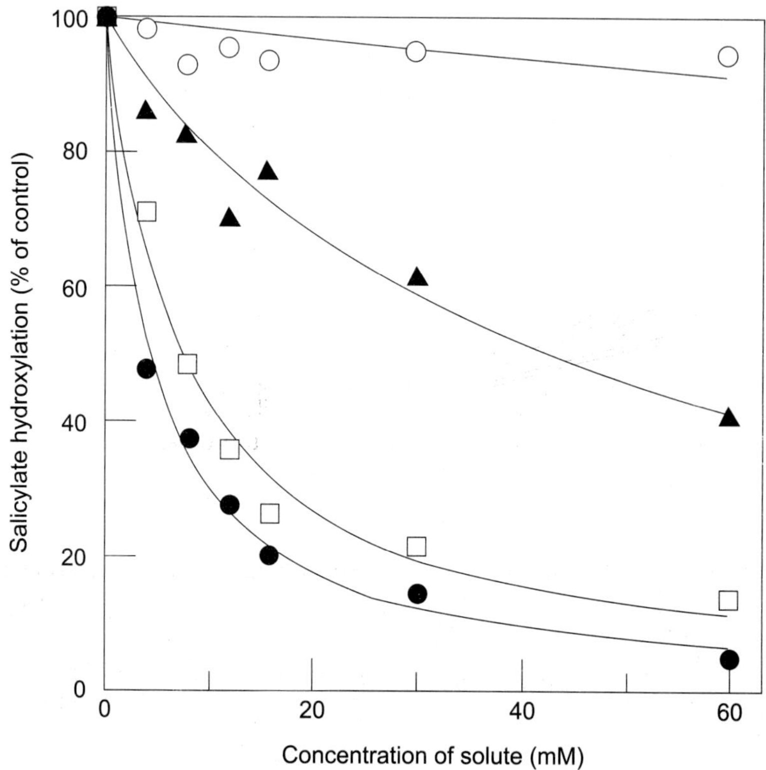

Fig. 11.2 Hydroxyl radical scavenging activities of citrulline and various compatible solutes. Citrulline (closed circles), mannitol (squares), proline (triangles) and glycinebetaine (open circles). (After Akashi et al., 2001).

indicating that citrulline is one of the most potent scavengers among the compatible solutes examined (Table 11.1). Thus the citrulline accumulated in watermelon leaves may increase the antioxidative potential of the cells, and protect the cells from oxidative damage. The function of citrulline as hydroxyl radical scavenger appears more important than even ascorbate and glutathione.

Table 11.1 Second-order rate constants for the reactions between hydroxyl radicals and various compounds (After Akashi et al., 2001).

Compound	ID_{50} (mM)	Rate constant (M^{-1} s^{-1})
Citrulline	6.6± 1.2	(3.9±0.82)´ 10^{-9}
Mannitol	13± 3.0	(2.1± 0.58)´ 10^{-9}
Proline	48± 9.0	(5.4± 0.94)´ 10^{-8}
Glycinebetaine	500± 280	(8.2± 0.31)´ 10^{-7}

Data represent the mean ± SEM (n = 3)

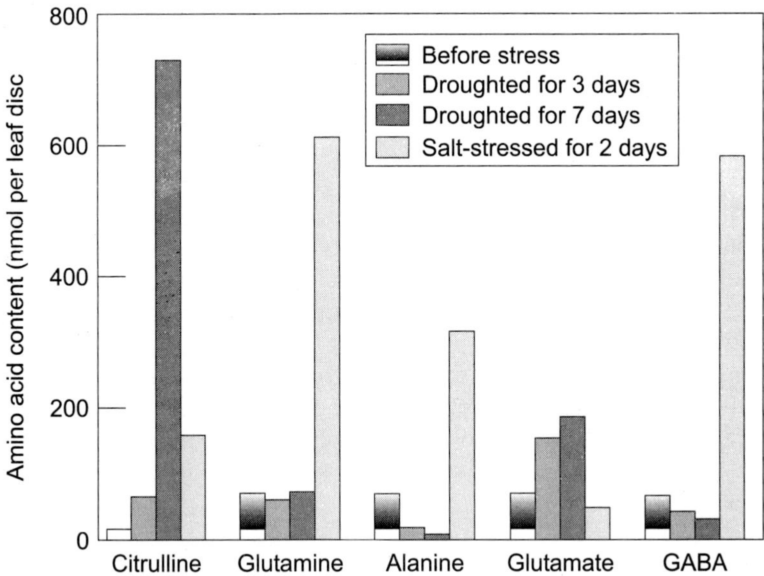

Fig. 11.3 Changes in the free-amino acid composition in the leaves of wild watermelon droughted for 3 and 7 days and treated with 150 mM NaCl for 2 days. (After Kawasaki et al., 2000).

MECHANISM OF ACTION

Arginine accumulated in the leaves of wild watermelon up to 0.1 M which functions as a hydroxyl radical scavenger. Citrulline has a similar chemical structure to arginine and is a good candidate as a radical scavenger. Citrulline effectively protects DNA and metabolic enzymes from oxidative injuries. Then in 2001, Akashi et al. also examined whether or not citrulline effectively protects DNA from ROS attacks. The ROS-induced DNA damages were found to be mediated by the interaction of hydroxyl radicals with C-4′ of deoxyribose, leading to cleavage of the phosphodiester. Hydroxy radicals caused cleavage of one strand of the supercoiled plasmid DNA and yielded the nicked form addition of citrulline at concentrations of 50-200 mM, strongly reduced the formation of nicked DNA demonstrating that citrulline effectively protected DNA from ROS attacks. Subsequently they also examined whether or not citrulline is capable of protecting metabolic enzymes from oxidative injuries, using pyruvate kinase as a model enzyme. Pyruvate kinase is readily inactivated by ROS. At 200 or 400 mM citrulline protected the enzyme from oxidative damage. Thus, citrulline accumulated in droughted wild watermelon maintains these ROS vulnerable components active and maintains their cellular potential to survive during prolonged drought. Further, Yokota et al. (2002) concluded that citrulline contributes to oxidative stress tolerance under drought conditions as a novel hydroxyl radical-scavenger. Akashi et al. (2005) further found that the activities of cytosolic metabolic enzymes are not inhibited by 600 mM citrulline, indicating that physiological concentration of citrulline does not interfere with cellular metabolisms. In order to address possible protective roles of accumulated citrulline, they examined the free radical scavenging activity of citrulline *in vitro* . Salicylate competitive trapping assay demonstrated that the second-order rate constant for the reaction of citrulline with hydroxyl radical was approximately 3×109 $M^{-1}s^{-1}$, suggesting that citrulline is one of the most effective scavengers among known compatible solutes. Moreover, the physiologically relevant concentration of citrulline exhibited significant protective effects on an enzyme and DNA from free radical attack. These data strongly suggest that citrulline, as the novel compatible solute, contributes to the tolerance against oxidative stress under drought conditions in wild watermelon plants.

Genetic basis

O'Reilly and Devine (1994) have reported that the citrulline biosynthetic operon *argC-F* located at 100 degrees on the *Bacillus subtilis* chromosome contains seven open reading frames which encode all the enzymes required for the biosynthesis of citrulline. The operon is transcribed as a single transcription unit. The second cistron of the operon is homologous to *ArgJ* (ornithine acetyltransferase) from *Bacillus stearothermophilus* and *Neisseria gonorrhoeae,* suggesting that the acetylation of glutamate and the deacetylation of acetylornithine are carried out by a single enzyme in a cyclical pathway. The *argF* gene is an orthologue of *argF* from *Pseudomonas aeruginosa* and a paralogue of *arcB* from *P. aeruginosa* and *argF/argI* from *Escherichia coli.*

Wild watermelon (*Citrullus lanatus* sp) has the ability to tolerate severe drought/high light stress conditions despite carrying out normal C_3-type photosynthesis. Akashi et al. (2004) employed mRNA differential display to isolate drought-responsive genes in the leaves of wild watermelon. One of the isolated genes, *CLMT2,* shared significant homology wth type-2 metallothionein (MT) sequences from other plants. The second-order rate constant for the reaction between a recombinant *CLMT2* protein and hydroxyl radicals estimated as 1.2×10^{11} M^{-1} s^{-1}, demonstrating that *CLMT2* has an extraordinary high activity for detoxifying hydroxyl radicals. Moreover, hydroxyl radical-catalyzed degradation of watermelon genomic DNA was effectively suppressed by *CLMT2 in vitro.* The results suggest that *CLMT2* induction contributes to the survival of wild watermelon under severe drought/high light stress conditions.

REFERENCES

Akashi, K.; C. Miyake and A. Yokota, 2001. Citrulline, a novel compatible solute in drought-tolerant wild watermelon leaves, is an efficient hydroxyl radical scavenger. FEBS Letters 508 (3): 438-42.

Akashi, K.; N. Nishimura; Y. Ishida and A. Yokota, 2004. Potent hydroxyl radical scavenging activity of drought-induced type-2 metallothionein in wild watermelon. Biochem. Biophys. Res. Commu. 323: 72-78.

Akashi, K.; C. Miyake; T. Kohchi and A. Yokota, 2005. Citrulline function as an efficient hydroxyl radical scavenger : implication for the drought tolerance of wild watermelon. Doi: 10.1071/SAO403460.

Kawasaki, S.; C. Miyake; T. Kohchi et al. 2000. Responses of wild watermelon to drought stress : Accumulation of an *ArgE* homologue and citrulline in leaves during water deficits. Plant Cell Physiol. 41: 864-73.

O'Reilly, M. and K.M. Devine, 1994. Sequence and analysis of the citrulline biosynthetic operon *argC-F* from *Bacillus subtilis.* Microbiology 140: 1023-25.

Sugawara, K.; Y. Yoshizawa; S. Tzeng; et al. 1998. Colorimetric determination of citrulline residues in proteins. Anal. Biochem. 265(1): 92-96.

Yokota, A.; S. Kawasaki; M. Iwano and C. Nakamura, 2002. Citrulline and *DRIP-1* protein (*ArgE* homologue) in drought tolerance of wild watermelon. Ann. Bot. 89: 825-32.

SUGGESTED READING

Akashi, K.; C. Miyake and A. Yokota, 2001. Citrulline, a novel compatible solute in drought-tolerant wild watermelon leaves, is an efficient hydroxyl radical scavenger. FEBS Letters 508 (3): 438-42.

Yokota, A. S. Kawasaki; M. Iwano and C. Nakamura, 2002. Citrulline and *DRIP-1* protein (*ArgE* homologue) in drought tolerance of wild watermelon. Ann. Bot. 89: 825-32.

12

ROS AND ANTIOXIDANTS

Introduction

Abiotic stresses – drought, cold, salinity—generally produce two types of stresses, viz. (i) water-deficit stress, and to counter that plants generate certain compatible osmolytes and other osmoprotectants (discussed in earlier chapters), and (ii) reactive oxygen species (ROS), and to counter that plants have developed both (a) enzymatic [superoxide dismutase (SOD), catalase, and peroxidase], and (b) non-enzymatic (ascorbate, glutathione, carotenoids, and tocopherols) defences. The above ROS degenerating/countering compounds generated in response to different types of stresses are not similar in quality or quantity, and are thus described separately for each type of stress.

ROS GENERATION AND ROS STRESS

Since oxygen has two unpaired electrons, needs activation as the two unpaired electrons have parallel spins. Activation of oxygen occurs by two different mechanisms : absorption of sufficient energy to reverse the spin on one of the unpaired electrons, or monovalent reduction (O_2^- , H_2O_2, OH^-). By simultaneous transfer of two electrons in divalent reduction, the singlet oxygen (O-O:) is much more reactive towards organic molecules than its triplet (.O-O.) counterpart. Superoxide (O_2^-) acts either as an oxidant or a reductant; it can oxidize sulfur, ascorbic acid or NADH, or reduce cytochrome C and metal ions. A dismutation reaction leading to the formation of H_2O_2 and oxygen can occur spontaneously or is catalyzed by the enzyme superoxide dismutase (SOD) . Hydrogen peroxide readily permeates membranes and is, therefore, not compartmentalized in the cell.

Oxidative stress (attack) on proteins results in site-specific amino acid modifications, aggregation of cross-linked reaction products, altered electrical charge and increased susceptibility to proteolysis. The sulfur-containing amino acids, and thiol groups specifically, are very susceptible sites. The ROS can abstract an H atom from cysteine residues to form a thiyal radical that will cross-link to a second thiyal radical to form disulfide bridges. Alternatively, oxygen can add to a methionine residue to form methionine sulfoxide derivatives. ROS and ROS generating agents such as ionizing radiation, induce numerous lesions in DNA that cause deletions, mutations and other lethal genetic effects. Cross-linking of DNA to protein is another consequence of hydroxyl radical attack on either DNA or its associated proteins.

One to three percent of all oxygen consumed by respiration is inevitably reduced to superoxide radicals and H_2O_2 . The main sources of ROS in plants under physiological conditions are respiration, photosynthesis, and N_2-fixation (Matamoras et al., 2003 – Table 12.1). Similarly, the term reactive nitrogen species (RNS) refers to nitrogen derived molecules with variable reactivity and includes free radicals (nitric oxide) and non-free radicals (peroxynitrite). Nitric oxide reacts with superoxide radicals to form peroxynitrite and probably with thiol compounds to form nitrosothiols (Matamoras et al., 2003).

Table 12.1 Production of ROS and RNS in plants (Adapted from Matamoras et al., 2003).

ROS and RNS	Cellular sources
Superoxide radical	Electron transport chains of mitochondria, chloroplasts, endoplasmic reticulum, peroxisomes, and plasma membrane. NADPH oxidase in membranes. Oxidation of leghaemoglobin in cytosol. Xanthine oxidase and membrane polypeptides in peroxisomes.
Hydrogen peroxide	Electron transport chains of mitochondria, chloroplasts, endoplasmic reticulum, and plasma membrane. CuZnSOD in cytosol and plastids, MnSOD in mitochondria, FeSOD in plastids, photorespiration, fatty acid β-oxidation, urate oxidase, and MnSOD in peroxisomes.
Organic and lipid peroxides	Nonenzymatic lipid peroxidation. Lipoxygenase.
Hydroxyl radical	Reaction of superoxide radical with H_2O_2 catalyzed by trace amounts of Fe or Cu.
Singlet oxygen	Photoinhibition in chloroplasts.
Nitric oxide	Nitrate reductase in leaves and other plant organs. Nitric oxide synthase (?) in peroxisomes.
Peroxynitrite	Reaction of nitric oxide with the superoxide radical.

The ROS is generated at least at four different sites within the chloroplast : (a) PS I can reduce oxygen, which is an important mechanism of oxygen activation in the chloroplast, (b) the photoactivated chlorophyll normally transfers its excitation energy to the PS reaction centers. After stress-induced stomatal closure, this energy excites oxygen from the triplet to singlet form, (c) the oxidizing side of PS II facilitates 4 single electron transfers from water to the PS II reaction center releasing triplet oxygen, and (d) leaks of electrons from this site to molecular oxygen, or release of partially reduced oxygen products make a relatively minor contribution to the ROS production.

Reactive oxygen species are formed as by-products in the electron transport chains of chloroplasts, mitochondria and plasma membrane. The lifetime of ROS within the cellular environment is determined by the antioxidative system, which provides crucial protection against oxidative damage. The ROS-induced senescence is related to the extent of gene *LSC 54*, metallothionein encoding protein (Navabpour et al., 2003). Further the effects of H_2O_2 and NO on gene expression are given in Fig. 12.1 by Neill et al. (2002).

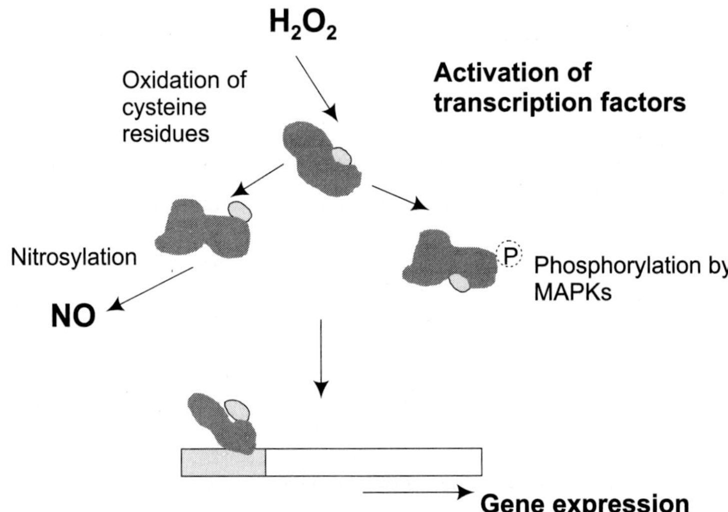

Fig. 12.1 H_2O_2, NO and gene expression. H_2O_2 and NO might regulate the activity of transcription factors directly via nitrosylation (NO) or oxidation of cysteine reduces (H_2O_2). Activation of transcription factors can also occur via activation of a mitogen-activated protein kinase (MAPK) cascade which leads to phosphorylation of the tranacription factor (After Neill et al., 2002).

THE PLANT DEFENCE MECHANISM : GENERATION OF ROS-DEGENERATING ENZYMES AND ANTIOXIDANTS

To control the level of ROS and to protect cells under stress conditions, plant tissues contain several enzymes scavenging ROS (superoxide diismutase, catalase, peroxides and glutathione peroxidase), detoxifying lipid peroxidation products (glutathione S-transferases, phospholipid hydroperoxide, glutathione peroxidase, and ascorbate peroxidase) and a network of low molecular mass antioxidants (ascorbate, glutathione, phenolic compounds and tocopherols). In addition, a whole array of enzymes is needed for the regeneration of the actiforms of the antioxidants (monodehydroascorbate reductase, dehydroascorbate reductase and glutathione reductase).

Enzymatic defence

Superoxide dismutase enzyme catalyzes dismutation of superoxide to hydrogen peroxide and oxygen. The activity of this enzyme determines the reactive proportions of the two constituents of the reaction that generates hydroxyl radicals. There are two distinct types of SOD classified on the basis of the metal cofactors : Cu/Zn-SOD, Mn-SOD and Fe-SOD isomers. Superoxide dismutase has a central role in defence against oxidative stress. Gupta et al. (1993) developed transgenic tobacco plants that express a chimeric gene that encodes chloroplast-localized Cu/Zn-SOD from *Pisum sativum.* To investigate whether increased expression of chloroplast targeted SOD could alter the resistance of photosynthesis to environmental stress, these plants were subjected to chilling temperatures and moderate (500 µmol of quanta m^{-2} s^{-1}) or high (1500 µmol of quata m^{-2} s^{-1}) light intensity. During exposure to moderate stress, transgenic SOD plants retained rates of photosynthesis $\approx$ 20 percent higher than the untransformed tobacco plants, implicating active oxygen species in the reduction of photosynthesis during chilling. Unlike untransformed plants, transgenic SOD plants were capable of maintaining nearly 90 percent of their photosynthetic capacity following exposure to chilling at high light intensity for four hours. These results demonstrate that SOD is a critical component of the active-oxygen-scavenging system of plant chloroplasts and indicate that modification of SOD expression in transgenic plants can improve plant stress tolerance.

Catalase, the heme-containing enzyme, catalyzes dismutation of H_2O_2 into water and oxygen. In maize plants, three isoforms of catalase termed *cat-1, cat-2* and *cat-3* located on separate chromosomes are differentially expressed and independently regulated. *cat-1* and *cat-2* are localized in peroxizomes and the cytosol, whereas *cat-3* is mitochondrial (Scandilios, 1990). Catalase is light sensitive, and salinity, heat shock or cold reduce catalase activity.

Relatively small increases or localized bursts of H_2O_2 influence only part of the network and modify gene expression in such a way as to strengthen plant defence responses. In contrast, large increases in H_2O_2 trigger a distinct local sequence of events in gene expression that leads inevitably to programed cell death (Foyer et al., 1997). Low concentrations of H_2O_2 inhibit *Cat-1, Cat-3* and *Cst-1* gene expression in maize, while the higher doses strongly induced these genes (Polidoros and Scandalios, 1999). The higher concentrations of H_2O_2 induced *Cat-1, Cat-2* and *Cat-3* gene expression to higher levels, and in lesser time, than the lower concentrations. Induction of *Cat-3* was superimposed on the circadian regulation of the gene. The results demonstrate a direct signaling action of H_2O_2 in the regulation of antioxidant gene responses in maize. However, Willekens et al. (1997) used transgenic tobacco with ~ 10 percent wild-type catalase activity to study the role of catalase and effects of H_2O_2 stress in plants. The catalase-deficient plants showed no visible disorders at low light, but at elevated light rapidly developed white nectrotic lesions on the leaves. An alternative H_2O_2 –scavenging mechanisms may have compensated for reduced catalase activity, as shown by increased ascorbate peroxidase and glutathione peroxidase levels. The catalase-deficient plants showed increased susceptibility to salt and ozone stress, but not to chilling.

Oxidants such as H_2O_2 interact with other signaling systems, particularly hormones (Table 12.2 – Pastori and Foyer, 2002). They also influence and modify the action of other secondary messengers such as Ca^{2+} and NO. Similarly, H_2O_2 has a strong regulatory influence on fluxes through Ca^{2+} channels and on Ca^{2+} concentrations in different cellular compartments. The lifetime of H_2O_2 *in vitro* is determined by the capacity of the two major oxidant buffers to the plant cell, ascorbate and glutathione, together with the antioxidant enzymes that use these antioxidants (Noctor and Foyer, 1998). Most plant cells contain very large quantities of ascorbate (10-100 mM), and most intracellular compartments, hence, have the capacity to deal with very high fluxes of H_2O_2 production (Noctor et al., 2000).

Table 12. 2. Ascorbate, glutathione, and H_2O_2 function as upstream/downstream components of hormone-mediated signal transduction (Adapted from Pastori and Foyer, 2002).

Hormones	Tissue	Agent	Function	References
Auxin	Root	H_2O_2	Gravitropism	Joo et al. (2001)
Abscisic acid	Leaves	H_2O_2	Stomatal closure	Murata et al. (2001)
	Aleurone	Antioxidants	Cell survival	Bethke and Jones (2001)
Gibberellic acid	Aleurone	H_2O_2	Programed cell death	Berthke and Jones (2001)

Together with catalase, *ascorbate peroxidase* controls the amount of H_2O_2 present within the plant cell (but not the apoplast) so that it rarely approaches the concentrations that inhibit metabolism and trigger cell death (Storozhenko et al., 1998). Peroxides and catalases play an important role in the fine regulation of ROS concentration in the cell through activation and deactivation of H_2O_2 .

Non-enzymatic defence

Ascorbic acid is abundantly found in plant tissues , which functions as a reductant for many free radicals, thereby minimizing damage caused by oxidative stress; in addition, ascorbate has some other functions too. Ascorbate can directly scavenge oxygen-free radicals with and without enzyme catalysts and can indirectly scavenge them by recycling tocopherol to the reduced form. Ascorbate protects critical macromolecules from oxidative damage, and is present in chloroplast, cytosol, vacuole and extracellular components of the cell. Ascorbate also occurs in cell wall where it is the first line of defence against ozone and ultraviolet.

Ascorbic acid exists mostly in the reduced form (90 percent of the ascorbate pool) in leaves and chloroplasts; and its intracellular concentration can build up to millimolar range (e.g 20 mM in cytosol and 20-300 mM in chloroplast stroma). The ability to donate electrons in a wide range of enzymatic and non-enzymatic reactions makes ascorbic acid the main ROS-destroying compound in the aqueous phase. Ascorbic acid can directly scavenge superoxide, hydroxyl radicals and singlet oxygen and reduce H_2O_2 to water via ascorbate peroxidase reaction. In chloroplasts, ascorbic acid acts as a cofactor of violaxanthin de-epoxidase thus sustaining dissipation of excess excitation energy. Ascorbic acid regenerates tocopherol from tocopheroxyl radical providing membrane protection.

Glutathione is a tripeptide whose antioxidant function is facilitated by the sulfhydryl group of cystine. Glutathione has a redox potential of –340 mV that enables glutathione to reduce dehydroascorbate to ascorbate or to reduce the disulfide bonds of proteins. Glutathione is found in most plant tissues, cells and subcellular components; its levels are higher in light than in dark. Its concentration is highest in chloroplasts (70- 90 percent) averaging between 1 and 4 mM, but significant quantities also accumulate in the cytosol. Glutathione reacts chemically with singlet oxygen, superoxide and hydroxyl radicals, and therefore, functions directly as a free radical scavenger. It also stabilizes membrane structure by lipid peroxidation reactions.

An initial stress response is related to changes in the glutathione redox state, whereas acclimation is marked by increased glutathione concentrations, increased related enzyme activities, and/or a more reduced redox state of glutathione. However, when the stress level increases, glutathione concentration drops and the redox state becomes oxidized, which marks the degradation of the system (Tausz et al., 2004). When the cellular production of ROS exceeds the cell's antioxidant capacity, cellular macromolecules such as lipids, proteins and DNA get changed. ROS also plays a role in gene regulation and signal transduction pathways, which may be involved in defensive mechanisms against oxidative stress. With experimental inactivation of glutathione peroxidase, peroxidase accumulation increases which may act as a second messenger and regulate expression of antiapoptotic genes and the glutathione peroxidase itself to protect against cell damage (Miyamoto et al., 2003).

Glutathione scavenges cytotoxic H_2O_2, and reacts non-enzymatically with other ROS : singlet oxygen, superoxide radical and hydroxyl radical. The central role of glutathione in the antioxidant defence is due to its ability to regenerate another powerful water soluble antioxidant, ascorbic acid, via ascorbate-glutathione cycle (Fig. 12. 2 – Blokhina et al., 2002).

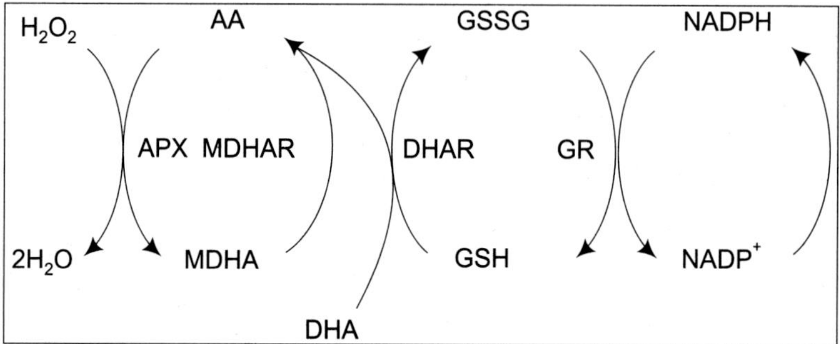

Fig. 12.2 Ascorbate-glutathione cycle (Halliwell-Asada pathway). APX = ascorbate-peroxidase; MDHAR= monodehydroascorbate reductase; DHAR= dehydroascorbate reductase; GR= glutathione reductase. (After May et al., 1998).

For increasing *in vivo* glutathione production (up to 4-fold higher) plants were transformed (Noctor et al., 2002). This process led to numerous insights into the control of glutathione synthesis or reduction (Fig. 12.3). The inter- or intra-cellular glutathione pools are linked by transport across membranes; it can cross the chloroplast envelope at rates similar to the speed of biosynthesis. Thus the control of glutathione concentration and redox state is due to a complex interplay between biosynthesis, utilization, degradation, oxidation/reduction, and transport. The level of glutathione also correlates with adaptation of plants to extremes of temperature. In addition, the size of reduced glutathione pool shows marked alterations in response to a number of environmental stresses.

The increased intracellular glutathione concentration does not prevent exogenous dehydroascorbate from inducing a cell cycle shift. Potters et al. (2004) thus concluded that together with a glutathione-driven dehydroascorbate reduction, a glutathione-independent pathway for dehydroascorbate also exists *in vivo*, and that both compounds act independently in growth control.

α-Tocopherol has antioxidant properties as a result of its ability to quench both singlet oxygen and peroxides (Fryer, 1992). Tocopherols and trocotrienols have antioxidant functions in membranes.

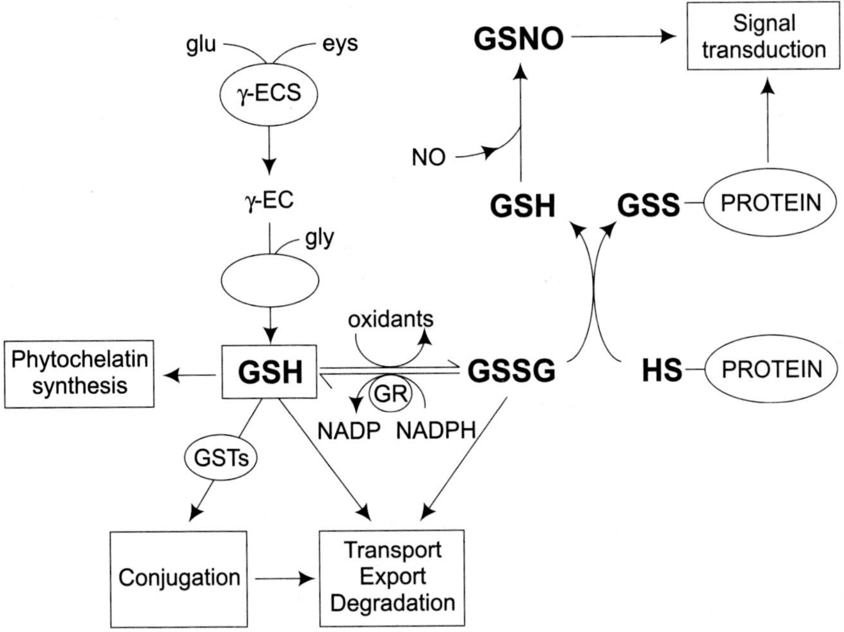

Fig. 12.3 Glutathione biosynthesis and interacting processes in plant cells. d-EC = d-glutamylcysteine; d-ECS = d-glutamylcysteine synthetase; GR = glutathione reductase; GSH = reduced glutathione; GSSH = glutathione disulfide; GSH-S = gluthione synthetase; GST = glutathione S-transferase. (After Noctor et al., 2002).

α-Tocopherol with its three methyl substituents has the highest antioxidant activity of tocopherols. However, α-tocotrienol is a better antioxidant than α-tocopherol in a membrane environment. The chloroplast membranes contain α-tocopherol as the predominant tocopherol isomer, and are hence well protected against the photooxidative damage.

Carotenoids are C40 isoprenoids and tetraterpenoids that are located in plastids of both photosynthetic and non-photosynthetic plant tissues. Carotenoids have the ability to detoxify various forms of activated oxygen and triplet chlorophyll that are produced as a result of excitation of the photosynthetic complexes by light. Carotenoids are of two types, (a) carotenes, and (b) xanthophylls. The main protective role of β-carotene in photosynthetic tissues is its direct quenching of triplet chlorophyll, which would prevent generation of singlet oxygen, and therefore, completely avoids oxidative stress.

Carotenoids are involved in protection of photosynthetic apparatus against photoinhibitory damage by singlet oxygen ($'O_2$), which is produced by the excited singlet state of chlorophyll. Carotenoids can directly deactivate $'O_2$ and can also quench the excited triplet state of chlorophyll, thus indirectly reducing the formation of 'O_2 species (Foyer and Harbinson, 1994). On the other hand, zeaxanthin is involved in the de-excitation of excess energy.

Flavonoids and isoflanoids with their polyphenolic structures have the ability to scavenge free radicals and to chelate transition metals, a basis for their potent antioxidant abilities (Arora et al., 2000). Further, they have the ability to stabilize membranes by decreasing membrane fluidity. The localization of flavonoids and isoflavonoids into the membrane interiors and their resulting restrictions on fluidity of membrane components could sterically hinder diffusion of free radicals and thereby decrease the kinetics of free radical reactions (Arora et al., 2000).

STRESS INDUCED ROS-DETOXIFICATION BY ANTIOXIDANTS

Several abiotic stresses like drought, salinity, cold, intense light, and toxic metals, and the biotic stresses like wounding, pathogenesis (not discussed), and chemical stresses (e.g. some herbicides – not discussed) produce ROS, viz. $O^{\cdot -}$, H_2O_2, OH^- which affect cytoplasmic macromolecules, lipids, proteins, and membranes, and cause considerable subcellular damage. ROS occur as by-products of regular cellular metabolism such as photosynthesis. However, under stress conditions, their formation is usually exacerbated. In general, ROS (particularly $O^{\cdot -}$ and OH^- radicals) are damaging to essential cellular components such as DNA, proteins and lipids. Lipid peroxidation disrupts the membrane integrity of the plant cell. As a result, essential solutes leak out of organelles and from the cell, causing disruption in membrane function and metabolic imbalances. DNA is the blueprint for both future form and function. Any damage to its integrity means that proteins that would have been essential for optimal function of the plant will not be synthesized. Similarly, denaturation of important proteins essential for biochemical reactions leads to the whole plant being negatively affected and unable to cope with the stress.

To prevent the damage caused by free radicals, plants have evolved complex protective mechanisms. The primary constituents include antioxidant enzymes such as superoxide dismutase, catalase, ascorbic peroxidase and glutathione reductase, and the free radical scavengers such as ascorbate, glutathione, carotenoids and tocopherol. ROS superoxide and singlet oxygen are produced in chloroplasts by photoreduction of oxygen and energy transfer from triplet excited chlorophyll to oxygen, respectively. Hydrogen peroxide and hydroxyl radicals are formed as a result of the reactions of superoxide. All these species are reactive and potentially damaging, causing lipid peroxidation and inactivation of enzymes. They are normally scavenged by a range of antioxidants and enzymes which are present in the chloroplast and other subcellular compartments. When CO_2 fixation is limited by water deficit, the role of ROS formation increases in chloroplasts as excess excitation energy, not dissipated by the photoreactive mechanisms, is used to form superoxide and singlet oxygen. However, non-lethal water deficits often result in increased activity of superoxide dismutase, glutathione reductase and monodehydroascorbate reductase. The tolerant cells recover upon rehydration and are able to reduce their glutathione pool. The non-tolerant species, however, continue to show further oxidative damage including lipid peroxidation (Smirnoff, 1993).

Kocsy et al. (2002) studied the effect of different abiotic stresses on the synthesis of glutathione and hydroxymethylglutathione, on the ratio of reduced to oxidized forms of these thiols (GSH/GSSG, hmGSH/ hmGSSG), and on the glutathione reductase activity in maize and wheat genotypes having different sensitivity to low temperature stress. Cold treatment induced a greater increase in total glutathione content and in glutathione reductase activity in tolerant genotypes of both species than in sensitive ones. The GSH/GSSG and hmGSH/hmGSSG ratios were increased by this treatment only in the frost-tolerant wheat variety. High-temperature stress increased the total glutathione content and the GSH/GSSG ratio only in the chilling-sensitive maize genotype, but the glutathione reductase activity was greater after this treatment in both maize genotypes. Osmotic stress resulted in a great increase in the total glutathione content in wheat and the glutathione reductase activity in maize. The amount of total hydroxymethyl-glutathione increased following all stress treatments. These results indicate the involvement of these antioxidants in the stress responses of wheat and maize.

Drought

Activities of superoxide dismutase, catalase, and peroxidase, as well as malondialdehyde contents and solute potentials, were studied in seedlings of seven wheat species (9 genotypes representing 3 ploidy levels; hexaploid, tetraploid, diploid) subjected to water stress for 4, 8, and 12 days by withholding water (Zhang and Kirkham,, 1994). Solute potentials of all genotypes were lowered by water stress. In most species, superoxide dismutase and catalase activities showed an increase or maintenance in the early phase of drought and then a decrease with further increase in magnitude of water stress. On the contrary, peroxidase activities and malondialdehyde contents greatly increased in response to water stress. Enzymatic activities partly recovered and malondialdehyde contents decreased with rewatering. Under drought, hexaploid wheats had higher peroxidase activities and malondialdehyde contents than tetraploid and diploid wheats; solute potentials and activities of superoxide dismutase and catalase, however, were similar among the three groups. These results suggest that water stress alters the equilibrium between free radical production and enzymatic defence reactions in wheat species and that hexaploid wheats have less efficient antioxidant systems (e.g., the ascorbate-glutathione cycle and the nonezymatic system) than tetraploid and diploid wheats. Further, McKersie et al. (1996) produced transgenic alfalfa expressing Mn-superoxide dismutase cDNA which reduced injury from water-deficit stress as determined by chlorophyll fluorescence, electrolyte leakage, and regrowth from crowns. A 3-year field trial indicated that yield and survival of transgenic plants were significantly improved, supporting the hypothesis that tolerance of oxidative stress is important in adaptation to field environments.

Both mesophyll and bundle sheath cells of maize (a C_4 species) leaves contained the enzymes ascorbate peroxidase and glutathione reductase which are involved in H_2O_2 detoxification (Brown et al., 1995). Glutathione reductase was unaffected by drought in whole leaf tissue and mesophyll cells, but did increase slightly in bundle sheath cells. Ascorbate peroxidase and glutathione reductase activity levels were similar in mesophyll cells, bundle sheath cells and in whole leaf tissue. The authors suggest that moderate drought has little effect on enzymes of the H_2O_2 scavenging system, and that mesophyll and bundle sheath cells may be exposed to similar levels of H_2O_2 .

Loggini et al. (1999) analyzed the antioxidative defences, photosynthesis and xanthophyll pigments in two wheat (a C_3 species) cultivars, Adamello (drought sensitive) and Ofanto (drought tolerant), during dehydration and rehydration to determine the difference in their sensitivities to drought and to elucidate the role of different protective mechanisms against oxidative stress. During dehydration, glutathione content decreased in both wheat cultivars, but only cv Adamello showed a significant increase in glutathione reductase and hydrogen peroxide-glutathione peroxidase activities (Figs. 12.4 and 12.5). After exposure to drought, cv Adamello showed a larger reduction in the actual PS II photochemical efficiency and a higher increase in non-radioactive energy dissipation than cv Ofanto. Although the differences in zeaxanthin content were not sufficient to explain the difference in drought tolerance between the two cultivars, zeaxanthin formation may be relevant in avoiding irreversible damage to PS II in the more sensitive cultivar.

Drought imposed in three contrasting wheat genotypes at two stages after anthesis resulted in an increased H_2O_2 accumulation and lipid peroxidation, and decrease in ascorbic acid content (Sairam et al. 1998). The activities of superoxide dismutase, ascorbate peroxidase and catalase

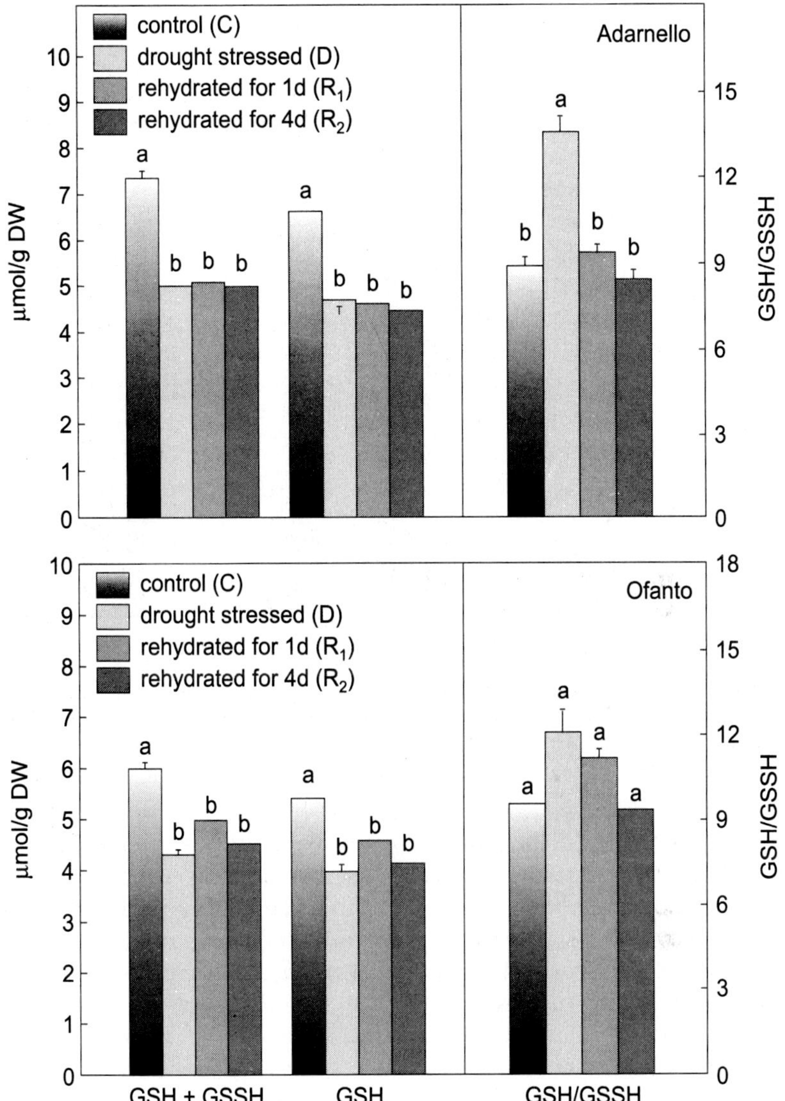

Fig. 12.4 Effects of dehydration and rehydration on GSSH and GSH contents in the wheat cvs Adamello and Ofanto. Bars represent the SE (n=10 repetitions from 3 independent experiments). One-way analysis of variance used for comparisons between the means. Bars with different letters are significantly different at $P \leq 0.01$. (After Loggini et al., 1999).

increased. The drought tolerant genotype C306 showed highest ascorbate peroxidase and catalase activities and ascorbic acid content and the lowest accumulation of H_2O_2 and lipid peroxidation as compared to the susceptible

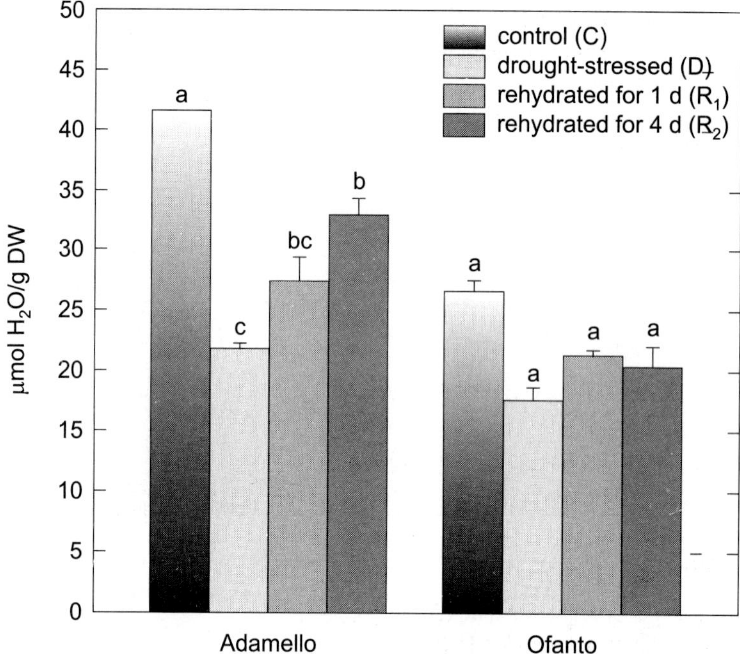

Fig. 12.5 Effects of dehydration and rehydration on hydrogen peroxide content in the wheat cvs Adamello and Ofanto. Bars represent the SE (n= 10 repetations from 3 independent experiments). One-way analysis of variance used for comparisons between the means. Bars with different letters are significantly different at $P \leq 0.01$ (After Loggini et al. 1999).

genotype HD2329. Further, Sairam and Saxena (2000) reported that the antioxidant enzymes ascorbate peroxidase, glutathione reductase and also a non-specific peroxidase increased in three contrasting wheat genotypes under drought stress. The genotype PBW175 having highest ascorbate peroxidase, glutathione reductase and peroxidase activities shows lowest lipid peroxidation, highest membrane stability, and the contents of chlorophyll and carotenoids, as against the susceptible genotype WH542. Genotype HD2402 showed intermediate values.

Accumulation of H_2O_2 serves as an important stress signal and promotes stomatal closure, whereas ascorbic acid is the major antioxidant that scavenges H_2O_2 . The level of H_2O_2 and the ascorbic acid redox state in guard cells and whole leaves are diurnally regulated such that the former increases during the afternoon, whereas the latter decreases (Chen and Gallie, 2004). Plants with an increased guard cell ascorbic acid redox state were generated by increasing dehydroascorbic acid redox expression, and these exhibited a reduction in the level of guard cell H_2O_2 . Guard cells with an increase in ascorbic acid redox state were less responsive to H_2O2 or

ABA signaling, and the plants exhibited greater water loss under drought conditions, whereas suppressing dehydroascorbic acid redox serves to maintain a basal level of ascorbic acid recycling in guard cells that is insufficient to scavenge the high rate of H_2O_2 produced in the afternoon, thus resulting in stomatal closure. As the ascorbic acid in guard cells varies diurnally, the level of glutathione reductase in cotton also varies seasonally, but not in response to water deficit (Mahan and Wanjura, 2005). The amount of ascorbate and the activity of ascorbate peroxidase increased under water stress ($\approx$ 5x and 1.5x, respectively). While the glutathione metabolism appears sufficient for oxidative stresses resulting from field water deficits, altered ascorbate metabolism may be a response to water deficit in the field.

Changes in carotenoids, tocopherol and diterpenes during drought and recovery were determined by Munne-Bosch and Alegre (2000). Under severe stress : (a) α-tocopherol increased 9-fold g^{-1} dry weight and 20-fold $unit^{-1}$ chlorophyll, (b) lutein and β-carotene contents decreased on a dry weight basis, but an 80 percent increase in lutein and constant levels of β-carotene were observed on a chlorophyll basis, (c) there were transient and sustained increases in the de-epoxidation state of the xanthophyll cycle, and (d) the highly oxidized abietane diterpene isorosmanol increased 8-fold as a result of the oxidation of carnosic acid. The photoprotection conferred by the xanthophyll cycle and the antioxidant function of tocopherols, lutein and diterpenes may help to avoid irreversible damage in severe drought, making possible the recovery of functional membranes after the autumn rainfalls. Besides, chlorophyll loss reduces the amount of photons absorbed by leaves, which enhances the photoprotective and antioxidant capacity of leaves per amount of photons absorbed, since the ratios of xanthophylls, α-tocopherol and abietane diterpenes to chlorophyll increase. Then three years later, Munne-Bosch and Alegre (2003) tried to assess the antioxidative protection by carnosic acid in combination with that of other low molecular weight antioxidants (α-tocopherol and ascorbate) in chloroplasts of three Labiatae species, differing in their carnosic acid contents, exposed to drought stress in the field. The three species showed significant increases in α-tocopherol, a shift of the redox state of α-tocopherol toward its reduced state, and increased ascorbate levels in chloroplasts under stress. These results suggest that carnosic acid, in combination with other low molecular weight antioxidants, helps to prevent oxidative damage in chloroplasts of water-stressed plants.

Salinity

Hernandez et al. (1993) studied the effect of salt stress on ROS production in leaf mitochondria from two NaCl-treated cultivars of *Pisum sativum* with different sensitivity to NaCl. In mitochondria from NaCl-sensitive plants, salinity brought about a significant decrease of MnSOD. Conversely, in salt-tolerant plants, NaCl treatment produced an increase in the mitochondrial MnSOD activity. The enhanced rates of superoxide production by mitochondria from the salt sensitive plants were concomitant with a strong decrease in the mitochondrial MnSOD activity, whereas the NaCl-tolerant plants appear to have a protection mechanism against salt-induced increase in active oxygen production by means of the induction of mitochondrial MnSOD activity. These results indicate that in the subcellular toxicity of NaCl in pea plants, an oxidative stress mechanism mediated by superoxide radicals is involved, and also imply a function for mitochondrial MnSOD in the molecular mechanisms of plant tolerance to NaCl. Later, Gömez et al. (2003) described the intrachloroplast localization and the changes that take place in the thylakoid and stroma-located SOD and ascorbate peroxidases, in response to long term NaCl stress in *Pisum sativum* cv Puget plants. The thylakoidal Fe-SOD activity was induced by NaCl concentration as low as 70 mM, while CuZn-SOD was induced at 90 mM. The NaCl stress also induced stomatic Fe-SOD and CuZn-SOD activities, although these inductions started at higher NaCl concentration (90 mM) and were significant at 110 mM NaCl. Significant increase in H_2O_2 content and a reduction in ascorbate level at 90 mM NaCl occurred, although the oxidized ascorbate pool at the highest NaCl concentration did not show significant changes. The authors suggest that the up-regulation of the above enzymes would contribute to the adaptation to moderate NaCl salinity.

Ruiz and Blumwald (2002) studied the role of sulfur-assimilation and the biosynthesis of cysteine and glutathione during the response to salt stress of wild-type and salt-tolerant transgenic *Brassica napus* (canola) plants overexpressing a vacuolar Na^+/H^+ antiporter. A 3-fold increase in cysteine and glutathione content was observed in wild-type plants exposed to salt stress, but not in the transgenic plants. The induction of cysteine and glutathione synthesis during salt stress in the wild-type plants suggests a possible protective mechanism against salt-induced oxidative damage. On the other hand, the salt-tolerant transgenic plants did not show significant changes in either cysteine or glutathione content, confirming the role of vacuolar Na^+ accumulation and ion homoestasis in salt tolerance.

Salinized tomato seedlings wilted rapidly but recovered if returned to nonsaline nutrient solution within 6 hours, and after 9 hours of salt treatment 100 percent of the seedlings remained wilted and died (Shabala and Neumann, 2001). Remarkably, addition of the antioxidant, 0.5 mM ascorbic acid, to the medium prior to and during salt treatment for 9 hours, facilitated the subsequent recovery and long term survival of about 50 percent of the wilted seedlings (Fig. 12.6). Salt stress increased accumulation in roots, stems and leaves, of lipid peroxidation products produced by interactions with damaging ROS. Additional ascorbic acid partially inhibited this response (Table 12.3).

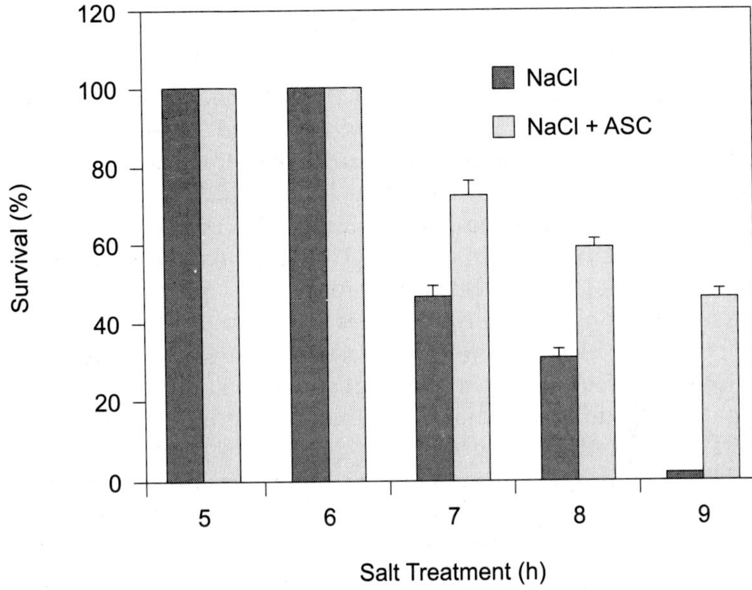

Fig. 12.6 Tomato seedling survival of NaCl-induced wilting is increased by treatment with exogenous ascorbic acid (seedlings were incubated with or without 0.5 mM ascorbic acid for 24 hours and then for an additioal 9 hours with or without 300 mM NaCl). (After Shabala and Neumann, 2001).

Table 12.3 Inhibitory effect of exogenous ascorbic acid (0.5 mM) on lipid peroxidation (TBARS)* in root, stem and leaf tissues of salinized (300 mM NaCl) tomato seedlings (After Shabala and Neumann, 2001).

Treatment	TBARS* (μmol g^{-1} fresh weight)		
	0 hour NaCl	6 hour NaCl	9 hour NaCl
Root	577± 40	710± 67	1084± 58
Root + ascorbate	497± 9	458± 11	520± 50
Stem	300± 23	445± 17	723± 42
Stem + ascorbate	250± 25	310 27	470± 16
Leaf	378± 22	577± 40	886± 19
Leaf + ascorbate	325± 20	454± 13	587± 42

- TABRS = Cellular accumulation of lipid peroxidation products in the form of thiobarbituric acid reactive substances.

Addition of exogenous ascorbic acid to the root medium remarkably increased seedling survival of treatments with 300 mM NaCl for 7, 8 or 9 hours. Thus, seedlings treated for 9 hours with 300 mM NaCl alone showed zero percent recovery. However, approximately 50 percent of seedlings treated for 9 hours with 300 mM NaCl and 0.5 mM ascorbic acid showed rapid shoot recovery after the return to non-saline medium. Recovered seedlings eventually produced new lateral roots and leaves.

Mittova and coworkers (2002; 3003 a, b; 2004) made a detailed study on the response of cultivated tomato and its wild salt-tolerant relative *Lycopersicon pennellii* to salt-dependent oxidative stress as indicated by the increased H_2O_2 and lipid peroxidation levels which were accompanied with increased contents of the oxidized forms of ascorbate and glutathione. The hydrogen peroxide level decreased, lipid peroxidation level slightly decreased and the levels of the reduced forms of ascorbate and glutathione increased in plastids of *L. pennellii* species in response to salinity stress. The better protection of *L. pennellii* root plastids from salt-induced oxidative stress is correlated with increased activities of superoxide dismutase, ascorbate peroxidase, guaiacol peroxidase, monodehydroascorbate reductase, glutathione peroxidase, glutathione-S-transferase and phospholipid hydroperoxide glutathione peroxidase (Mittova et al., 2002). In the plastids of both species, activities of superoxide dismutase, ascorbate peroxidase and guaiacol peroxidase could be resolved into several isozymes. In *L. esculentum* plastids, two Cu/ZnSOD isozymes are found whereas in *L. pennelli* an additional FeSOD-type is also detected. In response to salinity, activities of selected superoxide dismutase, ascorbate peroxidase and guaiacol peroxidase isozymes are increased in *L. pennellii*, while in *L. esculentum*- plastids the activities of most of the superoxide dismutase and guaiacol peroxidase isozymes decrease. Taken together, Mittova et al. (2002) suggest that plastids play an important role in the adaptation of *L. pennellii* roots to salinity.

Salinity increases reduced glutathione content and reduced : oxidized glutathione ratio in oxidative stress-tolerant *L. pennellii* but not in *L. esculentum*. These changes are associated with salt-induced up-regulation of δ-glutamylcysteine synthetase protein, an effect which is prevented by preincubation with buthionine sulfoximine. Salt treatment induces glutathione peroxidase and glutathione-S-transferase but not glutathione reductase actvities in *L. pennellii*. These results suggest a mechanism of coordinate up-regulation of synthesis and metabolism of reduced glutathione in *L. pennellii*, that is absent from *L. esculentum* (Mittova et al., 2003 a). Further, Mittova et al. (2003 b) report that oxidative stress is evident in *L. esculentum*-mitochondria as indicated by their raised levels of lipid peroxidation and H_2O_2 content whereas their reduced ascorbate and reduced glutathione contents decrease. Concomitantly, the superoxide dismutase activity decreases whereas the ascorbate peroxidase and guaiacol peroxidase activities remain at control level. In contrast, the mitochondria of salt-treated *L. pennellii* do not exhibit salt-induced oxidative stress. In their case salinity induces an increase in the activities of superoxide dismutase, ascorbate peroxidase, monodehydroascorbate reductase, dehydroascorbate reductase and glutathione-dependent peroxidase. The *L. pennellii* peroxisomes exhibit increased superoxide dismutase, ascorbate peroxidase, monodehydroascorbate reductase and catalase activity and their lipid peroxidation, and the H_2O_2 levels are not affected by the salt treatment. The activities of all these enzymes remain at control level in peroxisomes of salt-treated *L. esculentum* plants. The salt-induced increase in the antioxidant enzyme activities in *L. pennellii* plants confer cross-tolerance towards enhanced mitochondrial and peroxisomal reactive oxygen species production imposed by salicylhydroxamic acid and 3-amino-1,2,4-triazole, respectively.

Further, Mittova et al. (2004) studied the salinity-induced up-regulation of the antioxidative system in root mitochondria and peroxisomes of tomato species. Salt stress-induced oxidative stress in *L. esculentum* mitochondria, as indicated by the increased levels of lipid peroxidation and H_2O_2 . These changes are associated with decreased activities of superoxide dismutase and guaiacol peroxidases and contents of ascorbate and glutathione. By contrast, in the mitochondria of salt-treated *L. pennellii* plants both H_2O_2 and lipid peroxidation levels decrease while the levels of ascorbate and glutathione and activities of superoxide dismutase, several isoforms of ascorbate peroxidase, and guaiacol peroxidase increase. Similar to mitochondria, peroxisomes isolated from roots of salt-treated *L. pennellii* plants also exhibit decreased levels of lipid peroxidation and H_2O_2, and increased superoxide dismutase, ascorbate peroxidase and catalase activities. In spite of the fact that salt stress decreases activities of antioxidant enzymes in *L. esculentum* peroxisome, oxidative stress is not evident in these organelles.

Cold stress

Prasad et al. (1994) isolated cDNAs representing three chilling acclimation-responsive genes which were differentially expressed during acclimation. Identification of one of these genes at *cat3*, which encodes the mitochondrial catalase 3 isozyme, led to hypothesize that chilling imposes oxidative stress in maize seedlings. The H_2O_2 levels were elevated during both acclimation and chilling of nonacclimated seedlings. The levels of *cat 3* transcripts and the activities of catalase 3 and guaiacol peroxidase were elevated in mesocotyls during acclimation. Catalase 3 seems to be an important H_2O_2 – scavenging enzyme in maize seedlings. The control 3-day-old seedlings pretreated with H_2O_2 or menadione, a superoxide generating compound, at 27°C induced chilling tolerance. Prasad and coworkers suggested that peroxide has dual effects at low temperatures. During acclimation, its early accumulation signals production of antioxidant enzymes such as catalase 3 and guaiacol peroxidase. At 4°C, in non-acclimated seedlings, it accumulates to damaging levels in the tissues due to low levels of these, and perhaps other, antioxidant enzymes.

With a view to analyzing their protective function against chilling-induced injury, Kocsy et al. (2001) increased the pools of glutathione and its precursors, cysteine and δ-glutamylcysteine in the chilling-sensitive maize inbred line Penjalinan, using a combination of two herbicide safeners. Compared with the controls, great increase in pool size of the 3 thiols was detected in the shoots and roots when both safeners were applied at a concentration of 5 μM. This combination increased the relative protection from chilling from 50 – 75 percent. The increase in total glutathione level was accompanied by a rise in glutathione reductase activity. During chilling, the ratio of reduced : oxidized thiols first decreased independently of the treatments, but increased again to the initial value in safener-treated seedlings after 7 days at 5°C. The results show a linear relationship between glutathione and glutathione reductase, and a biphasic relationship between relative protection and glutathione reductase and glutathione, thus demonstrating the relevance of the glutathione levels in protecting maize seedlings against chilling-induced injury. Glutathione reductase activity is found only in the mesophyll fractions of leaves, with no glutathione reductase activity being detectable in bundle sheath extracts (Pastori et al., 2000). The post-transcriptional regulation prevents glutathione reductase accumulation in the bundle sheath cells of maize leaves. The resulting limitation on the capacity for regeneration of reduced glutathione in this compartment may contribute to the extreme chilling sensitivity of maize leaves. Further, Gömez et al. (2004) cloned genes encoding δ-glutamylcysteine synthetase (d-*Ecs*) and glutathione reductase from maize and specific antibodies produced. The chilling-induced increase in δ-*Ecs* transcripts was not accompanied by enhanced total leaf δ-*Ecs* protein or extractable activity. δ-*Ecs* and glutathione synthetase transcripts and proteins were found in both the bundle sheath and the mesophyll cells under optimal conditions. Chilling increased δ-*Ecs* transcript and protein in the bundle sheath but not in the mesophyll cells. Leaf total glutathione significantly increased only in the recovery period, when the reduced glutathione : glutathione disulfide ratio decreased 3-fold. Thus, while there was a specific increase in the potential contribution of the bundle sheath cells to glutathione synthesis during chilling, it did not result in enhanced leaf glutathione accumulation at low temperatures. Return to optimal temperatures allowed glutathione to increase, particularly glutathione disulfide, and this was associated with leaf chlorosis.

To test the hypothesis that enhanced tolerance to oxidative stress would improve winter survival, McKersie et al. (1999) transformed two clones of alfalfa (*Medicago sativa*) with a MnSOD targeted to the mitochondria or the chloroplast. After one winter, most transgenic plants had higher survival rates than control plants, with some at 100 percent. Similarly, some independent transgenic plants had twice the herbage yield of the control plants. Although many of the transgenic plants had higher winter survival rates and herbage yield, there was no apparent difference in primary freezing injury, and therefore, the trait is not associated with a change in the primary site of freezing injury of alfalfa plants.

Heat stress

Huang et al. (2001) determined whether oxidative stress is involved in leaf injury induced by high soil temperatures in two creeping bentgrass (*Agrotis palustris*) cultivars, heat-tolerant L-93 and heat-sensitive Penncross. Shoots and roots were exposed to four different temperature regimes : (i) 20/20°C (control), (ii) 20/35°C (high soil temperature), (iii) 35/20°C (high air temperature), and (iv) 35/35°C (high soil and air temperatures). Electrolyte leakage and malondialdehyde contents increased under high soil temperature alone or in combination with high air temperature regimes in both cultivars, but to a greater extent in Penncross than in L-93 (Fig. 12.7). Decreases in turf quality, and increases in electrolyte leakage and malondialdehyde were more pronounced at 20/35°C than at 35/20°C. The activities of superoxide dismutase and catalase decreased with prolonged periods of high temperatures and to a greater extent for Penncross than for L-93. The reduction in superoxide dismutase and catalase activities were more severe at 20/35°C than at 35/20°C. These results suggest that high soil temperature caused more severe oxidative damage to leaves than the high air temperature.

Dat et al. (1998) investigated changes in endogenous salicylic acid and antioxidants in relation to induced thermotolerance of mustard (*Sinapis alba*) seedlings. The reduced : oxidized ascorbate ratio was 5-fold lower than in the control, one hour after treatment but recovered by two hours after treatment. The glutathione pool became slightly more oxidized from 2 hours after treatment. Glutathione reductase activity was more than 50 percent higher during the first two hours. Activities of dehydroascorbate reductase and monodehydroascorbate reductase decreased by at least 25 percent during the first two hours, but were 20 – 60 percent higher than the control levels after 3 – 6 hours. One hour after heat acclimation ascorbate peroxidase activity increased by 30 percent. Young leaves appear better protected by antioxidant enzymes following heat acclimation than the cotyledons or stem. Sairam et al. (2000) also observed increased antioxidant

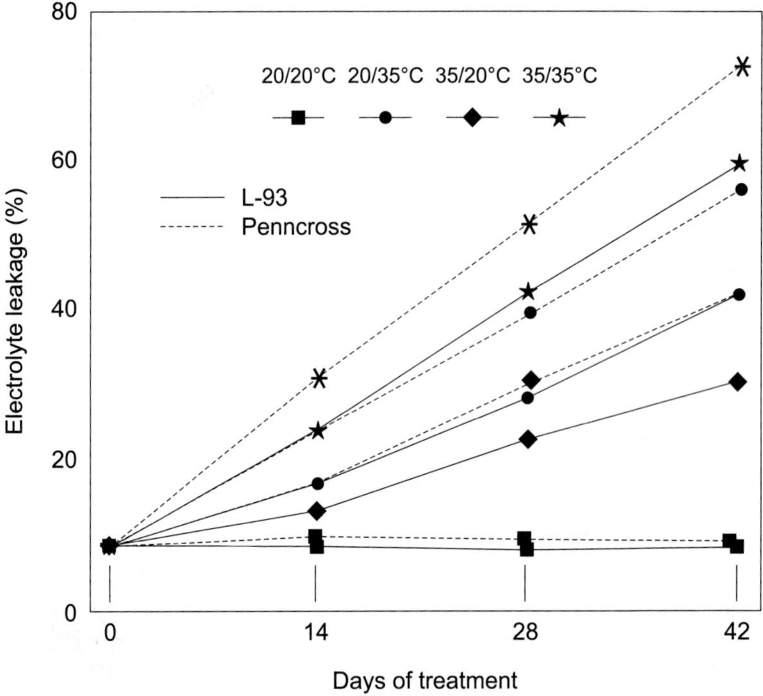

Fig. 12.7 Cell membrane permeability expressed as electrolyte leakage of two creeping bentgrass cultivars, L-93 (solid line and symbol) and Penncross (dotted line), in response to differential air/soil temperature. Vertical bars indicate LSD values (P= 0.05) for treatment and cultivar comparisons at a given day (After Huang et al., 2001).

activity under elevated temperature conditions of heat-tolerant wheat genotypes.

Hypoxia/Anoxia

Oxygen shortage/oxygen deprivation stress in plant cells is distinguished by three physiologically different states ; transient hypoxia, anoxia and reoxygenation. Generation of reactive oxygen species is characteristic for hypoxia and especially reoxygenation. Of the ROS, hydrogen peroxide and superoxide ($O_2^{\cdot-}$) are both produced in a number of cellular reactins, including the iron-catalyzed Fenton reaction, and by various enzymes such as lipoxygenases, peroxidases, NADPH oxidase and xanthine oxidase. The main cellular components susceptible to damage by free radicals are lipids (peroxidation of unsaturated fatty acids in membranes), proteins (denaturation), carbohydrates, and nucleic acids (Blokhina et al., 2001). Consequences of hypoxia-induced oxidative stress depend on tissue and/ or species (i.e. their tolerance to anoxia), on membrane properties, on endogenous antioxidant content and on the ability to induce the response in the antioxidant system. Effective utilization of energy resources (starch, sugars) and the switch to anaerobic metabolism and the preservation of the redox status of the cell are vital for survival (Blokhina et al., 2002). The formation of ROS is prevented by an antioxidant system : low molecular mass antioxidants (ascorbic acid, glutathione, tocopherols), enzymes generating the reduced forms of antioxidants, and ROS-interacting enzymes such as superoxide dismutase, peroxidases and catalases. In plant tissues many phenolic compounds (in addition to tocopherols) are potential antioxidants; flavonoids, tannins and lignin precursors may work as ROS-scavenging compounds. Antioxidants act as a cooperative network, employing a series of redox reactions. Interactions between ascorbic acid and glutathione, and ascorbic acid and phenolic compounds are well known. Such aspects as compartmentalization of ROS formation and antioxidant localization, synthesis and transport of antioxidants, the ability to induce the antioxidant defence and cooperation (and/or compensation) between different antioxidant systems are the determinants of the competence of the antioxidant system (Blokhina et al., 2002).

Tolerance of anoxia is greatly improved when pretreated with 2 – 4 hours of hypoxia. But anoxia followed by reoxygenation causes extensive damage to cellular components through generation of ROS. Anoxia pretreatment protects soybean against H_2O_2 concentrations that induces programed cell death in normoxic cells (Amora et al., 2000). H_2O_2 removal in anoxia pretreated soybean culture is faster. Protection is associated with increased action of alternative oxidase and peroxidases. Alternative oxidase inhibition abolishes the protective effect, while induction of alternative oxidase protects normoxic cells against H_2O_2 . Amora et al. propose that during anoxia, plant cells can prepare for reoxygenation injury by up-regulating their antioxidant capacity, and that alternative oxidase is involved in this process.

Biemelt et al. (1998) investigated the response of ascorbate-glutathione cycle in roots of young wheat seedlings that were deprived of oxygen either by subjecting them to root hypoxia or to entire plant inoxia and then re-aerated. Although higher total levels of ascorbate and glutathione were observed under hypoxia, only the total amount of ascorbate was increased under anoxia. Under both treatments a significant increase in the reduced form of ascorbate and glutathione was found, resulting in increased reduction states. Upon the onset of re-aeration the ratios started to decline rapidly, indicating oxidative stress. Hypoxia caused higher activity of ascorbate peroxidase, whereas activities of monodehydroascorbate reductase, dehydroascorbate reductase, and glutathione reductase diminished or were only slightly influenced. Under anoxia, activities of ascorbate peroxidase and glutathione reductase decreased significantly to

39 and 62 percent, respectively. However, after re-aeration of hypoxically or anoxically pretreated roots, activity of enzymes approached the control levels. This corresponds with the restoration of the high reduction state of ascorbate and glutathione within 16 to 96 hours of re-aeration, depending on the previous duration of anoxia. Apparently, anoxia followed by re-aeration more severely impairs entire plant metabolism compared with hypoxia, thus leading to decreased viability.

A large-scale investigation of monodehydroascorbate reductase and dehydroascorbate reductase activities, and ascorbic acid and glutathione contents in 11 species with contrasting tolerance to anoxia has revealed an increase in monodehydroascorbate reductase and/or dehydroascorbate reductase in the anoxia tolerant plants after several days of anoxia treatment (Wollenweber-Ratzer and Crawford, 1994). However, in the intolerant plants the activities were very low or without any changes. While glutathione decreased significantly during the post-anoxia period, ascorbic acid showed increased values in the tolerant species. Further, the investigation by Biemelt et al. (1998) on the antioxidative defence system in the roots of wheat seedlings under root hypoxia or whole plant anoxia revealed a significant increase in the reduced forms of ascorbate and glutathione. Nevertheless, a rapid decrease in the redox state of both antioxidants was observed during re-aeration. Ascorbate works in cooperation not only with glutathione, but also takes part in the regeneration of α-tocopherol, providing synergestic protection of the membranes (Thomas et al., 1992). Tocopherol is in direct interaction with reduced glutathione.

Heavy metal stress

Redox-active metals, such as iron, copper and chromium, undergo redox cycling whereas redox-inactive metals, such as lead, cadmium, mercury and others deplete cells' major antioxidants, particularly thiol-containing antioxidants and enzymes. Either redox-active or redox-inactive metals may cause an increase in production of reactive oxygen species such as hydroxyl radical, superoxide radical or hydrogen peroxide. Enhanced generation of ROS can overwhelm cells' intrinsic antioxidant defences, and result in a condition known as "oxidative stress" (Ercal et al., 2001). Maximum accumulation of cadmium occurs in roots followed by stems and leaves of *Pisum sativum* (Dixit et al., 2001). An enhanced level of lipid peroxidation and an increased tissue concentration of hydrogen peroxide in both roots and leaves indicate that cadmium causes oxidative stress in pea plants. *Cadmium*-induced enhancement in superoxide dismutase activity is more at 40 µM than at 4 µM in leaves. While catalase prominently increases in leaves both at 4 and 40 µM cadmium, ascorbate peroxidase shows maximum stimulation at 40 µM cadmium in roots. Enhancement in glutathione reductase activity is also more at 40 µM than at 4 µM cadmium in roots. While glutathione peroxidase activity decreases in roots and remains almost unmodified in leaves, glutathione-S-transferase shows pronounced stimulation in both roots and leaves of pea plants exposed to 40 µM cadmium.

Cadmium and some other metals cause a transient depletion of glutathione and an inhibition of antioxidative enzymes, especially of glutathione reductase (Schützendübel and Polle, 2002). Assessment of antioxidative capacities by metabolic modeling suggests that the reported diminution of antioxidants is sufficient to cause H_2O_2 accumulation. The depletion of glutathione is apparently a critical step in cadmium sensitivity since plants with improved capacities for glutathione synthesis displayed higher cadmium tolerance. Available data suggest that cadmium, when not detoxified rapidly enough, may trigger, via the disturbance of the redox control of the cell, a sequence of reactions leading to growth inhibition, stimulation of secondary metabolism, lignification, and finally cell death.

It has been observed that some plants in certain mycorrhizal associations are less sensitive to cadmium stress than non-mycorrhizal plants. Schützendübel and Polle (2002) suggest that it may be possible that mycorrhizal fungi provide protection via glutathione, since higher concentrations of this thiol are found in pure cultures of the fungi than in bare roots. The development of stress tolerant plant-mycorrhizal associations may thus be a promising new strategy for phytoremediation and soil amelioration measures.

Genetically engineered Indian mustard (*Brassica juncea*) plants were developed by Zhu et al. (1999) to overexpress *Escherichia coli gsh1* gene encoding δ-glutamylcysteine synthetase, targeted to plastids. The δ-glutamylcysteine synthetase transgenic seedlings showed increased tolerance to cadmium and had higher concentrations of phytochelatins, δ-glutamylcysteine, glutathione, and total nonprotein thiols compared with wild-type seedlings. When tested in a hydroponic system, δ-glutamylcysteine synthetase in mature plants, accumulated more cadmium than the wild-type plants; shoot cadmium concentrations were 40 to 90 percent higher. In spite of their higher tissue cadmium concentration, the δ-glutamylcysteine synthetase plants grew better in presence of cadmium than the wild-type. Zhu et al. concluded that overexpression of δ-glutamylcysteine synthetase increases biosynthesis of glutathione and phytochelatins, which in turn enhances cadmium tolerance and accumulation. Thus, the overexpression of δ-glutamylcysteine synthetase appears a promising strategy for the production of plants with superior heavy metal phytoremediation capacity.

Later, Freeman et al. (2004) reported that increased glutathione biosynthesis plays a role in *nickel* tolerance in *Thlaspi* (*Brassicaceae*) nickel hyperaccumulation. *Thlaspi* species hyperaccumulate nickel up to 3 percent of shoot dry weight. They observed that concentrations of glutathione, in shoot tissues, is strongly correlated with the ability to hyperaccumulate nickel in various *Thlaspi* hyperaccumulators collected from serpentine soils. Further, analysis of the Australian nickel hyperaccumulator *T. goesingense* revealed that the high concentration of glutathione observed in this hyperaccumulator coincides with constitutively high activity of glutathione reductase. The change in glutathione metabolism also coincides with the ability to both hyperaccumulate nickel and resist its damaging oxidative effects. Freeman et al. concluded that the elevated glutathione concentrations driven by constitutively elevated serine acetyltransferase activity, are involved in conferring tolerance to nickel-induced oxidative stress in *Thlaspi* nickel accumulators.

Aluminum tolerance mechanism, together with oxidative stress tolerance was investigated in an Al-tolerant cell line (*ALT301*) and the parental Al-sensitive line (*SL*) of tobacco (Devi et al., 2003). During Al exposure in a simple calcium solution for 24 hours, Al triggered the evolution of ROS in Al-sensitive cell line much higher than in *ALT301*. *ALT301* contained ascorbate and glutathione levels that were higher than *SL* under normal growth conditions. Under these oxidant exposures, *ALT301* contained lower levels of intracellular H_2O_2 or lipid peroxides, and maintained higher amounts of ascorbate and glutathione than the *SL*. Taken together, Devi and coworkers concluded that the accumulation of aluminum in cells enhances peroxidation of lipids exclusively under growing conditions, and that the higher content of ascorbate and glutathione in *ALT301* than in *SL* seems to be in part responsible for the tolerance mechanism of *ALT301* to aluminum by protecting cells from either lipid peroxidation of H_2O_2 commonly enhanced by aluminum or other oxidants. Involvement of ascorbate system in the response of pumpkin (*Cucurbita pepo*) roots to aluminum stress was further studied by Dipierro et al. (2005). Treatment of 5-day-day-old pumpkin seedlings with 50 mM aluminum sulfate resulted in approximately 60 percent inhibition of root growth within 48-60 hours of treatment, while aluminum accumulated in the roots reaching a maximum within 48 hours. During the same period, the hydrogen peroxide content of the roots was strongly enhanced. The increased level of hydrogen peroxide was matched by both increased ascorbate peroxidase activity and ascorbate free radical reductase activity, while dehydroascorbate reductase and glutathione reductase did not change. The levels of ascorbate in the roots were also increased by the aluminum treatment. It was concluded that an oxidative burst is probably involved in the toxicity of aluminum in pumpkin roots and that plants react to the enhanced production of reactive oxygen species by expressing higher levels of scavenging systems such as ascorbate-ascorbate peroxidase system.

REFERENCES

Amora, Y.; M. Chevionb and A. Levinea, 2000. Anoxia pretreatment protects soybean cells against H_2O_2 –induced cell death : possible involvement of peroxidases and of alternative oxidase. FEBS Lett. 477: 175-80.

Arora, A.; T.M. Byrem; M.G. Nair and G.M. Strasburg, 2000. Modulation of liposomal membrane fluidity by flavonoids and isoflavonoids. Arch. Biochem. Biophys. 373: 102-09.

Bethke, P.C. and R.L. Jones, 2001. Cell death of barley aleurone protoplasts is mediated by reactive oxygen species. Plant J. 25: 19-29.

Biemelt, S.; U. Keetman and G. Albrecht, 1998. Re-aeration following hypoxia or anoxia leads to activation of the antioxidative defence system in roots of wheat seedlings. Plant Physiol. 116: 651-58.

Blokhina, O.B.; T.V. Chirkova and K.V. Fagerstedt, 2001. Anoxic stress leads to hydrogen peroxide formation in plant cells. J. Exp. Bot. 52: 1-12.

Blokhina, O.B.; E. Virolainen and K.V. Fagerstedt, 2002. Antioxidants, oxidative damage and oxygen deprivation stress : a review. Ann. Bot. 91: 179-94.

Brown, P.S.; D.P. Knievel and E.J. Pell, 1995. Effects of moderate drought on ascorbate peroxidase and glutathione reductase activities in mesophyll and bundle sheath cells of maize. Physiol. Plant. 95: 274-80.

Chen, Z and R. Gallie, 2004. The ascorbic acid redox state controls guard cell signaling and stomatal movement. Plant Cell 16: 1143-62.

Dat, J.F.; C.H. Foyer and I.M. Scott, 1998. Changes in salicylic acid and antioxidants during induced thermotolerance in mustard seedlings. Plant Physiol. 118: 1455-61.

Devi, S.R.; Y. Yamamoto and H. Matsumoto, 2003. An intracellular mechanism of aluminum tolerance associated with high antioxidant status in cultivated tobacco cells. J. Inorg. Biochem. 97: 59-68.

Dipierro, N.; D. Mondelli; C. Paciolla; et al. 2005. Changes in the ascorbate system in the response of pumpkin (*Cucurbita pepo* L.) roots to aluminum stress. J. Plant Physiol. 162: 529-36.

Dixit, V.; V. Pandey and R. Shyam, 2001. Differential antioxidative response to cadmium in roots and leaves of pea (*Pisum sativum* L. cv Azad) J. Exp. Bot. 52(358): 1101-09.

Ercal, N.; H. Gurer-Orhan and N. Aykin-Burns, 2001. Toxic metals and oxidative stress. Part I. Mechanisms involved in metal-induced oxidative damage. Curr. Top. Med. Chem. 1 (6): 529-39.

Foyer, C.H. and J. Harbinson, 1994. Oxygen metabolism and the regulation of photosynthetic electron transport. pp 1-42 In : Causes of Photooxidative Stress and Amelioration of Defence System in Plants (eds) C.H. Foyer and P.M. Mullineaux, CRC Press, Boca Raton, FL. .

Foyer, C.H.; H. Lopez-Delgado; J. Dat and I. Scott, 1997. Hydrogen peroxide- and glutathione-associated mechanisms of acclimatory stress tolerance and signaling. Physiol. Plant. 100: 241-54.

Freeman, J.L.; M.W. Persans; K. Nieman; et al. 2004. Increased glutathione biosynthesis plays a role in nickel tolerance in *Thlaspi* nickel hyperaccumulators. Plant Cell 16: 2176-91.

Fryer, M.J. 1992. The antioxidant effects of thylakoid vitamin E (α-tocopherol). Plant Cell Environ. 15: 381-92.

Gömez, J.M.; A. Jimenez; E. Olmos and F. Sevilla, 2003. Location and effects of long-term NaCl effects on superoxide dismutase and ascorbate peroxidase isozymes of pea (*Pisum sativum* cv Puget) chloroplasts. J. Exp. Bot. 55(394): 119-30.

Gömez, L.D.; H. Vanacker; P. Buckner; et al. 2004. Intercellular distribution of glutathione synthesis in maize leaves and its response to short-term chilling. Plant Physiol. 134: 1662-71.

Gupta, A.S.; H. Heinen; A.S. Holaday; et al. 1993. Increased resistance to oxidative stress in transgenic plants that overexpress chloroplastic Cu/Zn superoxide dismutase. Proc. Nat. Acad. Sci. 90: 1629-33.

Hernandez, J.A.; F.J. Corpas; M. Gömez; et al. 1993. Salt-induced oxidative stress mediated by activated oxygen species in pea leaf mitochondria. Physiol. Plant. 89: 103-10.

Huang, B.; X. Liu and Q. Xu, 2001. Supraoptimal soil temperatures induced oxidative stress in leaves of creeping bentgrass cultivars differing in heat tolerance. Crop Sci. 41: 430-35.

Joo, J.H.; Y.S. Bae and J.S. Lee, 2001. Role of auxin-induced reactive oxygen species in root gravitropism. Plant Physiol. 126: 1055-60.

Kocsy, G.; P. van Ballmoos; A. Rüegsegger; et al. 2001. Increasing the glutathione content in a chilling-sensitive maize genotype using safners increased protection against chilling-induced injury. Plant Physiol. 127: 1147-56.

Kocsy, G.; G. Szalai and G. Galiba, 2002. Induction of glutathione synthesis and glutathione reductase activity by abiotic stresses in maize and wheat. Scientific World J. 2: 1699-705.

Loggini, B.; A. Scartazza; E. Brugnoli; et al. 1999. Antioxidative defence system, pigment composition, and photosynthetic efficiency in two wheat cultivars subjected to drought. Plant Physiol. 119: 1091-1100.

Mahan, J.R. and D.F.. Wanjura, 2005. Seasonal patterns of glutathione and ascorbate metabolism in field grown cotton under water stress. Crop Sci. 45: 193-201.

Matamoras, M.A.; D.A. Dalton; J. Ramos; et al. 2003. Biochemistry and molecular biology of antioxidants in the *Rhizobia*-legume symbiosis. Plant Physiol. 133: 499-509.

May , M.J.; T. Vernoux; C. Leaver; et al. 1998. Glutathione homeostasis in plants : implications for environmental sensing and plant development. J. Exp. Bot. 49: 649-67.

McKersie, B.D.; S.R. Bowley; E. Harjanto and O. Leprince, 1996. Water deficit tolerance and field performance of transgenic alfalfa overexpressing superoxide dismutase. Plant Physiol. 111: 1177-81.

McKersie, B.D.; S.R. Bowley and K.S. Jones, 1999. Winter survival of transgenic alfalfa overexpressing superoxide dismutase. Plant Physiol. 119: 839-48.

Miyamoto, Y.; Y.H. Koh; Y.S. Park; et al. 2003. Oxidative stress caused by inactivation of glutathione peroxidase and adaptive responses. Biol. Chem. 384 (4): 567-74.

Mittova, V.; M. Guy; M. Tal; et al. 2002. Response of the cultivated tomato and its wild salt-tolerant relative *Lycopersicon pennellii* to salt-dependent oxidative stress increased activities of antioxidant enzymes in root plastids. Free Radic. Res. 36: 195-202.

Mittova, V.; F.L. Theodoulou; G. Kiddle; et al. 2003 a. Coordinate induction of glutathione biosynthesis and glutathione metabolising enzymes is correlated with salt tolerance in tomato. FEBS Lett. 554: 417-21.

Mittova, V.; M. Tal; M. Volokita; et al. 2003 b. Up-regulation of the leaf mitochondrial and peroxisomal antioxidative systems in response to salt induced oxidative stress in the wild salt-tolerant tomato species *Lycopersicon pennellii*. Plant Cell Environ. 26(6): 845-56.

Mittova, V.; M. Guy; M. Tal; et al. 2004. Salinity up-regulates the antioxidative system in root mitochondria and peroxisomes of the wild salt-tolerant tomato species *Lycopersicon pennellii*. J. Exp. Bot. 55 (399): 1105-13.

Munne-Bosch, S. and L. Alegre, 2000. Changes in carotenoids, tocopherols and diterpenes during drought and recovery, and the biological significance of chlorophyll loss in *Rosmarinus officinalis* plants. Planta 210: 925-31.

Munne-Bosch, S. and L. Alegre, 2003. Drought-induced changes in the redox state of α–tocopherol, ascorbate, and the diterpene carnosic acid in chloroplasts of Labiatae species differing in carnosic acid contents. Plant Physiol. 131: 1816-25.

Murata, Y.; Z.M. Pei; I.C. Mori; et al. 2001. Abscisic acid activation of plasma membrane Ca^{2+} channels in guard cells requires cytosolic NAD(P)H and is differentially disrupted upstream and downstream of reactive oxygen species production in *abi 1-1* and *abi 2-1* protein phosphatase 2C mutants. Plant Cell 13: 2513-23.

Navabpour, S.; K. Morris; R. Allen; et al. 2003. Expression of senescence-enhanced genes in response to oxidative stress. J. Exp. Bot. 54(391): 2285-92.

Neill, S.; J.R. Desikan; A. Clarke; et al. 2002. Hydrogen peroxide and nitric oxide as signaling molecules in plants. J. Exp. Bot. 53(372): 1237-47.

Noctor, G. and C. Foyer, 1998. Ascorbate and glutathione : keeping active oxygen under control. Annu. Rev. Plant Physiol. Plant Mol. Biol. 49:249-79.

Noctor, G.; S. Veljovic-Jovanovic and C.H. Foyer, 2000. Peroxide processing in photosynthesis : antioxidant coupling and redox signaling. Philos. Trans. R. Soc., Lond. 355: 1465-75.

Noctor, G.; L. Gomez; H. Vanacker; et al. 2002. Interactions between biosynthesis, compartmentation and transport in the control of glutathione homeostasis and signaling. J. Exp. Bot. 53 (372): 1283-1304.

Pastori, G.M.; P.M. Mullineaux and C.H. Foyer, 2000. Post-transcriptional regulation prevents accumulation of glutathione reductase protein and activity in the bundle sheath cells of maize. Plant Physiol. 122: 667-76.

Pastori, G.M. and C.H. Foyer, 2002. Common components, networks, and pathways of cross-tolerance to stress. The central role of "Redox" and abscisic acid-mediated controls. Plant Physiol. 129: 460-68.

Polidoros, A.N. and J.G. Scandalios, 1999. Role of hydrogen peroxide and different classes of antioxidants in the regulation of catalase and glutathione S-transferase gene expression in maize (*Zea mays* L.). Physiol. Plant. 106: 112-20.

Potters, G.; N. Horemans; S. Bellone; et al. 2004. Dehydroascorbate influences the plant cell cycle through a glutathione-independent reduction mechanism. Plant Physiol. 134: 1479-87.

Prasad, T.K.; M.D. Anderson; B.A. Martin; et al., 1994. Evidence for chilling induced oxidative stress in maize seedlings and a regulatory role for hydrogen peroxide. Plant Cell 6: 65-74.

Ruiz, J.M. and E. Blumwald, 2002. Salinity induced glutathione synthesis in *Brassica napus*. Planta 214: 965-69.

Sairam, R.K.; P.S. Deshmukh and D.C. Saxena, 1998. Role of antioxidant systems in wheat genotype tolerance to water stress. Biol. Plant. 41: 387-94.

Sairam, R.K. and D.C. Saxena, 2000. Oxidative stress and antioxidants in wheat genotypes : possible mechanism of water stress tolerance. J. Agron. Crop Sci. 184 : 55.

Sairam, R.K.; G.C. Srivastava and D.C. Saxena, 2000. Increased antioxidant activity under elevated temperatures : a mechanism of heat stress tolerance in wheat genotypes. Biol. Plant. 43: 245-51.

Scandalios, J.G. 1990. Response of plant defence genes to environmental stress. Adv. Genet. 28: 1-41.

Schützendübel, A. and A. Polle, 2002. Plant responses to abiotic stresses : heavy metal-induced oxidative stress and protection by mycorrhization. J. Exp. Bot. 53 (372): 1351-65.

Sgherri, C.L.M. and F. Navari-Izzo, 1995. Sunflower seedlings subjected to increasing water deficit stress : oxidative stress and defence mechanisms. Physiol. Plant. 93: 25-30.

Sgherri, C.L.M.; B. Loggini; A. Bochicchio; et al. 1994 a. Antioxidant system in *Boea hygroscopica:* changes in response to desiccation and rehydration. Phytochemistry 37: 377-81.

Sgherri, C.L.M.; B. Loggini; S. Puliga; et al. 1994 b. Antioxidant system in *Sporobolus stapfianus*: changes in response to desiccation and rehydration. Phytochemistry 33:561-65.

Shabala, A. and P.M. Neumann, 2001. Exogenous ascorbic acid (vitamin C) increases resistance to salt stress and reduces lipid peroxidation. J. Exp. Bot. 52 (364): 2207-11.

Smirnoff, N. 1993. The role of active oxygen in the response of plants to water deficit and desiccation. New Phytol. 125: 27-58.

Storozhenko, S.; P. de Pauw; M. van Montagu; et al. 1998. The heat shock element is a functional component of the *Arabidopsis APX 1* gene promoter. Plant Physiol. 118: 1005-14.

Tausz, M.; H. Sircelj and D. Grill, 2004. The glutathione system as a stress marker in plant ecophysiology : is a stress-response concept valid ? J. Exp. Bot. JXB Advance Access published online on July 2, 2004.

Thomas, C.E.; L.R. McLean; R.A. Parker; et al. 1992. Ascorbic acid and phenolic antioxidant interactions in prevention of liposomal oxidation. Lipids 27: 543-50.

Willekens, H.; S. Chamnogpol; M. Davey; et al. 1997. Catalase is a sink for H_2O_2 and is indispensable for stress defence in C_3 plants. The EMBO J. 16(16): 4806-16.

Wollenweber-Ratzer, B. and R.M.M. Crawford, 1994. Enzymatic defence against post-anoxic injury in higher plants. Proc. Royal Soc. Edinbergh 102B: 381-90.

Zhang, J.X. and M.B. Kirkham, 1994. Drought-stress-induced changes in activities of superoxide dismutase, catalase, and peroxidase in wheat species. Plant Cell Physiol. 35: 785-91.

Zhu, Y.L.; E.A.H. Pilon-Smits; A.S. Tarun; et al. 1999. Cadmium tolerance and accumulation in Indian mustard is enhanced by overexpressing δ-glutamylcysteine synthetase. Plant Physiol. 121 : 1169-77.

SUGGESTED READINGS

Blokhina, O.; E. Virolainen and K.V. Fagerstedt, 2002. Antioxidants, oxidative damage and oxygen deprivation stress : a review. Ann. Bot. 91: 179-94.

Loggini, B.; A. Scartazza; E. Brugnoli and F. Navari-Izzo, 1999. Antioxidative defense system, pigment composition, and photosynthetic efficiency in two wheat cultivars subjected drought. Plant Physiol. 119: 1091-1100.

Matamoras, M.A.; D.A. Dalton; J. Ramos; M.R. Clemente; M.C. Rubio and M. Becana, 2003. Biochemistry and molecular biology of antioxidants in the *Rhizobia*- legume symbiosis. Plant Physiol. 133: 499-509.

Shabala, A. and P.M. Neumann, 2001. Exogenous ascorbic acid (vitamin C) increases resistance to salt stress and reduces lipid peroxidation. J. Exp. Bot. 52 (364): 2207-11

APPENDIX

Measurement of osmotic adjustment

Osmotic adjustment is calculated as the difference in osmotic potential at full turgor between well-watered and drought-stressed plants. In practice, the relative water content is monitored during the stress period. When relative water content reaches a value where wilting occurs in the test plants, plants are rehydrated and samples are collected after the plants regain full turgor (Zhang et al., 1999). The rehydration method is more suitable for extensive measurement of osmotic adjustment when evaluating a large number of genotypes. Measurement of osmotic adjustment requires that water deficit should develop slowly so that full expression of solute accumulation potential occurs before plants wilt completely.

Cell turgor measurement

A micro-pipette is used which is fabricated using a double-pull method from borosilicate tubing with an internal filament. A subminiature pressure transducer is housed adjuscent to the pressure-probe micro-pipette in a small brass holder connected to a micrometer-driven piston by thick-walled teflon tubing (1.59 mm outer diameter × 0.254 mm inner diameter).The piston, tubing, holder, and micropipette are filled with silicon oil. The flexible tubing minimizes mechanical vibration, which can damage the cell, especially during long-term measurements. The micropipette is fabricated using a double-pull protocol to achieve a final diameter of about 1.5 µm. To avoid cellular damage during implacement and, thus, improve chances of observing turgor regulation during long-term monitoring (40-50 min) of cell turgor pressure in a single epidermal cell. In fact, after implacement with the small aperture pressure probe, cells maintain normal cytoplasmic streaming and exhibit normal electrical properties. However, the small aperture (and relatively large silicone oil reservoir) slows fluid movement through the micropipette tip. The restricted flow means that pressure readings must be made after the meniscus has stopped moving (requiring about 5-10 s) to ensure that hydraulic equilibrium is achieved. Once filled, the micropipette is introduced into the APW bath solution. Capillary action causes the bath solution to enter the tip of the micropipette. This is offset by applying pressure (0.31 ± 0.06

MPa, *n*=17) to bring the oil/APW meniscus to the tip of micropipette. The baseline offset is monitored for a minute or so to ensure there is no significant drift both before and after cell turgor measurements. Turgor pressure measurements are referenced to the baseline offset, that is, the baseline offset is used as "zero" pressure in cell turgor measurements. Pressure is monitored to a digital oscilloscope as voltage output from the transducer using standard electronics.

The micropipette tip is localized in the vacuole upon implacement. Immediately after implacement, the oil/vacuolar sap interface moves back into the micropipette due to the cell turgor. Pressure is applied to bring the meniscus to the pipette tip. The cell turgor pressure is then calculated from the voltage readings based on calibrations performed with pressurized air and corresponds to the cell turgor pressure.

Membrane potential measurement

Membrane potentials of the mesophyll cells are measured with glass microelectrodes before and 40 minutes after the onset of hyperosmotic stress using the MIFE electrometer. Electrodes have a tip diameter < 1 μm and are backfilled with 0.5 M KCl.

Osmotic potential measurement

The osmotic potential of expressed leaf sap is measured with a vapor pressure osmometer using potassiun chloride solutions as standards at 25°C. The chamber is equilibrated for different time periods (usually 1 – 3 min) before taking measurement, and the stability of the instrument during the measurement period is tested in each experiment.

Leaf osmolality

Leaf discs (1.1 cm diameter) are weighed and freeze-dried. After determination of the dry weight they are homogenized in 0.5 mL of distilled water. Samples are boiled in a water bath for 5 minutes and centrifused for 30 minutes at 16,000 g_n . Osmolality of the supernatants is determined in a freezing-point depression osmometer. Samples are taken from the same leaves that are used to determine freeze-thaw damage. Samples from at least three different plants are taken from every independent treatment set grown under either acclimating or non-acclimating conditions. Significance levels for the measured differences in leaf osmolality are evaluated with an unpaired Student's test.

Patch-clamp method

Double-barreled microelectrodes are prepared using a double-pull protocol with intermediate twist. The electrodes are backfilled with 200 mM KCl and connected by AgCl electrodes to IE-251 electrometers. Before electrodes are implanted in the root hair, the absence of cross talk is confirmed by injecting 1 nÅ of current through one electrode and checking for significant voltage deflection in the other electrode.

Current-voltage measurements are performed using an operational amplifier controlled by a data acquisition board via a compiled C program using the current injection capability of the electrometer. The current injected through one of electrodes is measured via the electrometer and sampled by the data acquisition board after filtering at 200 Hz with an eight-pole Bessel filter. Both current and voltage traces are displayed on an oscilloscope and printed as a hard copy. The experimental protocol during voltage-clamp measurements involve regular (every 2-3 minutes) clamping of the MP at different values (-3000 to 0 mV range) for 40 to 50 seconds. In most cases, a bipolar staircase of voltage clamp (alternative clamps above and below the resting potential) is used, each clamp followed by 1 to 2 minutes of no clamping.

In voltage-clamp experiments, net ion fluxes are measured from the surface of a young root hair cell. The ion-selective K^+ vibrating microelectrode is located at a distance of 4 to 6 µm from the root hair surface. The double-barreled microelectrode is implanted from the opposite side of the root hair, and the voltage (resting potential) is monitored for about 1 minute. Once the reading is stable, flux measurements are commenced. Net K^+ fluxes and clamping currents are averaged for each voltage-clamp treatment.

Membrane permeability and electrolyte leakage

Leaves or leaf segments of heat-treated plants are cut into 1 cm^2 pieces and placed in test tubes with 20 ml deionized distilled water (0.5-0.8 g fresh leaf tissue per sample). After vortexing the samples for 3 seconds, the initial electrical conductivity (EC_0) for each sample is measured. The samples are stored at 4°C for 24 hours, and conductivity (EC_1) is measured again. Samples are then autoclaved for 15 minutes, cooled to room temperature, and conductivity (EC_2) is measured for the third time. The relative permeability of cell membranes is calculated as under :

$$\text{Relative permeability } (\%) = [(EC_1 - EC_0)/(EC_2 - EC_0) \times 100$$

The tolerance of leaf cell membranes to high temperature also is estimated by continuously monitoring the electrical conductivity of the incubation solution while the temperature of the system is slowly increased. Approximately 3 gram of leaf tissue is vacuum infiltrated with distilled deionized water for 10 minutes. Leaf pieces are then transferred to a glass beaker with 1 liter of distilled deionized water and placed on a stirring hot plate. Controls are adjusted to provide gentle stirring and to increase the solution temperature from 25 to 70°C at a rate of 0.3-0.4°C min^{-1} . Electrical conductivity and temperature are monitored using a conductivity meter connected to a computer, data are logged at 20s intervals.

Isolation of chloroplasts (After Nelson et al., 1998)

Five gram of leaf tissue from NaCl treated plants is ground in 125 mL of buffer (1 mM $MgCl^{-}_{2}$, 5 mM ascorbic acid, 20 mM Hepes, pH 7.5, 0.5 M sorbitol, and 0.25% BSA) by using a polytron at low speed. The extract is filtered through Miracloth and centrifused at -2000 *g* for 3 minutes. The pillet is suspended in 10 ml of grinding buffer and again filtered. The extract is layered on top of a Percoll gradient consisting of 80% Percoll in grinding buffer and 40% Percoll in grinding buffer. After centrifugation (8 min at 4000 rpm), the layer of intact chloroplasts is collected, diluted 20-fold in a grinding buffer, and centrifused at 4000 rpm for 3 minutes. The pillet is extracted with 80% acetone and centrifused. The supernatant is kept for chlorophyll determination, and the pillet is resuspended in buffer for SDS-PAGE.

Collection of phloem exudates and measurement of solutes (After Nelson et al., 1998)

Side shoots bearing two pairs of 8-week-old plants are excised 4 hours after initiation of the light period. The cut surface is submerged in 2 ml of 10 mM EDTA, pH 8.0, contained in liquid scintillation vials. After a 2-hour period to wash out solutes released directly at the cut surface, the side shoots are placed in fresh media for an additional 4 hours. The buffer is then lyophilized for determination of sucrose, inositol, and ononitol levels.

Ion flux measurements

Net fluxes of K^{+} , Cl^{-} , and Na^{+} are measured non-invasively using ion selective vibrating microelectrodes. Micropipettes are pulled from borosilicate tubing and then salinized. The tips are broken to an outer diameter of about 3 to 5 µm. Commercially available ionophore cocktails (Fluka catalog nos. 60031 for potassium, 24902 for chloride, and 71178 for sodium) are used to fill the tips after backfilling with 0.2 M KCl (K^{+} selective) or 0.5 M NaCl (Na^{+} and Cl^{-} selective). The electrodes are calibrated in sets of standards before and after use. Electrodes with a response of less than 50 mV per decade are discarded. The probe excursion distance is either 15 or 25 µm, the movement frequency is 0.1 Hz, and the sampling rate is 15 Hz.

The Petri dish containing 5- to 7-day-old test tissue is filled with APW solution (0.1 mM KCl, 0.1 mM $CaCl_2$, 0.1 mM $MgCl_2$, and 0.5 mM NaCl;

unbuffered pH approximately 5.6) and the tissue is conditioned for 2 to 3 hours. The solution is periodically replaced (every 20-30 min) by fresh solution to prevent formation of depletion zones and provide aeration. Twenty minutes before measurements, the Petri dish is transferred onto the microscope stage, and electrodes are positioned 20 µm above the tissue surface, with their tips separated by 2 to 3 µm and aligned parallel to the surface. All measurements are performed in the mature fully elongated zone 6 to 10 mm the root apex.

Ion fluxes are measured in the steady state for 5 to 10 minutes then the hyperosmotic treatment (100/100 mM mannitol/sorbitol made up in APW) is given. About 30 ml of solution is replaced (8 to 9 times of the chamber volume), and net ion fluxes are measured for another 50 to 60 minutes. The time required for solution replacement and establishment of a diffusion gradient (unstirred layer) is about 3 minutes. The interval is later discarded from the analysis and appears as a gap in most figures.

Direct Na^+ and K^+ measurements in the transpiration stream (Adopted from Watson et al., 2001).

The xylem-feeding insect *Philaenus spumarius* is used to analyze sodium and potassium fluxes in the xylem of intact, transpiring plants. With this method for example, two wheat cultivars, salt-excluding and non-excluding, can be compared for this trait. The reduced ion accumulation is specific to sodium; accumulation of K^+ is unaffected by NaCl in salt-excluding cultivars, such that it develops a much lower leaf Na^+ / K^+ ratio than the non-excluding cultivars. In the 150 mol m^{-3} NaCl treatment, sodium levels in the leaf xylem reaches only 2-3 mol m^{-3} in the salt-excluding cultivar, compared with 8-10 mol m^{-3} in the non-excluding cultivar. The lower leaf xylem content alone is thus sufficient to account for the reduced accumulation of sodium in leaves of the salt excluding cultivar.

Water soluble carbohydrate extraction and chromatography

Water soluble carbohydrates are extracted (six to 8 plants/extraction) from finely cut plant material. The cut tissue is heated at 65 - 70°C in dehydrated ethanol. The ethanol extract is siphoned off and stored on ice. Distilled water (20 mL) is added to the tissue and the tissue is ground in a Polytron on medium speed for three minutes. The water extract is centrifused for 10 minutes at 900 × *g*, and the supernatent is siphoned off and combined with the ethanol. Four additional water extractions are conducted. All supernatents are combined, concentrated under vacuum at ~ 40°C, and oven dried at 60°C to constant weight. The weight of the remaining tissue pillet (dried to a constant weight at 60°C) plus the weight of the dried extract is taken to be the total plant dry weight. The dried extract is rehydrated in water. A 0.03 M $Pb(NO_3)_2$ solution is added (dropwise until no precipitation occurs) to the extract to precipitate ions that could damage the HPLC column. The treated, diluted extracts are mixed, centrifused for 5 minutes at 900 × *g*, and filtered through a 0.45-µm filter. Carbohydrates in the extract are separated by HPLC. A Bio-Rad Aminex HPX-87P column is used at 84°C with degassed HPLC grade water at the mobile phase flowing at 0.4 mL/minute. A differential refractometer connected to a recorder is used to detect simple sugars (sucrose, glucose, and fructose). Peak heights relative to standard curves are used to quantify sugar levels. Fructan is also separated with HPLC. Sucrose is used as the standard for fructan (sucrose yields a refractive index of 1.2 times that of fructan). Retention times in minutes are as follows : fructan, 10.8; sucrose, 15.4; glucose, 18.6, and fructose, 25.2. The response factors are 0.09, 0.08, and 0.03 kg m^2 $Å^{-1}$ s^{-3} mg^{-1} for sucrose, glucose, and fructose, respectively.

Lipid peroxidation

The comparative rates of lipid peroxidation are assayed by determining the levels of malondialdehyde in 1 g aliquots of root, stem (hypocotyl tissue from below the cotyledons) or leaf tissue. Malondialdehyde is a product of lipid peroxidation and is assayed by the thiobarbituric acid reaction. Addition of extra ascorbic acid or NaCl to the extraction medium does not affect the measurements and significant readings are not obtained without addition of the reactive thiobarbituric acid. Results of roots, stems and leaves are expressed as means ± SE for separate experiments with batches of 10 or more plants per treatment.

Salt-induced membrane damage

The roots of intact seedlings are gently rinsed for 5 minutes in large volumes of 0.5 mM $CaCl_2$ and then with water prior to incubation under dim light in small vials containing 5 ml of deionized water. After incubation for 1 hour the electrical conductivity of the bathing solution is measured using a conductivity meter. The roots are then killed by immersion in 5 mL of fresh water at 95°C for 5 minutes, cooled and incubated for 1 hour more, prior to an additional measurement of electrical conductivity. The electrical conductivity resulting from ion leakage from live roots is expressed as a percentage of total conductivity (electrical conductivity live root plus electrical conductivity killed root). The resultant value is taken as a comparative measure of treatment effects on root cell-membrane integrity.

Analysis of mannitol in plants (After Shen et al., 1997 a)

Leaf mannitol content is measured by HPLC analysis. Intact chloroplasts are prepared from mesophyll protoplasts. Tobacco leaves (2 – 4 g) are cut into sections that are incubated in 40 ml of enzyme solution containing 1.2% cellulase, 0.4% macerozyme R10, and 0.45 M sucrose in a B5 medium. After 4 – 6 hours of incubation, the cell suspension is filtered through a 100μm nylon mesh and is centrifused at 60 g for 5 minutes. Intact protoplasts are floated on a 0.45 M sucrose medium and then washed with a B5 medium containing 0.45 M glycinebetaine. Chloroplasts are released from the protoplasts by passing the protoplasts through a 25μm nylon mesh three times and are collected by centrifugation at 100 g for 1 minute. The amount of mannitol in the pillets of protoplasts and chloroplasts is measured as described by Tarczynski et al. (1992).

Hydroxyl radical production in plant cells (After Shen et al., 1997 a)

Production of hydroxyl radicals in mesophsyll cells is determined using DSMO (an ideal molecular probe for trapping hydroxyl radicals in plant cells) as a probe. The mesophyll cells containing approximately 300 μg of chlorophyll are incubated with different concentrations of DSMO and 100 μM MV (methyl viologen) at 4°C in darkness for 30 minutes. After dark incubation, cells are placed in a water bath at 25°C and illuminated (500 μmol m^{-2} s^{-1}) for 1 hour. The reaction is terminated by adding 500 μL of an extraction buffer (methanol : chloroform : water, 12: 5 : 3, v/v). The extract is centrifused at 10,000 *g* for 5 minutes and the supernatant is dried in a vacuum drier at room temperature. The resulting pellet is suspended in 1 mL of water and passed through a Sep-Pak column (Millipore) to remove hydrophobic interferences. MSA (methane sulfinic acid) in samples is determined. The recovery of standard MSA using this procedure is more than 85 percent.

Estimation of free proline

Fully expanded leaves are selected and homogenized in 3 percent sulfosalicylic acid. The filtered homogenate is allowed to react with acid ninhydrin and glacial acetic acid for one hour at 100°C. The reaction mixture is extracted with toluene. The chromophore containing toluene is aspirated from the aqueous phase and the absorbance read at 520 nm. Proline concentration is determined from a standard curve and calculated on a fresh weight basis.

Δ^1-Pyrroline-5-carboxylate synthetase (P5CS) assay (From Kavi Kishore et al.,1995)

Leaves are homogenized in an extraction buffer (100 mM Tris-Cl, pH 7.5,

10 mM β-mercaptoethanol, 10 mM $MgCl_2$, and 1 mM PMSF). The extracts are centrifused at 10,000 *g* for 15 minutes. Ammonium sulfate (15 – 35% saturation) precipitation is carried out and the resulting pillets are collected and dissolved in the extraction buffer. This fraction is dialyzed against 10 mM TrisCl (pH 7.5) containing 5 mM $MgCl_2$ for 4 h with three changes of the buffer. P5CS enzyme assay is performed in the presence of ^{14}C-glutamate. This assay is carried out in a mixture containing 50 mM Tris, 20 mM $MgCl_2$, 10 mM ATP, 5 mM NADPH, 0.1 μCi ^{14}C-glutamate (Du Pont/NEN), leaf extract (10 μg of protein) in a final volume of 15 μL and final pH of 7.0 to 7.5. A blank reaction (-ATP) is performed in a mixture containing leaf extract of the transgenic line and other components except ATP. P5CS enzyme is partially purified from *Escherichia coli* by 35 percent ammonium sulfate precipitation and DEAE-cellulose chromatography. The reaction containing 0.1μg of the parcially purified P5CS serves as a positive control. After incubation at 35°C for 15 min, the mixture is chilled on ice. The products (2 μL of the mixture) are resolved by TLC on a silica gel. Glutamine, glycine, proline, and P5C (1 μg of each) are used as reference markers. ^{14}C-glutamate and ^{14}C-proline are also used as standards. TLC gel is developed with a mobile solution (phenol : water : acetic acid, 75:25:5) [w/v/v] containing 0.3% (w/v) ninhydrin in a saturated chamber. After development, the gel is dried at 65°C for 15 min or until the amino acid spots appear. The gel is wrapped with Saran Wrap and analyzed on a Phosphor Image or exposed to X-ray film.

Determination of glycinebetaine content

Samples consisting of leaves or aliquots of solution are frozen in liquid nitrogen immediately after collection. Glycinebetaine is quantified using NMR methods. Thawed leaf samples are used to express sap. Ten microliter of the extract is dried under a stream of nitrogen in a desiccator, and the final volume is made up to 1 mL with D_2O. Glycinebetaine is quantified using NMR spectrometer operating at 11.75 T (499.869 MHz for ^1H NMR) and spectra are measured at 30°C using a 5 mm triple-resonance inverse detection probe. One dimensional proton spectra are required with 16 transients, using 19.2 K data points, a pulse repetition rate of 2.0 s^{-1}, a flip angle of 60°, and a spectral width of 8,000 Hz centered on the water peak. The free induction decays are zero filled to 64 K prior to Fourier transformation using Varian NMR.

Betaine extraction and quantification

Oven-dried leaf material (10-40 mg) is ground in methanol: chloroform: water (12:5:1) and d_9–betaine (500 nmol) is added as an internal standard. The aqueous phase is fractionated by Dowex-1-OH^- and Dowex-50-H^+ ion-exchange chromatography. The betaine fraction is eluted with 6 M NH_4OH, dried under a stream of N_2 at 45°C, and dissolved in 1 mL of distilled water. Liquid chromatography/continuous flow secondary-ion mass spectrometry is used to obtain spectral data.

Assay of choline oxidase activity

Choline oxidase (COX) activity is determined spectrophotometrically by a BADH-coupled enzymatic reduction of NAD^+ at 22°C. BADH activity is measured independently by betaine aldehyde-specific reduction of NAD^+ at 22°C. One unit of BADH equals 1 nmol NAD^+ reduced min^{-1} mg^{-1} protein. The reactions are carried out in a final volume of 1 mL containing 50 mM HEPES-KOH, pH 8.0, 10 mM EDTA, 1 mM NAD^+ , 130 units of *E. coli* BADH, 20 mM choline, and protein extract. One unit of COX activity equals 1 nmol NAD^+ reduced min^{-1} mg^{-1} protein.

HPLC analysis of polyamines (After Hennion and Martin-Tanguy, 2000)

Seedlings or any plant tissues are frozen in liquid nitrogen, then lyophilized. Shoots and roots are separated after lyophilization. For each treatment, shoots or roots are pooled for amine analysis. In order to extract free amines, aerial parts and roots are ground respectively in 1 mL or in 0.5 mL of M HCl. After extraction for 1 hour, samples are pelleted by centrifugation at 18 000 *g* for 30 minutes, and the supernatent, containing free amines, is stored frozen at –20°C. HPLC and spectrofluorimetry are used to separate and quantify amines prepared as their dansyl derivatives. Aliquots (100 µL or 200 µL of samples, depending on sample dry weight) of the supernatent are added to 200 µL of saturated sodium carbonate and 600 µL of 27.8 mM dansyl chloride in acetone in a 5 mL tapered vial. After brief vortexing, the mixture is incubated in darkness at room temperature for 16 hours. Excess dansyl reagent is removed by reaction with 300 µL of 1.3 M proline and incubation for 30 minutes in the dark. Dansylated amines are extracted in 1 mL ethyl acetate. The organic phase is collected and then evaporated to dryness. The residues, containing the free amines fraction, is dissolved in methanol and stored in glass vials at –20°C. The amine fraction is analysed by HPLC using an LKB 2152 plus LKB 2150 chromatography system with an HPLC column packed with reverse phase spherisorb ODS-2 (particle size 5 mm; 4.6 × 250 mm. Portions (20 µl) of the amine fractions are applied to the column and eluted with a programmed methanol : water solvent gradient, changing from 60 percent to 95 percent over 23 minutes at a flow rate of 0.8 mL per minute. Elution is complete after 7 minutes. For detection of dansyl amines, an excitation wavelength of 365 nm is used with an emission wavelength of 510 nm. The results are standardized with equimolar (0.1 nmol) mixtures of dansylated amines.

Detection of trehalose by HPLC with ELSD (Zhou et al., 2001)

As a method for detecting low concentration of trehalose in transgenic plants was not available, Zhou et al. (2001) developed the high performance liquid chromatograph (HPLC) with evaporative light-scatting detector (ELSD) using water : methyl cyanide (1: 2.6 v/v) as mobile phase. An ODS column Zorbax RX-SIL is employed. The trehalose detection limits of ELSD is 5 mg L^{-1}.

Trehalase assay (After Jang et al., 2003).

Crude enzyme extracts are obtained by grinding frozen plant material in extraction buffer containing 50 mM Tris-HCl (pH 7.5), 250 mM sucrose, 1 mM EDTA (pH 8.0), and 10 mM phenylmethylsulfonyl fluoride. The suspension is incubated for at least 2 hours at 0°C and centrifused (5,000 rpm for 5 minutes). The supernatent is used for the enzyme activity assay. Trehalase activity is measured by estimating both the glucose produced by hydrolysis of trehalose and trehalose reduced using HPIC with a Carbo-Pac PA 1 column (4 × 250 nm) using the DX500 HPIC system (Dionex 500). The reaction mixture containing 30 mM trehalose (Sigma) is incubated at 37°C for 1, 2, and 3 hours and stopped by boiling for 2 minutes. Soluble protein is determined with the Bradford method (Bradford, 1976).

Analysis of ectoine

For the identification of intracellular ectoine, 5-day-old suspensions of BY2 cells are harvested, and then 1 gram fresh weight of the cells is transferred to a 15-mL centrifuge tube, suspended in 5 mL of extraction buffer (ethanol : chloroform : water, 12 :5 : 2, v/v), and sonicated. The cell extract is separated from the cell pellet by centrifugation at 3,000 *g* for 5 minutes. The pellet is re-extracted twice by the same method and all the cell extracts are pooled in a 50 mL centrifuge tube. After the addition of 10 mL of chloroform and 5 mL of water and centrifugation at 3,000 *g* for 5 minutes, the aqueous layer is collected in a 15 mL centrifuge tube and evaporated at 80°C. The residue is then dissolved in 5 mL of water and filtered through a 1.2-µm-pore syringe filter. The filtered extract is then passed through an ion-exchange column containing AG50W-X8 (H^+ form, Bio-Rad), washed with two bed volumes of water, and eluted with 3 N NH_4OH. The eluate is evaporated at 80°C, and then the residue is dissolved in 1 mL of water. The dissolved solution is filtered through centrifugal filter units (0.2-µm pore) and subjected to liquid chromatography/electrospray ionization mass spectroscopy. The samples are loaded onto a column (2×250 mm) at 40°C and eluted over 30 minutes at a flow rate of 0.2 mL per minute with 0.1 percent (v/v) aqueous formic acid. The effluent is fed directly to the electrospray interface of the mass spectrometer. Ions are detected throughout the entire LC step over a *m/z* (mass-to-charge ratio) range of 100 to 300. Authentic ectoine is purified from the cells of *H. elongata* OUT30018 by a method described by Ono et al. (1998).

Colorimetric determination of citrulline

Sugawara et al. (1998) described a method for the direct colorimetric determination of citrulline residues in proteins based on the reaction with diacetylmonoxime in the presence of lower concentrations of sulfuric acid. The reduced sensitivity due to the lower acid concentration is overcome by the addition of ferric chloride which also contributes to the color stabilization. Insoluble proteins or proteins resulting in turbidity are analyzed following partial hydrolysis of those with enzyme or acid. The molar absorption coefficient (epsilon) for citrulline at 464 nm is 2.8×10^{-4}. The method enables to determine low levels of protein-bound citrulline which are beyond the limitations of conventional methods using an amino acid analyzer.

Hydrogen peroxide determination

Hydrogen peroxide concentration is evaluated as described by Sgherri et al. (1994 a). This method is very sensitive and reproducible, and excludes the interference of other peroxides (except for a small effect of lipid peroxide). A standard curve in the 0- to 350-μM range is used.

GSH and GSSH determination

Fresh leaf tissue (0.5 g) is homogenized in ice-cold 5% sulfosalicylic acid (w/v), centrifused at 12,100 *g* for 15 minutes, and the supernatent is used for total and GSSH determinations by the 5,5′-dithio-bis-(2-nitrobenzoic acid)/GSSH reductase recycling procedure, as described by Sgherri and Navari-Izzo (1995). GSSH is determined after removal of GSH by 2-vinylpyridine derivatizations. Changes in absorbance of the reaction mixture are measured at 412 nm at 25°C, and the contents of total glutathione and GSSH are calculated as described by Sgherri et al. (1994 b). GSH is determined by subtraction of GSSH (as GSH equivalents) from the total glutathione content.

Ascorbic acid

The effect of exogenous supply on tissue levels of ascorbic acid is determined as under: Seedlings are incubated with or without 0.5 mM ascorbic acid for 24 hours and then for an additional 9 hours with or without 300 mM NaCl. After rinsing in 0.5 mM $CaCl_2$, root, stem and leaf tissues are frozen with liquid nitrogen, and ground in a mortar and pestle. Ground tissue is mixed with 10 mL of 10% TCA and the supernatent obtained after centrifugation at 18,000 *g* for 15 minutes at 4°C is analyzed for total ascorbic acid. Briefly, any dehydroascorbic acid is reduced to ascorbate with β-mercaptoethanol and excess mercaptoethanol is complexed with *N*-ethylmaleimide. Total ascorbate is then assayed under acid conditions by measuring the reduction of Fe^{2+} to Fe^{3+} and the formation of a pink colored bipiridyl-Fe complex at 525 nm.

REFERENCES

Bradford, M. 1976. A rapid and sensitive method for the quantitation of microgram quantities of protein utilizing the principle of protein-dye binding. Anal. Biochem. 72: 248-54.

Hennion, F. and J. Martin-Tanguy, 2000. Amines of the subantarctic crucifer *Pringlea antiscorbutica* are responsive to temperature conditions. Physiol. Plant. 109: 232-43.

Jang, I.C.; S.J. Oh; J.S. Seo; et al. 2003. Expression of a bifunctional fusion of the *Escherichia coli* genes for trehalose-6-phosphate synthase and trehalose-6-phosphate phosphatase in transgenic rice plants increases trehalose accumulation and abiotic stress tolerance without stunting growth. Plant Physiol. 131: 416-24.

Kavi Kishore, P.B.; Z. Hong; G.H. Miao; et al. 1985. Overexpression of D^1-pyrroline-5 carboxylate synthetase increases proline production and confers osmotolerance in transgenic plants. Plant Physiol. 108: 1387-94.

Loggini, B.; A. Scartazza; E. Brugnoli; et al. 1999. Antioxidative defence system, pigment composition, and photosynthetic efficiency in two wheat cultivars subjected to drought. Plant Physiol. 119: 1091-1100.

Nelson, D.E.; G. Rammesmayer and H.J. Bohnert, 1998. Regulation of cell specific inositol metabolism and transport in plant salinity tolerance. Plant Cell 10: 753-64.

Ono, H.; M. Okuda; S. Tongpim; et al. 1998. Accumulation of compatible solutes, ectoine and hydroxyectoine, in a moderate halophyte, *Halomonas elongata* KS3 isolated from dry salty land in Thailand. J. Ferment Bioeng. 85: 362-68.

Sgherri, C.L.M.; B. Loggini; A. Bochicchio; et al 1994 a. Antioxidant system in *Boea hygroscopica:* changes in response to desiccation and rehydration. Phytochemistry 37: 377-81.

Sgherri, C.L.M. and F. Navari-Izzo, 1995. Sunflower seedlings subjected to increasing water deficit stress : oxidative stress and defence mechanisms. Physiol. Plant. 93: 25-30.

Shen, B.; R.G. Jensen and H.G. Bohnert, 1997 a. Increased resistance to oxidative stress in transgenic plants by targetting mannitol biosynthesis to chloroplasts. Plant Physiol. 113: 1177-83.

Sugawara, K.; Y. Yoshizawa; S. Tzeng; et al. 1998. Colorimetric determination of citrulline residues in proteins. Anal. Biochem. 265(1): 92-96.

Tarczynski, M.C.; R.G. Jensen and H.J. Bohnert, 1992. Expression of a bacterial *mtlD* gene in transgenic tobacco leads to production and accumulation of mannitol. Proc. Nat. Acad. Sci. 89: 2600-04.

Watson, R.; J. Pritchard and M. Malone, 2001. Direct measurement of sodium and potassium in the transpiration stream of salt-excluding and non-excluding varieties of wheat. J. Exp. Bot. 52(362): 1873-81.

Zhang, J.; H.T. Nguyen and A. Blum, 1999. Genetic analysis of osmotic adjustment in crop plants. J. Exp. Bot. 50(332): 291-302.

Zhou, J.; B. Yang and X. Dai, 2001. [Detection of trehalose in transgenic tobacco by HPLC with ELSD]. Wei Sheng Wu Xue Bao 41 (3): 378-80 (In Chinese).

AUTHOR INDEX

Abebe, T. 84, 85
Acevedo, E. 7
Ackerson, R.C. 6
Aeschbacher, R.A. 150
Akashi, K. 3, 5, 190, 191, 192, 193, 194
Akiyama, T. 139
Albrecht, G. 163, 173
Alegre, L. 209
Ali, R.M. 136
Alia, K.Y. 105, 107
Allard, F. 107, 113
Allen, G. 18
Allen, R.D. 82
Altman, A. 160
Amiard, V. 163, 169
Amora, Y. 217
Anderson, A.D. 7
Anderson, S.E. 142
Angeles Martinez-Cordero, M. 15
Arakawa, K. 105
Arora, A. 204
Arrabaca, M.C. 48, 51
Audran, J.C. 97, 98
Avonce, N. 152, 158

Bachman, M. 75
Bailly, C. 56, 58
Bancal, A. 171
Barka, E.A. 97, 98
Basnayake, J. 8
Basra, R.K. 139
Bellaloui, N. 81
Ben-Hayyim, G. 130
Bertand, A. 173
Bethke, P.C. 200
Bhagwat, K.A. 109, 110, 111, 112
Biemelt, S. 217, 218
Blackman, S.A. 54, 56
Blatt, M.R. 33
Blokhina, O.B. 202, 216, 217
Blumwald, E. 39, 210
Bornman, C.H. 97
Bouchereau, A. 128
Bourot, S. 107
Brandford, M. 234
Bray, E.A. 9
Bremer, E. 182
Bright, S.W.J. 97
Brown, P.H. 81
Brown, P.S. 206
Bruni, F.B. 43
Buitink, J. 44
Buris, J.S. 45
Buschmann, P.H. 24

Caimi, P.G. 168, 175
Cairns, A.J. 167, 176
Cakmak, I. 22
Canovas, D. 181, 182, 183
Capell, T. 128, 133, 138, 142
Caplan, A. 92, 99
Carden, D.E. 31
Castonguay, Y. 45, 47
Cerda, A. 13, 17, 31
Chattopadhyay, M.K. 134, 135
Chen, Y. 45
Chen, Z. 208
Colmer, T.D. 118
Cowley, T. 141
Crawford, R.M.M. 218
Cress, W.A. 3, 88, 91
Crowe, J.H. 81
Cuin, T.A. 32
Cunnigham, S.M. 76, 77

Dat, J.F. 215
Davenport, R. 32
De, B. 25, 26, 27
Deane, C.R. 98
Deguchi, M. 80
Delanuey, A.J. 89, 90
Demel, R.A. 163, 168, 171
de Roover, J. 169
da Silva, 48

Devaux, C. 62, 64
Devi, S.R. 220
Devine, K.M. 194
Dhindsa, R.S. 36
Dietrich, P. 22, 37
Dionne, J. 46, 171, 172
Dipierro, N. 220
Dixit, V. 218
Dörffling, K. 97
Downie, B. 59, 83
Dube, R.S. 59
Duncan, D.R. 97

Eastmond, P.J. 148
Edelman, J. 164
Erdei, L. 137
Ercal, N. 218

Fellman, J.K. 72
Fernando, M. 23
Fernendez-Ballester, G. 31
Foyer, C. 200
Foyer, C.H. 199, 200, 204
Francelchetti, M. 129
Freeman, J.L. 220
Fricke, W. 32
Fryer, M.J. 202
Fukushima, E. 59

Galiba, G. 7, 44, 48, 50, 52, 53, 57, 98, 137, 143, 153
Galinski, E.A. 181
Gallie, R. 208
Garcia, A.B. 155
Garg, A.K. 4, 147, 153, 154, 155, 156, 157, 158
Gassmann, D. 20
Gilbert, G.A. 76
Goddijn, O.J. 153
Golladack, D. 37
Gömez, J.M. 210
Gömez, L.D. 214
Gong, M. 33
Gonzales, E.M. 18
Gonzalez de Mejia 138
Grabov, A. 33
Grewal, J.S. 35
Grote, E.M. 117
Guenzi, A.C. 85
Gupta, A.S. 199
Guy, C.L. 46
Handa, S. 88, 96
Harbinson, J. 204
Hare, P.D. 3, 88, 91, 106
Havaux, M. 115
Hebert, R.R. 6
Hendry, G.A. 163
Hennion, F. 131, 233
Henson, C.A. 170
Hernandez, J.A. 209
Hincha, D.K. 47, 48, 49
Hoekstra, F.A. 44, 45, 50, 53, 57
Holmström, K.O. 107, 108, 109, 110
Hong, Z. 99
Hu, Y. 14, 17
Huang, B. 15, 16, 17, 34, 35, 215
Huang, J. 13, 37, 108, 215, 216
Hummel, I. 131, 132, 133, 134, 139

Igarashi, Y. 96
Ilan, N. 33
Ishitani, M. 105, 120
Itai, C. 96
Iyer, S. 92, 99

Jang, I.C. 153, 154, 156, 234
Jansson, E. 97
Jebbar, M. 5, 180, 184, 186
Jefford, T.G. 164
Jha, A.B. 59
Jiang, Y. 15, 16, 17, 34, 35
Jolivert, Y. 104
Jones, M.M. 6
Jones, R.L. 200
Jonker, H.H. 167
Joo, J.H. 200

Kafkafi, U. 34, 35
Karami, E. 6
Karakas, B. 78, 80
Kasinathan, V. 137
Kasukabe, Y. 133
Kavi Kishore, P.B. 3, 95, 96, 97, 231
Kawakami, A. 171
Kawasaki, S. 190, 191, 193
Kaya, C. 37
Keller, F. 75
Kerepesi, T. 174
Kirkham M.B. 205
Kishitani, S. 112
Kiyasue, T. 97
Koch, K. 21
Koch, K.E. 43, 45, 62
Kocsy, G. 205, 214

Koops, A.J. 167
Kosmas, S.A. 158
Krall, J.P. 104
Krapp, A. 45
Krishnamurthy, R. 109, 110, 111, 112
Kueh, J.S.H. 97
Kuhlmann, A.U. 182
Kumar, S. 122
Kunte, H.J. 184
Kuo, T.M. 70

Ladyman, J.A.R. 109, 117
Lafata, A.M. 61
La Rosa, P.C. 95
Lee, S.B. 159
Leopold, A.C. 43, 44
Lepri, O. 129
Leslie, S.B. 13
Lew, R.R. 13, 14
Li, T. 100, 101
Li, W. 19
Li, Y. 31
Lilius, G. 109
Lilley, J. M. 9
Lin, P. 31
Lin, W. 141
Lin, X.H. 38
Liu, J. 27, 134
Liu, K. 130
Liu, Y.L. 134
Livingston III. D.P. 170
Loescher,W.H. 72
Loggini, B. 206, 207, 208
Lorenzen, J.H. 61
Louis, P. 181
Lu, C.A. 63, 68
Lu, C.M. 107
Luan, S. 28
Ludlow, M.M. 2
Maathuis, F.J.M. 25
Madore, M.A. 80
Mahan, J.R. 209
Maiale, S. 135
Majore, I. 18
Malmberg, R.L. 140
Mamta, Rai, 188
Mamzanera, M. 181
Marten, I. 20
Martin-Tanguy, J. 131, 233
Martinez, C.A. 93, 94
Matamoras, M.A. 197
Matsuura-Endo, C. 47
May, M.J. 202
Mc Kersie, B.D. 206, 215
Mc Neil, S.D. 106, 108, 120
Mehta, R.A. 142
Menendez, C.M. 46
Mengel, K.K. 21
Min-Yu, L. 185
Mittova, V. 212, 213
Miyamoto, Y. 202
Mo, H. 134
Moghaieb, R.E.A. 186, 187
Monroy, A.F. 36
Morcuende, R. 174
Moran, J.F. 83
Morgan, J.M. 7, 8, 23
Muchow, R.C. 2
Müler, J. 148, 150, 151, 152, 153
Munne-Bosch, S. 209
Murata, N. 104, 105, 108, 110, 123
Murata, Y. 200

Naas, H.G. 171
Nagata, S. 185
Nagy, Z. 58
Naidu, B.P. 113
Nakamura, T. 106, 121
Nakayama, H. 5, 182, 185
Narang, R. 170
Navabpour, S. 198
Navari-Izzo, F. 235
Neill, S. 198
Nelson, D.E. 71, 79, 228
Newman, I.A. 36
Newmann, P.M. 210, 211
Noctor, G. 200, 202, 203
Nomura, M. 110
Noury, M. 141
Nuccio, M.L. 118, 119

Ono, H. 182, 186, 235
Onraedt, A. 185
O'Reilly, M. 194

Paleg, L.G. 96
Papageorgiou, G.C. 105
Park, E.J. 122
Park, K.Y. 122, 142
Parvanova, D. 115
Pastori, G.M. 200, 214
Pattanagul, W. 80
Paul, M. 147, 149
Peng, Y.H. 32

Peterbauer, T. 70
Pharr, D.M. 72, 73
Phillips, G.C. 134
Pillai, M.A. 139
Pilon-Smits, E.A.M. 163, 164, 169
Pitman, M.G. 19
Polidoros, A.N. 200
Polle, A. 58, 219
Potters, G. 202
Prasad, T.K. 213, 214
Premchandra, G.S. 16
Pua, E.C. 134
Pubela, A.E. 163, 171
Purcell, P.C. 66

Quan, R. 108

Rajashekar, C.B. 113, 114
Rajendrakumar, C.S.V. 105, 107
Rantsch, D. 92
Raschke, K. 14
Rascio, A. 37, 95
Rathinasabapathi, B. 106
Rawson, H.M. 6
Raymond, M.J. 92
Redmann, R.E. 13, 37
Rhodes, D. 8, 88, 96, 110, 117
Rich, P.J. 117
Richter, A. 69
Ritsema, T. 167
Roberts, S.K. 18, 19, 20, 28
Rontein, D. 2, 120
Roxas, V.P. 10
Roy, D. 94
Roy, M. 142
Rubio, F. 25
Rubio, L. 38
Ruiz, J.M. 210
Russell, B.L. 109

Sadiqov, S.T. 28
Sairam, R.K. 206, 208, 215
Sakamoto, A. 104, 108, 110, 123
Saneoka, H. 116, 118
Santa-Cruz, A. 136
Santoiani, C.S. 170
Savitch, L.V. 59, 60
Saxena, D.C. 208
Scandilios, J.G. 199, 200
Schachtman, D.P. 25
Schioacke, R. 93
Schmidhalter, U. 14, 17
Schroeder, J.I. 20, 25
Schützendübel, A. 219
Scoccianti, V. 130
Seo, H.S. 153, 154
Serrano, R. 10
Sevenier, R. 168, 175
Sgherri, C.L.M. 235
Shabala, A. 210, 211
Shabala, L. 17, 36
Shabala, S. 13, 14, 17, 18, 20, 29, 30, 33, 36
Sharmila, P. 91
Sharp, R.E. 92, 96
Shen, B. 3, 81, 82, 83, 231
Shen, W. 138
Sheveleva, E.V. 73, 74, 79, 84
Shinozaki, K. 10
Singh, S.N. 35
Sionit, N. 7
Siripornadulsil, S. 91
Smeekens, S.C. 153, 165, 166, 167
Smirnoff, N. 75, 92, 205
Snowman, B.N. 28
Spalding, E.P. 14
Sprenger, N. 167, 175
Steponkus, P.L. 6
Storozhenko, S. 200
Stitt, M. 45
Streeter, J.C. 151, 154
Su, H. 15
Subbarao, G.V. 25
Sugawara, K. 235
Sumaryati, S. 98
Suresh, P. 26
Suzuki, M. 171
Swapan, K.D. 74

Tabaei-Aghdaei, S.R. 47
Takabe, T. 108
Talibart, R. 180
Tan, M.K. 8
Tangpremsri, T. 2
Tarczynaski, M.C. 8, 73, 78, 231
Tausz, M. 201
Tester, M. 32
Tetsch, L. 184
Thomas, C.E. 218
Tiburcio, A.F. 129
Tillberg, J.E. 174
Triboi, E. 17

Umashankar, R. 26
Upadhyay, T. 135

Urano, K. 143

Vander Ende, W. 167
Van der Meer, I.M. 175
van Dun, K. 153
van Smeekens, K. 153
Van Swaaij, A.C. 97
Verbruggen, N. 96
Vereyken, I.J. 4, 164, 168, 169
Verma, D.P.S. 90
Verslues, R.E. 92
Vijn, I. 165, 166, 167, 175, 176
Vinocur, B. 160
Voetberg, G.S. 96
Walker, D.J. 23
Wallace, R.K. 163
Walters, D.R. 141
Wang, C. 174
Wang, H.L. 45, 48, 51, 54, 55
Wang, T.B. 23, 24
Wang, Y.B. 185
Wanjura, D.F. 209
Watson, M.B. 140
Watson, R. 229
Wegner, I.N. 14
Weinstein, L.H. 140
Wellfare, D. 19
Wi, S.T. 142
Widholm, J.M. 97
Wiemken, A. 167
Willekens, H. 200
Williams, W.P. 115
Williamson, J.D. 66
Wingler, A. 137, 149, 153
Wollenweber-Ratzer, B. 218
Wu, R. 142
Wu, S.J. 31

Xiong, L. 12

Yamaguchi-Schinozaki, K. 10
Yang, G. 113, 118
Yang, J. 45, 51
Yang, W.J. 8, 105, 107, 116, 117
Yang, X. 32, 115
Yang, X.H. 107
Yeo, E.T. 159
Yokota, A. 191
Yonamine, I. 99
Yoshida, M. 171
Yoshida, Y. 89, 90, 95, 97
Yu, B. 136
Yu, S. 62, 63
Yu, S.M. 62, 63, 65

Zabotina, O.A. 45
Zhang, H.X. 38
Zhang, J. 2, 9, 225
Zhang, J.X. 205
Zhao, Y. 104, 105
Zheng, Y.Z. 100, 101
Zhou, J. 234
Zhu, J.K. 27, 31
Zhu, Y.L. 219

SUBJECT INDEX

Abscisic acid (ABA) 10, 18, 19, 20, 26, 28, 31, 89, 96, 100, 101, 113, 158
Acclimation 36, 45, 46, 47, 48, 59, 60, 61, 112, 113, 114, 116, 171, 172, 201, 213, 214, 215
Accumulater 104, 119
Acidification, cytosolic 33
Adaptation 1, 9, 10, 19, 25, 37, 45, 62, 94, 160, 173, 180, 184, 202, 206, 210
  mechanism 31, 43
Adaptive features 1
  process 50, 57
Agmantine 132, 138
Alkalization, mesophyll 37
α-Amylase 59, 63
Antioxidants 5, 196-220
  defence 58, 59
Arginine 128, 134, 135, 140, 142, 188, 190, 191, 193
  biosynthesis 191
Ascorbate (ascorbic acid) 5, 82, 192, 196, 199, 201, 202, 204, 208, 209, 211, 212, 217, 218, 220
  determination 235-236
  peroxidase 5, 56, 58, 199, 200, 206, 207, 208, 209, 210, 212, 213, 217, 218, 220, 221
  reductase 218

Back-crossing 9, 118
Betaine extraction 232, 233

Cadmium 91, 140
Carbohydrate depletion 64
  metabolism 46, 47, 48, 61
  reserve 62
Carotenoids 5, 203, 204, 208, 209
Catalase 5, 56, 58, 198, 199, 217
Cell turgor 225-226
Chloroplast isolation 228
Choline oxidase activity assay 233
Citrulline 4, 5, 190-194
  bioassay 191, 194
  determination 235
  protective role 194
Compartmentation 17, 18, 104, 105, 196, 205, 217
Compatible solutes (osmolytes) 1, 2, 13, 14, 17, 62, 79, 88, 92, 94, 96, 104, 181, 192, 194, 196
Crassulacean acid metabolism 106
Crop improvement 9, 17
  productivity 115
Cryoprotective 46
Cyclitol 69, 70, 71, 79, 83, 84
Cysteine biosynthesis 210
Cytosolic calcium 34

Dehydroascorbate reductase 218

Ectoine 4, 5, 180-188
  analysis 234-235
  biosynthesis 181, 182, 183, 185
  biosynthetic gene 184, 186
Electrical gradient 14
Electrolyte leakage 49, 206, 215, 216, 227-228
Enhanced tolerance 115, 123, 134, 160, 215
Energy conservation 22
Enginnering genes 160
  glycine betaine 116
Ethylene 33
  biosynthesis 143
Excitation energy 197

Fructan 4, 5, 8, 50, 53, 58, 61, 163-177
  accumulators 163, 169, 176
  biosynthesis 60, 61, 164, 166, 167, 176, 177
  graminan-type 167, 175
  inulin-type 168
  producing 109
  producing genes 167
Galactinol 56, 69, 70, 71, 75, 76, 80, 170
Gene encoding 15, 153, 154, 155, 160, 175, 185

expression 37, 47, 62, 63, 79, 91, 110, 128, 153, 170, 199
regulation 65, 66, 201
transfer 10
Genetic enginnering 8, 96, 104, 115, 116, 122
variation 46
Genetically modified 85
Genotypic variability (see variability)
Gibberellin 63
Glutamate 88, 188
Glutathione 5, 82, 83, 91, 199, 200, 204, 210, 212, 217, 218, 220
biosynthesis 203, 219, 220
metabolism 209
peroxidase 198, 199, 200, 202, 212
pool 215
reductase 5, 56, 205, 206, 208, 209, 215, 217, 219, 220
-S-transferase 212
Guaiacol peroxidase 212, 213, 214
Glycinebetaine 6, 7, 8, 93, 104-123, 180, 187, 191, 199
accmulating 116, 120
biosynthesis 106, 108, 119, 120, 122
deficient 8
precursor 107
Hardening 6, 45, 47, 170
Hardiness 47, 48
Heavy metal 218
Heritability 109
High-affinity uptake 15, 23, 24, 25
Hydrogenperoxide (H_2O_2) 199, 200
determination 235
detoxification 206
Hydroxyl radical production 231
Hyperosmotic 28, 30
Hyperpolarization 33

Improved growth 84, 180
photosynthesis 147, 149
survival 154
tolerance 7, 157, 186
Improvement 3, 9, 10, 37, 38, 44, 76, 83, 98, 99, 108, 110, 113, 116, 118, 122, 141, 159, 175, 185, 217, 219
Indoleacetic acid (IAA) 28
Induced tolerance 160
Inheritance 7, 8, 117, 118
Inorganic ions 62
Inosital 69, 70, 71, 74, 75, 79, 80, 83
Invertase, acid 59, 60
vacuolar 47
Ion flux measurement 228-229
Ion uptake 27

Lipid peroxidation 230

Macromoleules 3, 44, 61, 73, 81, 88, 89, 104, 105, 201
Mannitol 5, 8, 13, 14, 27, 28, 29, 30, 38, 63, 66, 71, 72, 73, 78, 80, 81, 82, 83, 84, 85, 98, 134, 137, 182, 185 161
analysis 231
Membranes 18, 44, 88, 109, 185, 202, 203, 204
barrier 168
damage 2, 230
efficiency 153
fluidity 152
fusion 48
integrity 3, 104, 105, 113, 139, 204
lipids 36, 164
permeability 37, 185, 216, 227
phospholipids 43
potential measurement 20, 226
preservation 164
properties 216
protecting role 167, 201
stabilization 43, 45, 46, 127, 208
transport 13, 36
Metabolic adjustment 92
engineering 2, 5, 9, 10, 78, 118, 186
Metallothionein 194, 198
Mineral deficiency 91, 174
Myo-inositol 170

Nitric oxide (NO) 197, 198, 200
Nitrogen fixation 149, 150, 197
metabolism 21, 22
starvation 174
Nodulation 150
Nodules 151, 152
Nutrient acqusition 17
stress 81, 174
uptake 2, 20
Nutritional quality 142
Nutritive value 142

Ononitol 8, 69, 70, 71, 74, 79
Ornithine 88, 89, 127, 128, 131, 138, 140, 142, 190
Osmolality 32, 47, 48, 184, 226
Osmolytes 1, 5, 6, 43, 45, 73, 80, 89, 94, 95, 185, 190, 191
Osmoprotectants 1, 2, 4, 10, 43, 45, 48, 63, 88, 96, 106, 107, 115, 116, 118, 133

Osmoregulation 2, 3, 4, 6, 7, 14, 23, 24, 31, 91, 104
Osmolarity 113, 182
Osmotica, major 37
Osmotic adjustment 1, 2, 3, 4, 5, 7, 8, 9, 13, 14, 16, 17, 18, 31, 35, 37, 45, 51, 78, 80, 83, 84, 88, 96, 105, 118, 186, 225
Osmotic potential measurement 226
Osmotolerance 14
Over-expressing 8, 82, 108, 129, 210, 219
-expression 95, 99, 141
-producing 3, 96, 153
Oxidative damage 7, 192, 193, 194, 205, 209, 210, 215
metabolism 18
stress 5, 58, 82, 99, 143, 191, 196, 203, 206, 212, 213, 217, 220
Partitioning 32
Patch-clamp method 19, 22, 226, 227
Permeability 14, 19
Peroxidase 217
Phloem exudate collection 228
Phloem loading 20, 48
Phosphorus starvation 1274
Photoinhibition 109
Photo-oxidative damage 35, 148, 158
Photoprotection 209
Pinitol 69, 70, 74, 75, 79, 84, 150
Plasma membrane 13, 15, 18, 19, 20, 24, 31, 32, 33, 36, 115, 130, 198
Polyphenolics 204
Potassium deficiency 140
level 3, 7
measurement 229
starvation 24
uptake 15, 24
Preconditioning 6, 15, 16, 17, 34, 74, 84
Pretreatment 24, 25, 33, 139, 217
Proline 2, 5, 8, 25, 26, 28, 75, 88-101, 115, 135, 185, 188, 191
biosynthesis 90, 92, 95, 97, 99
estimation 231
overproduction 97
Polyamines 127-143
analysis 233
biosynthesis 128, 129, 131, 138, 143
Promoter 133, 141, 142, 159, 182, 184
Protectants 4
Protection 4, 109, 110, 114, 140, 203, 209, 214
Protective mechanism 204, 206, 210
Purescine biosynthesis 131, 141

Quality (turf) 215

Raffinose 47, 53, 56, 57, 69, 71, 75, 77, 80, 83, 169, 170
Raffinose family oligosaccharides 75, 76, 78, 80
Reactive nitrogen species (RNS) 197
Reactive oxygen species (ROS) 22, 58, 81, 82, 190, 193, 196-220
detoxification 204
Recessive gene 117
Relative humidity 54, 56

Selection 9, 10, 99, 116, 122
Sorbitol 53, 55, 80, 81, 138
Stachyose 47, 53, 54, 56, 57, 69, 70, 75, 77, 80
Sucrose biosynthesis 44, 53, 55
hydrolysis 51, 54
levels 46
metabolism 45, 51, 58
phosphate synthetase 46, 59, 60
starvation 62, 66
starvation-induced genes 65
synthase 51, 61
Superoxide 196, 197, 204, 210
dismutase 56, 58, 82, 196, 198, 199, 205, 212, 213, 215, 217, 218
Stomatal closure 2, 18, 197, 209
conductance 80
movement 18, 19

Tocopherol 199, 201, 202, 203, 204, 209, 217, 218
Transgenic lines 157, 158
plants 4, 38, 58, 109, 110, 115, 119, 120, 121, 122, 128, 129, 134, 138, 142, 143, 147, 154, 155, 156, 159, 169, 175, 176, 186, 187, 188, 199, 206, 210, 215
Transporters 14, 15, 20, 21, 25, 28, 31, 32, 36, 93
Trehalase assay 234
biosynthesis 150
Trehalose 4, 53, 147-160, 168
biosynthesis 158
biosynthetic genes 147
detection 234
metabolism 148
overproduction 4

Up-regulation 23, 45, 47, 119, 142, 152, 210, 212, 213
UV-B radiation 141

Variability 23, 109, 116, 117, 206, 207

Zeaxanthin 204, 206